PREVENTIVE FORCE

Preventive Force

Drones, Targeted Killing, and the Transformation of Contemporary Warfare

Edited by
Kerstin Fisk and Jennifer M. Ramos

NEW YORK UNIVERSITY PRESS
New York

NEW YORK UNIVERSITY PRESS
New York
www.nyupress.org

© 2016 by New York University
All rights reserved

References to Internet websites (URLs) were accurate at the time of writing. Neither the author nor New York University Press is responsible for URLs that may have expired or changed since the manuscript was prepared.

Library of Congress Cataloging-in-Publication Data
Names: Fisk, Kerstin, editor of compilation. | Ramos, Jennifer M., editor of compilation.
Title: Preventive force : drones, targeted killing, and the transformation of contemporary warfare / edited by Kerstin Fisk and Jennifer M. Ramos.
Other titles: Drones, targeted killing, and the transformation of contemporary warfare
Description: New York : New York University Press, [2016] | Includes bibliographical references and index.
Identifiers: LCCN 2016001633 | ISBN 9781479857531 (hbk. : alk. paper) | ISBN 9781479857654 (pbk. : alk. paper)
Subjects: LCSH: Drone aircraft—Moral and ethical aspects. | Drone aircraft—Government policy—United States. | Preemptive attack (Military science) | Targeted killing—Government policy. | Terrorism—Prevention. | Air warfare—Government policy—Western countries. | United States—Military policy. | War (International law)
Classification: LCC UG1242.D7 P75 2016 | DDC 355.4—dc23
LC record available at http://lccn.loc.gov/2016001633

New York University Press books are printed on acid-free paper, and their binding materials are chosen for strength and durability. We strive to use environmentally responsible suppliers and materials to the greatest extent possible in publishing our books.

Manufactured in the United States of America

10 9 8 7 6 5 4 3 2 1

Also available as an ebook

CONTENTS

Acknowledgments vii

List of Figures and Tables ix

1. Introduction: The Preventive Force Continuum 1
   Kerstin Fisk and Jennifer M. Ramos

   PART I. UNDERSTANDING PREVENTIVE WAR

2. Preventive Force: The Logic of Costs and Benefits 33
   Jennifer Taw

3. Preventive War and Its Domestic Politics 58
   Miroslav Nincic

4. Sovereignty and Preventive War in the Twenty-First Century: A Retrospective on *Eve of Destruction: The Coming Age of Preventive War* 88
   Thomas M. Nichols

   PART II. PERSPECTIVES ON PREVENTIVE DRONE STRIKES

5. Why Drones Are Different 115
   Stephan Sonnenberg

6. The Drone: It's in the Way That You Use It 142
   David Glazier

7. Drones and the Law: Why We Do Not Need a New Legal Framework for Targeted Killing 170
   Daphne Eviatar

8. Studying Drones: The Low Quality Information Environment of Pakistan's Tribal Areas 199
   C. Christine Fair

PART III. THE FUTURE OF PREVENTIVE FORCE

9. The Contemporary Practice of Self-Defense: Evolving Toward the Use of Preemptive or Preventive Force? 229
   Avery Plaw and João Franco Reis

10. Restricting the Preventive Use of Force: Drones, the Struggle against Non-State Actors, and *Jus ad Vim* 257
    John Emery and Daniel R. Brunstetter

11. Drones and Dirty Hands 283
    Ben Jones and John M. Parrish

12. Beyond Preventive Force: Just Peace as Preventive Non-Intervention 313
    Deen Chatterjee

13. Conclusions 341
    Jennifer M. Ramos and Kerstin Fisk

About the Contributors 349

Index 353

ACKNOWLEDGMENTS

In spring 2014, the Bellarmine College of Liberal Arts at Loyola Marymount University sponsored a conference on preventive force and drone technology. The chapters herein reflect the substance of those conversations, and we are grateful to the authors for their contributions to both the conference and the volume. In addition, we were pleased to have the support of Dean Mike O'Sullivan and Associate Dean Jeffrey Wilson. Our colleagues within the Political Science Department were also supportive, particularly our chair, Richard Fox. It is safe to say that without his initial prodding, we would not have had the wonderful experience of the conference, nor this compilation of essays. We would also like to thank others who were instrumental in facilitating the conference: David Glazier, John Radson, Michelle Larson, President Dave Burcham, Jerry Green, Robert Williams, Justin Connelly, Peter Warren, Tom Plate, and Robert Greenwald. And we send a special shout-out to our students: Talin Bagdassarian, Alexia Barbaro, Loana Benjamin, Rubyann Park, Chris Eaton, Jeffrey Michels, Alfredo Hernandez, and Hannah Gioia.

As for the manuscript, we are grateful to Caelyn Cobb at New York University Press for her consistent attention and encouragement throughout this process. We dedicate this book to Kerstin's grandfather, Roman, and Jennifer's brother, Nate.

LIST OF FIGURES AND TABLES

Figures

3.1 "Very Important" U.S. Foreign Policy Goals, 1991	64
3.2 "Very Important" U.S. Foreign Policy Goals, 2002	64
3.3 Attitudinal Correlates of the Iraq War, 2003–2010	71
3.4 Why Bush Went to War, 2005 and 2006	75
3.5 Public Support for Preventive War, 2003–2006	77
3.6 Partisanship and the Iraq War	78

Tables

2.1 Types of Threat and Criteria for Preventive Response	43
2.2 Preventive Action: Costs, Benefits, and Alternative Actions	47
3.1 Domestic Responses to Initiation of Two Iraq Wars	68
3.2 Expectations Regarding War Outcomes	69
3.3 When the United States Is Justified in Overthrowing Foreign Governments	70
3.4 Iraq War as "Mistake"	72
3.5 Partisanship and Preventive War	78
3.6 The Iraq War and Congressional Election Issues	79
3.7 The Iraq War and Presidential Election Issues	80
11.1 Comparison of the Criteria Governing the Use of Force	298

1

Introduction

The Preventive Force Continuum

KERSTIN FISK AND JENNIFER M. RAMOS

Preventive force is a security strategy defined along a continuum. While the scale of preventive force varies, the strategic aim is the same: to thwart the development of possible future threats from suspected ill-willed actors. At one end of the spectrum are extreme forms of preventive force, including preventive nuclear strikes and preventive wars. The 2003 war in Iraq is one such example of a preventive war: the United States went to war based on concerns that Saddam Hussein was attempting to acquire nuclear weapons, which were considered threatening to the future security of the United States. Closer to the other end of the spectrum are smaller-scale applications of preventive force, including drone strikes[1] outside of legally recognized war zones, which target and kill individuals regarded as potential security threats.[2] Although preventive targeted killings entail a relatively more limited use of force compared to war, the dramatic expansion of the drone program is consistent with the United States' enduring commitment to preventive action.

There is much to debate with regard to the security ramifications of preventive force, as well as its political, legal, and ethical implications. For instance, does using force before a threat fully materializes complicate future security? Are the perceived benefits worth the potential costs? Some worry that expanding the customary understanding of a rightful response to a potential threat paves the way for other states— allies and adversaries alike—to follow the United States' lead in using force preventively (Fisk and Ramos 2014). As noted throughout this volume, there are also increasing concerns that preventive targeted killing will not only become a trend for other states in the international system, but also will create a "slippery slope" to war.³ For example, some claim that drones in particular may make preventive attacks against non-imminent threats easier, and thus "create an escalation risk insofar as they may lower the bar to enter a conflict" ("Recommendations and Report of the Task Force on U.S. Drone Policy," 2014, 31).

We focus in particular on drones because they represent *seemingly* ideal weapons in the war on terrorism: relatively low-cost, low-risk tools with disproportionately large benefits, especially for states that face "liberal complaisance" (Schweller 1992, 243). When faced with a non-imminent but likely lethal threat, "liberal complaisance often leaves states militarily unprepared for defensive action, much less large-scale preventive action" (Schweller 1992, 244). While there are other types of limited preventive force, including special operations and cyber-attacks, we concentrate on drone strikes not only because of the timeliness of the issue, but also because of the ethical, legal, and strategic considerations these strikes embody. We believe the international community is at a turning point in what it deems acceptable regarding the use of lethal force. And this has implications not only for fighting terrorism, but also for how states deal with other types of adversaries, including other state leaders and political challengers. An estimated nine states (including the United States, Israel, Britain, Iran, and China) have developed armed drone technology as of this writing, and this number is expected to grow in the near term (New America Foundation 2015a). Compounding this trend is the recent news of the Obama administration's policy change to allow armed drones—rather than unarmed drones—to be sold to America's allies (Ryan 2015). It has therefore become increasingly important to contemplate the costs and benefits of even smaller-scale uses of preventive force in both material and normative terms, and in relation to broader effects on the overall stability of the international system.

With these considerations in mind, this volume seeks to offer a comprehensive resource that speaks to the contours of preventive force as a security strategy. Our goals are threefold. First, we seek to clarify the pros and cons of U.S. drone policy by situating drone strikes within a broader preventive force framework. Doing so recognizes a long tradition of academic scholarship on preventive motivations and helps provide the theoretical underpinnings for analyzing and contextualizing drone policy, something that is missing from the current literature on targeted killings and drones. Second, we attempt to ascertain the long-term impact of the preventive use of force on the stability and security both of states and of the international system. Third, we endeavor to build upon existing literature that debates the legality and morality of targeted killings by advancing legal and ethical guidelines for preventive action, including, in particular, a regulatory framework for the use of drones.

Previous work on preventive force has applied this security concept in a useful, but somewhat limited way, often thinking of it only in its extreme form—war. Some scholars in the realist tradition, for instance, argue that anticipated shifts in the distribution of power in the international system—real or perceived—drive the declining leader to initiate war (see, for example, Thucydides 1972; Morgenthau 1948; Gilpin 1981; Copeland 2000; Betts 1982; Levy 1987; Van Evera 1999; Ripsman and Levy 2007; Weisiger 2013). The bargaining approach conceptualizes preventive war as the result of credible commitment and information asymmetry problems (as in Fearon 1995; Powell 2006; Lake 2010) when capabilities are expected to shift (for instance, due to acquisition of weapons of mass destruction [WMD]) and affect bargains between states at a future time. According to this view, states cannot credibly commit not to use future power in their own favor. This generates the incentive to strike first in order to secure a better deal than one expects to receive down the road. In essence, states adopt "better now than later" logic to curtail a threat expected to arise at an indeterminate point in the future.

Yet not all cases of systemic power shift result in preventive war. Thus, other scholars look to the domestic level for answers. For example, Schweller (1992) argues that democracies are less likely than other regimes to engage in preventive war (or do not do so at all). The logic is straightforward: democratic publics are averse to costly wars (especially ones justified on such shaky grounds); thus the projected domestic political costs lead political elites to reason that preventive war is too large a political risk. Other researchers disagree with this assertion, however. For instance, Jack Levy (2008) investigates several cases in which he argues that democratic publics were in favor of preventive military force (Israel in 1956 and 1981, and the United States during the Gulf War, for example). While Levy concedes that democracies are certainly more constrained than non-democracies when considering the preventive war option, the recent U.S. invasion of Iraq serves to further his point that regime type is only one factor in determining whether or not a state will wage a preventive war.

In terms of changing beliefs about preventive war, scholars of U.S. foreign policy have explored why and how state leaders have shifted away from the once dominant perspective that preventive war is illegitimate, immoral, and illegal and have, in turn, adopted a more benign view of prevention (Renshon 2006; Silverstone 2007). Thomas M. Nichols, in his 2008 book, *Eve of Destruction: The Coming Age of Preventive War*, argues that the international community as a whole is in a "new age of prevention" and therefore must find ways to protect and maintain international order by acting multilaterally and recommitting to the United Nations (also see Dombrowski and Payne 2006). An array of studies has also focused on international legal justifications for preventive war (Doyle 2008) and the ethical issues surrounding preventive war in the context of just war theory (Totten 2010; Chatterjee 2013; Colonomos 2013).

Others have examined key ethical and legal questions that accompany more limited uses of preventive force, such as targeted killing, in the context of the war on terrorism (including Plaw 2008; Otto 2011; Finkelstein, Ohlin, and Altman 2012; Guiora 2013; Kaag and Kreps 2014).[4] For example, while some assert that drones offer a more "humane" weapon of war (Lewis 2013; Strawser 2013), others focus on the moral hazard they present: "To say that we *can* target individuals without incurring troop casualties does not imply that, we *ought* to" (Kaag and Kreps 2012, emphasis in original; see also Brunstetter and Braun 2011, 2013). Legal analysts debate which body of law governs drone strikes, from domestic law to international human rights law to international humanitarian law, as well as their subsequent legality (Brooks 2004; Bakircioglu 2009; Dinstein 2010; Kreimer 2003; Lubell 2010; O'Connell 2012; Solis 2010; Van Schaack 2014). In essence, these studies ask us to reconsider the traditional definitions, limits, and guidelines of war and examine how and to what extent these apply (or do not) to the use of armed drones.

Our volume is distinct from these earlier works in two primary ways. First, it merges together essays authored by scholars and practitioners from a wide variety of backgrounds, including political science, law, philosophy, human rights, and defense. Second, it gathers these distinct but related perspectives into a broader, unified theme that situates targeted killings and drone strikes within the context of preventive force, as part of a larger and longer-term shift in the manner in which force is applied. In this way, it highlights the continued U.S. policy of preventive force; the narrative of the Iraq War in 2002–2003 is still in use today, albeit elsewhere.

Historical Trends in U.S. Preventive Logic

Anticipatory logic is nothing new in U.S. foreign policy. Throughout its relatively short history, the government of the United States has consistently maintained that the anticipatory use of force is both legal and legitimate, as well as in line with customary international law.[5] The criteria for determining the legitimate anticipatory use of military action were first established in a letter from U.S. Secretary of State Daniel Webster to Lord Ashburton following a British attack on an American ship called the *Caroline* in December 1837. In the event now commonly referred to as the "*Caroline* affair," the United States charged Britain with an unjustified attack on its ship near the U.S.-Canadian border. Webster's letter to Ashburton, dated July 1842, argued that the following conditions must hold in order for the initial use of force by a state to be considered justified: the first use of force should be "(i) 'overwhelming' in its necessity; (ii) leaving 'no choice of means'; (iii) facing so imminent a threat that there is 'no moment for deliberation'; and (iv) proportional" (quoted in Doyle 2008, 12). The resulting "Caroline standard" established that anticipatory force is acceptable so long as a state demonstrates that the threat of attack is imminent; the threat must be "instant, overwhelming, leaving no choice of means, and no moment for deliberation" (Webster 1842, quoted in Moore 1906). In other words, in order to be legitimate, the first use of force must be *preemptive*—demonstrably necessary as well as proportional to the overall goal of self-defense (for a more detailed description of circumstances leading up to the event, see Glazier, Chapter 6).

In contrast, *preventive* force—which refers to the anticipatory use of force in order to eliminate a non-imminent, perceived potential threat before it has a chance to manifest—has been considered immoral and illegitimate by the United States and the broader international community for nearly a century. For instance, in line with members of his administration who attributed World War II to the preventive motivations of

Adolf Hitler, President Truman labeled preventive wars "aggressive" and "the weapon of dictators" (Truman 1950). The preventive war option was also rejected by the Eisenhower administration, as was evident when Secretary Dulles told the press, "Any idea of preventive war is wholly out of the question as far as the United States is concerned" and "is not, and as far as I can forecast, never will be any part of the United States foreign policy" (quoted in Silverstone 2007, 84). Scott Silverstone (2007) maintains that the United States fully considered the preventive attack option at several points during the Cold War[6] but ultimately discarded it based on established norms, in particular the view that preventive force was immoral and inconsistent with American ideals.

A related concern was how Americans would be perceived by the rest of the world (Silverstone 2007). These worries over ethical culpability were confirmed following Israel's 1981 preventive strike on Iraq's Osirak nuclear facility, as the United Nations Security Council unanimously rebuked the move as aggressive and in violation of international law. In fact, the United States and Iraq worked together on the Security Council resolution that condemned Israel for acting "in clear violation of the United Nations Charter and the norms of international conduct" (Silverstone 2007, 160).[7]

U.S. Preventive Logic into the Twenty-First Century

The greater the threat, the greater the risk of inaction—and the more compelling the case for taking anticipatory action to defend ourselves, even if uncertainty remains as to the time and place of the enemy's attack.
—"National Security Strategy of the United States," 2002[8]

The condition that an operational leader present an "imminent" threat of violent attack against the United States does not require the United States to have clear evidence that a specific attack on U.S. persons will take place in the immediate future.
—U.S. Department of Justice, *White Paper*[9]

Scholars have developed various explanations for the United States' preventive war in Iraq in 2003, ranging from George W. Bush's personality

(Renshon 2006) to a "neoconservative moment" (Schmidt and Williams 2008, 194). For some, the turning point that paved the way for the preventive war came earlier. Silverstone (2007), for instance, maintains that heightened fear over nuclear proliferation in the 1990s provided the context for a normative shift favoring preventive force. The threat of nuclear proliferation to "rogue" states, including North Korea and Iraq, led the Clinton administration to seriously contemplate preventive strikes on nuclear facilities.

From this perspective, the preventive war in Iraq was not so much the result of a sudden, post-9/11 shift in U.S. views on the permissibility of preventive military force as some argue,[10] but rather the outgrowth of an "anti-proliferation imperative" for rogue regimes established a decade earlier. Unlike other scholars who looked to particular characteristics of the Bush foreign policy team (such as Moore and Slater 2004) or the psychological traits of Bush himself (Renshon 2006) in order to explain why the United States went to war with Iraq, Silverstone's logic predicted that U.S. proclivity to use preventive force would not likely disappear with the Bush administration. Silverstone's explanation instead suggested that as long as "rogues" attempt to acquire weapons of mass destruction, the United States would continue to embrace the logic of preventive military force.

Indeed, in our view, the preventive motive for war with Iraq should not be viewed in a vacuum or as a historical anomaly. Although he famously campaigned on the folly of the "Bush Doctrine" during his 2008 run for the presidency, President Obama has not sought to restore the qualitative distinctions between preemptive and preventive force. Instead, preventive drone strikes have become the tactical cornerstone of the Obama administration's security strategy—and routinely so in cases less severe than what Silverstone's "anti-proliferation imperative" would require.

Perhaps the clearest indicator of the Obama administration's acceptance of preventive force logic is its justification for expanding the U.S. drone program, in terms of both the overall number of strikes and the geographical scope of strikes. According to the New America Foundation's database on drone strikes in Pakistan and Yemen, there were forty-nine drone strikes during the Bush presidency, with only one strike occurring in Yemen in 2002. Under President Obama (from January 2009 to December 2014), the United States carried out 441 total drone strikes inside Pakistan (343) and Yemen (98). In June 2011, President Obama also expanded drone strikes into Somalia, which took place on ten separate occasions between 2011 and 2014 according to the Bureau of Investigative Journalism (2015).

The expansion of the drone program becomes even more pronounced when we consider the array of groups now targeted in the strikes. The Bush administration reportedly carried out strikes against three organizations in Pakistan, including al-Qaeda, the Taliban, and the Haqqani network. To these three previously targeted organizations, the Obama administration added the Islamic Movement of Uzbekistan (IMU) operating in Pakistan, al-Qaeda in the Arabian Peninsula (AQAP) in Yemen, Ansar al-Sharia in Yemen, and al-Shabaab in Somalia. Thus, the Obama administration further shifted the focus away from the al-Qaeda core in the years between 2009 and 2014. While the New America Foundation recorded 31 strikes on the al-Qaeda core under Obama during this period, it found that the majority targeted a combination of Taliban insurgents (107 strikes), the Haqqani network (33 strikes), and AQAP (approximately 84 strikes) (New America Foundation 2015b, 2015c).

An additional, important development coinciding with the increase in drone strikes under the Obama administration is reduced transparency surrounding *who* was actually targeted in the strikes (Shane 2015). According to New America Foundation data, sixteen of President Bush's drone strikes in Pakistan were against "unclear" targets, meaning that although major news outlets reported the occurrence of a drone strike, the target of the strike was not identified in the reports. In contrast, during the first six years of the Obama presidency, there were 153 strikes in Pakistan targeting individuals or organizations whose identities were reportedly not clear. Thus, when looking at the percentage of all drone strikes carried out in Pakistan under each administration, approximately 33 percent of Bush's targets were unclear in news reports, compared to 45 percent of Obama's targets (New America Foundation 2015b).

While the number and scope of targets pursued in accordance with the Bush-initiated "war on terror" have substantially increased under the Obama administration, the legal challenges inherent in applying the 2001 Authorization to Use Military Force (AUMF) to justify the ongoing war on terror are also growing as the number of years since 9/11 pass. For instance, in 2013, Representative Adam Schiff (D-California) introduced an amendment that proposes to "sunset" or repeal the AUMF, arguing, "The 2001 AUMF was never intended to authorize a war without end, and it now poorly defines those who pose a threat to our country. That authority, and the funding that goes along with it should expire concurrent with the end of our combat role in Afghanistan" ("Rep. Schiff Offers Amendment Sunsetting Authorization for Use of Military Force" 2013). Former White House Legal Adviser Harold Koh nevertheless maintains that repealing the AUMF would not undermine the legal authority of the president to continue to carry out targeted killings against al-Qaeda and its affiliates. Instead, the position of the Obama administration has been that targeted killings against al-Qaeda and the above "affiliates," in reaction to the events of September 11, 2001, are legitimate under the law of self-defense as outlined in Article 51 of the UN Charter. If and when the AUMF is eventually repealed, Koh argues there will be no "legal gap" in carrying out targeted killings since the legal authority derives from the right to self-defense against a "continuing and imminent threat" under international law (Koh 2014).

We argue, however, that "imminence" does not mean the same thing today as it did pre-9/11 or even pre-2008. The Obama administration has developed a significantly revised standard of imminence in line with its asserted authority (under the law of self-defense) to carry out targeted killings against ambiguous threats. John Brennan, the president's counterterrorism advisor turned CIA director, maintains that the imminence standard must be further relaxed given terrorists' "modern-day capabilities, techniques, and technological innovations" (Brennan 2011). These assertions clearly echo those of the Bush administration's 2002 National Security Strategy, which stated that the United States "must adapt the concept of imminent threat to the capabilities and objectives of today's adversaries" (Bush Administration 2002, 15). This standard is, in turn, used to justify the Obama administration's preventive, "signature"[11] drone strikes against suspected militants whose identities often are not confirmed before they are killed.[12] Targets of signature strikes are hit because their "patterns of life," such as proximity to known terrorists, signal a latent threat—they *might be* terrorists planning attacks against the United States or its interests abroad.

A related example of the shifting meaning of imminence is the targeting of low-level fighters. Peter Bergen of the New America Foundation estimates that the vast majority–approximately 94 percent—of militants killed prior to 2011 in Pakistan were low-ranking fighters, or "foot soldiers" (Bergen 2011). In 2013, senior U.S. officials admitted to conducting "dozens" of signature strikes in Yemen in which "none of the about three dozen militants killed . . . were 'household names'" or targets they considered high-level threats. Their logic was that "they might not be big names now . . . but these were the guys that would have been future leaders" (Schmitt 2013).

Critically, drone strikes outside of designated combat zones that target individuals based on speculation about suspicious behavior, as well as those that target low-level militants who "do not have the capacity to plot effectively against the United States," do not meet imminence criteria as conceived under customary international law (Bergen 2013, 4). The revised concept of imminence purportedly legitimizes the killing of a suspected militant who might do something harmful, someday. It requires neither substantiated evidence nor immediacy of the threat. The logic behind attacks on these potential threats is thus preventive logic.

As the U.S. government continues down its current path of relaxing the imminence standard and embracing preventive force to protect against unconventional threats including WMD and terrorism, Americans appear to be supportive. Recent polls indicate that approximately 58% approve of drone operations, with Republicans (74%) and Independents (56%) slightly more supportive than Democrats (52%) (Pew Research Center 2015).[13] Less than one-third are very concerned about the legality of or retaliation for drone strikes (31% and 29%, respectively (Pew Research Center 2015).[14] Thus, American public opinion does not reflect the very tense policy and scholarly debates surrounding preventive force.

Preventive Force Dilemmas

By conceptualizing preventive force along a continuum, with the scale of force employed along this continuum including preventive war and preventive uses of force short of war, we can see how U.S. strategy post-2001 contains the same underlying logic. This cross-continuum logic, formally incorporated into U.S. security strategy in 2002, is based on the asserted nature of two actor categories: rogue states and terrorists.

The rhetorical equivalence of the two was evident in the 2002 National Security Strategy, which asserted that "rogue states and terrorists do not seek to attack us using conventional means. They know such attacks would fail. Instead, they rely on acts of terror and, potentially, the use of weapons of mass destruction" (Bush Administration 2002, 15). President Bush's 2003 State of the Union address further communicated the perceived urgency: "Today, the gravest danger in the war on terror, the gravest danger facing America and the world, is outlaw regimes that seek and possess nuclear, chemical, and biological weapons. These regimes could use such weapons for blackmail, terror, and mass murder. They could also give or sell these weapons to terrorist allies, who would use them without the least hesitation." Threats from either of these actors are considered unpredictable and irrational, and assessed as being so grave that they must be treated as imminent. Purportedly, both deterrence and containment would be inadequate. The conceptual conflation of rogue states and terrorist organizations helped to set the stage for preventive force as a dual counterterrorism and counterproliferation strategy. It is not surprising, then, that the same strategic, legal, and ethical arguments are made across the preventive force spectrum.

Preventive Force: Strategic Controversies

Strategic arguments for and against the 2003 Iraq War as well as ongoing preventive drone strikes outside of declared theaters of war are very similar, because the underlying logic for counter-proliferation and counterterrorism is nearly identical. Proponents of the Iraq War and proponents of U.S. drone policy, while they may not overlap, both contend that the nature of today's threats requires a revised imminence standard. Since these threats cannot be deterred, "wait and see" strategies are untenable. Supporters of the preventive war in Iraq argued that Saddam Hussein was untrustworthy, irrational, and a risk-taker bent on one day attacking the United States. The United States had to take action because, in Condoleezza Rice's words, "We don't want the smoking gun to become a mushroom cloud" (Peterson 2001). The perceived inadequacy of deterrence is further pronounced in reference to non-state actors. For instance, Mark Totten argues that terrorists are risk-takers who have nothing to lose; they seek to cause as much devastation as possible, they have the capability to enact this devastation, and they are difficult to locate and identify. Waiting for evidence that fulfills imminence criteria thus "becomes a demand to take the first blow" (2010, 30–31, 32). For Totten, the global threat of terrorism "demands a shift toward strategies of prevention"; legitimizing anticipatory self-defense based on the imminence standard "asks too much in an age of al Qaeda" (2010, 32, 34). Russell Christopher similarly contends that the imminence requirement for legitimate self-defense "unduly distorts and restricts an actor's moral right to effective self-defense when necessary" (2012, 284).[15]

While many focus on the strategic benefits of relaxing the imminence standard to authorize the preventive use of force, others maintain that there are a number of negative strategic consequences that will result from erasing the distinction between preemption and prevention. For instance, the late U.S. Senator Edward Kennedy argued that the 2002 National Security Strategy illegitimately lowered the threshold for what is considered the rightful use of force in self-defense, thereby threatening future U.S. security. Kennedy demanded that we consider "the consequences of a policy that produces more preventive wars" (2006, 69). Similar arguments regarding the creation of a slippery slope toward wider conflict are apparent in the counterterrorism context as well. For instance, David Whetham contends that "'he [the terrorist] will plant the IED [improvised explosive device] in the morning' leads to 'He will give a lift to the UAV operator at the weekend' leads to 'He is planning something in the next couple of years' leads to 'He did it once, we're sure he'll do it again'" (2013, 78). At the same time, others argue that drone attacks work at cross-purposes with long-term U.S. foreign policy goals. Just as the United States is investing resources in stabilizing the governments of Pakistan and Somalia, it is also delegitimizing those governments with unilateral American drone strikes on their territory (Boyle 2013).

Critics also contend that preventive force risks alienating friends and allies, resulting in an overall loss of international support. John Lewis Gaddis asserts that the Iraq War "provoked complaints that great power was being wielded without great responsibility, followed by an unprecedented collapse of support for the United States abroad. From nearly universal sympathy in the weeks after September 11, Americans within a year and a half found their country widely regarded as an international pariah" (2005, 6). Concerns over reputational costs have echoed throughout the Obama administration, although now primarily in reference to drone strikes. According to a June 2014 report from a bipartisan task force on U.S. drone policy:

> UAV [unmanned aerial vehicle] strikes by the United States have also generated a backlash in countries not directly affected by the strikes, in part due to the perception that such strikes cause excessive civilian deaths, and in part due to concerns about sovereignty, transparency, accountability and other human rights and rule of law issues. . . . In February 2014, for instance, the European Parliament voted 534–49 for a resolution condemning US drone strikes. . . . National officials, parliamentarians and thought leaders in numerous allied countries and at the United Nations have questioned or condemned US targeted strikes. ("Recommendations and Report of the Task Force on U.S. Drone Policy," 30)

In fact, in all but three of the forty-four countries surveyed by the Pew Research Center in 2014 (the United States, Israel, and Kenya), the vast majority of citizens said they disapprove of U.S. drone strikes targeting extremists in Pakistan, Yemen, and Somalia. This includes majority opposition from the publics of Russia (78%) and Venezuela (92%) as well as traditional security allies like South Korea (75%), the United Kingdom (59%), and Germany (67%) (Pew Research Center 2014). Such public condemnation of U.S. drone policy could undermine U.S. cooperation with international partners, including access to valuable intelligence and other forms of assistance.

A third argument concerning strategic costs is that this U.S. strategy is doing more to heighten the threat of terrorism than to reduce it. In 2003, Jessica Stern argued that the Iraq War "has taken a country that was not a terrorist threat and turned it into one" and created "precisely the situation the administration has described as a breeding ground for terrorists" (Stern 2003). Similarly, with regard to drone strikes in Pakistan and Yemen, Michael Boyle maintains that the strikes can increase the threat from terrorism because they "corrode the stability and legitimacy of local governments, deepen anti-American sentiment and create new recruits for Islamist networks aiming to overthrow these governments" (2013, 3). Boyle points to evidence that the ranks of the Pakistani Taliban, or TTP, have grown substantially in Pakistan and that membership in al-Qaeda in the Arabian Peninsula, or AQAP, has grown substantially since drone strikes began in Yemen.

Empirical evaluation of the effects of drone strikes on terrorism in Yemen, Somalia, and Pakistan is currently limited, and the findings that do exist are mixed. For instance, Patrick Johnston and Anoop Sarbahi (2015) note that drone strikes in Pakistan are associated with a short-term reduction in the rate and lethality of terror attacks both in and immediately surrounding the areas where the strikes take place. However, a study carried out by Megan Smith and James Igoe Walsh uncovers no evidence that drone strikes have impaired al-Qaeda's ability to create and spread propaganda and argue that this effect points to the organization's "resilience and activity" (2013, 324). Moreover, to our knowledge, previous studies have not specifically examined the effects of targeting *low-level* operatives through signature strikes. Doing so is key because, as Leila Hudson, Colin Owens, and David Callen (2012) note, the population is more likely to oppose strikes against those who may be unaffiliated with terror groups, even if they are generally unopposed to strikes that target known leaders of terrorist organizations, or high-value targets.

Finally, critics of preventive force fear that the United States is setting a precedent for other states, and that these states will assert their own right to use force on preventive grounds to eliminate potential threats. For example, following the initiation of the Iraq War, Kegley and Raymond (2003) argued that other states would be emboldened to follow the lead of the United States and thus would increasingly wage preventive wars to deal with those considered threats. Likewise, with regard to U.S. drone policy, a predominant concern is that "the U.S. use of force in sovereign nations whose consent is questionable or nonexistent may encourage other states to follow suit with their own military platforms" and that therefore these states "may soon find themselves tempted to deploy lethal UAVs in similar fashion" (Abizaid and Brooks 2014, 10, 31).[16]

Indeed, in our previous work (Fisk and Ramos 2014), in which we examined the adoption of preventive force logic in the cases of India, Russia, China, and Germany in the wake of the 2003 Iraq War, we found that political and military elites in these countries referenced the "successful" U.S. example when discussing offensive shifts in their military strategies, as well as when emphasizing UAV, or drone, acquisition. We concluded that a "preventive force norm" has begun to cascade through the international system as these countries witness preventive uses of force by the United States in its prosecution of the war on terror and begin to shift their own strategies in this direction. Yet just because a behavior is gaining popularity or becoming part of the normative fabric of the international system does not mean that it is beyond ethical reproach.

Ethical Considerations

There are several approaches to thinking about the ethics of preventive force. We begin with two: deontological ethics and consequentialist ethics. In the context of preventive force, a deontological perspective focuses on whether or not "striking first" itself is ethical, whereas a consequentialist view centers on evaluating if the benefits of "striking first" outweigh the drawbacks. Following this discussion, we consider an ethical approach that was specifically formulated for matters of war: the just war tradition.

From a deontological perspective, the good or the bad of preventive force hinges on its intrinsic value. Whitley Kaufman, for instance, argues that "preventive war is legitimate for the very same reason that wars of self-defense are legitimate" (2005, 28). Therefore, within limits that guard against misuse of preventive war (for instance, convincing evidence of threat must be shown), states have a right to protect themselves from harm. Not doing so, from a leader's perspective, would be both irresponsible and a dereliction of duty (Kaufman 2005). This justification certainly aligns with two recent examples: Bush's Iraq War and Obama's "drone wars" (as an extension of Bush policy). Presumably, both leaders not only thought they had a legal right to pursue U.S. security in this way, but also thought that they were ethically justified in doing so.

On the other hand, there are plenty of observers who cite the decision to invade Iraq as unethical in that there was no immediate threat that required such extreme action. Indeed, it was difficult to convince the UN Security Council that there was sufficient evidence to demonstrate that Iraq possessed WMD. There was simply not enough indication that invading Iraq was justified. In a similar vein, critics could cite drone attacks directed at low-level militants, who themselves carry only a "suspected" label, to object to a preventive strike policy.

A similar controversy surrounding preventive force arises from the consequentialist argument, although in this case it centers on whether or not the means are worth the ends. That is, even if we have to sully ourselves, is doing so worth such minor costs if we can achieve benefits for the greater good? In thinking about preventive force, the consideration in this view is whether the result—not the action—is ethical. Some might argue that preventive war is not ethical because it could increase instability in the international system, in part because the misuse of preventive war will increase the overall number of wars and thus create more death and destruction (Lee 2005).

One could make the same argument about drone strikes outside of recognized theaters of war (many of which are part of a broader strategy of preventive force, given that they target suspected militants that might pose a threat), even though they tend to be individually less destructive due to their smaller scale. Because drone strikes offer an arguably cost-effective (in terms of both human and material resources) alternative to war, increased reliance on this foreign policy tool may destabilize the international system by encouraging more states to acquire and use weaponized drones without the robust checks and ethical restraints that typically govern the use of force. The point here is not to compare the ethical choice between preventive drone strikes and preventive war, but to highlight their similar ethical considerations (though clearly allowing for context).

Those arguing for preventive force from a consequentialist perspective might, on the other hand, point to the fact that it is better to strike now because doing so will save more lives than if one waited for the attack (Luban 2004; Yoo 2004; McMahan 1996). Certainly, one could argue that a preventive war against rising Germany in the 1930s may have saved many more lives than appeasement. The argument for signature drone strikes runs parallel: kill the suspected terrorists to save those in harm's way, even if the danger is not immediate or certain.

We thus far have reviewed two general ethical approaches and demonstrated how they might be used to consider the ethics of preventive force. We now turn to the just war tradition, which identifies general criteria by which a war can be judged to be "just" and has heavily influenced the development of modern international law by narrowing the lawful criteria for legitimately waging and executing war.[17] Just war principles, including target discrimination and proportionality, therefore cover not only a state's conduct during war, but also when it is "just" to go to war. The specific application of just war criteria is not without controversy, however.

While the invocation of just war principles pervades the rhetoric of recent U.S. administrations, philosophers and theorists differ on their interpretations of just war. Some assert that the just war tradition absolutely forbids the preventive use of force (Wills 2004; Rengger 2008). As Alberico Gentili argues, "A just cause for fear is demanded; suspicion is not enough" ([1612] 2003, 62). Yet others propose that such action is possible under certain conditions (Tuck 1999; Barnes 1982; Gray 2007). For example, in *On the Law of War and Peace* Hugo Grotius ([1625] 2010) argues that a preventive attack is permissible when there is "moral certainty" of a future enemy attack.

Though the debate on the permissibility of just war is yet to be settled, the advent of armed drones has pushed scholars to think about adaptations of the just war paradigm in the context of preventive use of force short of war. While Michael Walzer (2006), perhaps the most well-known modern just war theorist, argues that lethal force is morally wrong, he does concede that lethal force short of war is less troubling than war. Current debates focus on *jus ad vim* (the just use of force) and whether or not its application to drone strikes adequately reflects traditional just war principles. Since technology in the form of drones, for example, lowers the risk to the attacker, some believe it contributes to the lowering of barriers to the use of force (Lin 2010), and thus the application of *jus ad vim* could potentially be a convenient first choice, rather than last (Coady 2008).

As the above discussion makes clear, contemplating preventive force involves searching for the answers to a number of complex legal, strategic, and ethical questions. Formulating effective foreign and security policies entails thinking deeply and comprehensively about these questions. We thus believe that it is critical to engage in a fuller and more careful examination of preventive force, and this is what our volume sets out to achieve.

Organization of the Volume

The chapters in this volume are organized in three parts: "Understanding Preventive War"; "Perspectives on Preventive Drone Strikes"; and "The Future of Preventive Force." Within each of these parts, each chapter responds to one or more of the following key questions:

- In terms of practical considerations, what are the costs and benefits of employing preventive force?
- What are the political causes and consequences of using preventive force?
- What are the legal and ethical implications of these actions?

Part I focuses on preventive force in general as well as on the extreme case—preventive war. It begins within this context in order to elucidate the pros and cons of preventive force in a relatively clear, albeit extreme, situation. Here the authors deliberate the costs and benefits of preventive war in achieving the security objectives of a state. These analyses include not only the external strategic calculations, but also the domestic political considerations that leaders necessarily must consider when engaging in preventive war.

Part I begins with Jennifer Taw's chapter on "Preventive Force: The Logic of Costs and Benefits" (Chapter 2), in which she considers the types of threats and conditions that might warrant a state's use of preventive operations. Her cost-benefit analysis demonstrates that preventive action is only rarely worth the cost, given the alternatives, though it may be preferred in limited scenarios (as in the American drone program).

One of the key cost calculations, of course, in considering any use of military force, including preventive force, is casualties. In Chapter 3, "Preventive War and Its Domestic Politics," Miroslav Nincic focuses on the importance of projected casualties for public support of preventive war and demonstrates that public opinion restrains leaders in their decisions to wage preventive war more than it does in other types of war scenarios. Concurring with Taw, he suggests that lesser-scale uses of preventive force may be more likely to be initiated if the benefits are perceived as sufficiently high and few to no casualties are expected.

Nincic's chapter is followed by Thomas M. Nichols' "Sovereignty and Preventive War in the Twenty-First Century: A Retrospective on

Eve of Destruction: The Coming Age of Preventive War," a reflection by the author on his book of 2008. Whereas both Taw and Nincic maintain that preventive wars will continue to be rare, Nichols argues that preventive wars are likely to become more frequent due to the post–Cold War rise of global threats (humanitarian disasters, rogue states and terror organizations, and weapons of mass destruction) and the related decline of the sovereignty norm. Drawing on recent cases in Libya, Syria, Georgia, and Ukraine, Nichols maintains that the norm of preventive force is gaining ground in the international system, with foreseeably dangerous consequences.

Together, these three chapters provide insight into states' cost-benefit calculations when considering preventive military action, and in particular urge us to consider how drones might alter these calculations, the subject of the next section. In Part II, "Perspectives on Preventive Drone Strikes," the contributors reflect on the preventive use of force short of war. These chapters offer contrasting views about the impacts of drone strikes on civilian populations as well as the extent to which current drone policy operates within the constraints of established legal principles.

Stephan Sonnenberg leads this section with "Why Drones Are Different" (Chapter 5). He argues that drones differ qualitatively from other weapons and have had revolutionizing effects on warfare in the twenty-first century.[18] Consequently, Sonnenberg argues that drones require a new regulatory framework under the law of armed conflict. Based on his research experience analyzing the impact of preventive drone strikes on civilians in Pakistan, he suggests some steps forward in the global regulation of drones.

In contrast, David Glazier asserts that current domestic and international laws are sufficient; it is the policy that is lacking. In Chapter 6, "The Drone: It's in the Way That You Use It," Glazier acknowledges that drones have many positive attributes, including, theoretically, the ability to better distinguish between legitimate military targets and civilians than piloted aircraft. Whereas he agrees with Sonnenberg that most U.S. preventive drone attacks are legally problematic, Glazier argues that no new regulations are needed; drones can be advantageous tools in the war on terrorism so long as their use complies with existing domestic and international humanitarian laws.

In Chapter 7, "Drones and the Law: Why We Don't Need a New Legal Framework for Targeted Killing," Daphne Eviatar continues with this line of argument, concurring with Glazier that drone strikes can be useful counterterrorism tools if the requirements of international humanitarian law are met. Eviatar argues, however, that drone strikes on some groups, including the Taliban in Pakistan and AQAP in Yemen, do not meet the criteria for noninternational armed conflict (NIAC), and thus should be regulated under International Human Rights Law (IHRL). This requires the United States to recognize, in line with most other countries in the international system, that IHRL is binding beyond U.S. borders.

Part II concludes with C. Christine Fair's reminder of the necessity to be critical consumers of information. In her chapter "Studying Drones: The Low Quality Information Environment of Pakistan's Tribal Areas," Fair asserts that much of the critique of the U.S. drone program in Pakistan relies upon erroneous assumptions about the program, whom it targets, and with what outcomes, as well as the degree to which the Pakistani government has been complicit in the program. Fair problematizes the recent advocacy-driven research on the drone program, paying particular attention to the methods of data collection and analysis employed by various authors, and offers suggestions as to how organizations can better conduct empirical work in this key area.

The final part on "The Future of Preventive Force" aims to draw out ethical guidelines for a legitimate policy on the use of preventive force, with an eye toward targeted killing in particular. Given that targeted killing may become a more common tool of states' foreign policies, with preventive war still an exception (though see Nichols's earlier chapter), these chapters address how states might satisfy ethical imperatives while thwarting legitimate security threats. While several of the chapters in this section address this in terms of guidelines concerning the use of force short of war, the last chapter provocatively shifts the focus from preventive force to preventive nonintervention as the ultimate means to achieve states' security objectives.

Part III begins with Avery Plaw and João Franco Reis, who consider the basis upon which the legitimate use of force can be crafted (Chapter 9). In "The Contemporary Practice of Self-Defense: Evolving Toward the Use of Preemptive or Preventive Force?" Plaw and Reis argue for a law, or a legal interpretation of current law, that permits limited, episodic exercises of force by states in response to sustained patterns of attack by non-state actors. This approach, deemed "Caroline plus" by the authors, will allow states the necessary flexibility to confront immediate threats while avoiding costs associated with military adventures designed to eliminate future threats before they manifest.

"Restricting the Preventive Use of Force: Drones, the Struggle against Non-State Actors, and *Jus ad Vim*" (Chapter 10) considers the limited use of preventive force and outlines guidelines by which such force could be used. There, John Emery and Daniel Brunstetter offer an ethical framework that combines the standards of law enforcement—that is, indictment (in abstention) of suspected terrorists—with a circumscribed notion of imminence to define the rare circumstances under which the preventive use of lethal drones may be legitimate. Furthermore, the authors argue for the need to restrict the number of strikes to a minimum of isolated acts of preventive force, instead of pursuing the dangerous policy of a broader (preventive) drone war.

Ben Jones and John M. Parrish contribute a philosophical inquiry regarding "Drones and Dirty Hands," noting that the "War on Terror" has prompted a revival of interest in the idea of moral dilemmas and the problem of "dirty hands" in public life. They question the use of the "dirty hands" moral dilemma to characterize U.S. tactics in the war against al-Qaeda and affiliates, arguing that targeted killing should not be considered within this framework, in which the least evil choice is deemed both morally required and yet morally forbidden. Instead, Jones and Parrish maintain that such policies, if justified at all, must ordinarily be justified under the more exacting standards of just war theory and its provisions for justified killing—in particular the requirement that (with limited and defined exceptions) noncombatants be immune from intentional violence.

This part then concludes with a thought-provoking piece by Deen Chatterjee. "Beyond Preventive Force: Just Peace as Preventive Non-Intervention" shines a reflective light on all the previous chapters by investigating an alternative approach to preventive self-defense. Given the moral hazards of military responses to today's threats and consequences for international law, justice, and human rights norms, Chatterjee proposes "preventive nonintervention" to secure the peace, thereby moving away from the just war approach and instead prioritizing "just peace." He asserts that "just peace" as preventive nonintervention is better suited for today's global challenges than preventive self-defense.

We conclude the volume with a brief chapter that reviews the contributors' collective wisdom on preventive force. Our discussion is guided by the three main, and interrelated, questions we asked the authors to address. These questions stem from what we see as the key issues in preventive action: political (including costs and benefits), legal, and ethical. We then consider the future of preventive force as a security strategy.

In its entirety, this volume seeks to extend our knowledge and understanding of the preventive use of force from preventive war to preventive drone strikes. As noted earlier, the proliferation of drone technology and its continuing technological advances signify the indisputable staying power of drones. Yet drones are only one tool of preventive self-defense strategy. This is why a discussion of drones and drone strikes cannot take place within a vacuum; it should be contextualized within the broader policy spectrum of preventive force.

NOTES

1 We use the term "drone" here as it is widely adopted; however, we fully recognize that there are strongly preferred alternatives, including "remotely piloted aircraft" (RPA) and "unmanned aerial vehicle" (UAV).
2 The far end of the spectrum represents nonlethal forms of preventive force used abroad, including, for instance, detentions, raids and abductions, and attacks on infrastructure. For a more thorough description of the types of actions across the continuum, see Sofaer (2005).
3 For instance, a bipartisan task force on U.S. drone policy is concerned over "a slippery slope leading to continual or wider wars" ("Recommendations and Report of the Task Force on U.S. Drone Policy," 10–11).
4 The groundbreaking book *Wired for War* by Peter W. Singer (2009) and in-depth investigative reports from the *New Yorker*'s Jane Mayer and the *New York Times*'s Mark Mazzetti have likewise fueled popular debate on these issues.
5 As is detailed in later chapters in this volume (see especially Chapters 6 and 9), Article 51 of the UN Charter later established more restrictive criteria for the first use of force. It holds that states have an inherent right to use force in self-defense without first obtaining Security Council authorization if an armed attack has already taken place. Nevertheless, the United States maintained the traditional *Caroline* criteria for determining rightful self-defense as embodied in customary law.

6   A similar argument is advanced by Marc Trachtenberg (2007), who maintains that "preventive war thinking" was influential in U.S. foreign policy as early as 1941.
7   For more on the historical invocation of imminence criteria to denounce or justify the anticipatory use of force, see Totten (2010).
8   Bush Administration, 2002, "The National Security Strategy of the United States of America" (Washington, DC: The White House, September), http://nssarchive.us/NSSR/2002.pdf.
9   U.S. Department of Justice, undated, *White Paper: Lawfulness of a Lethal Operation Directed against a U.S. Citizen Who Is a Senior Operational Leader of Al-Qa'ida or an Associated Force* (Washington, DC: U.S. Department of Justice). http://msnbcmedia.msn.com/i/msnbc/sections/news/020413_DOJ_White_Paper.pdf.
10 For instance, Totten argues, "After the attacks of 9/11, the nation that birthed Webster's Rule suddenly disowned it" (2010, 29).
11 Signature strikes in Yemen have reportedly been given a new name by U.S. officials; the U.S. government now refers to these strikes on targets whose identity is unknown as "TADS," or terrorist attack disruption strikes (Becker and Shane 2012).
12 Anonymous sources have indicated that "many, or even most" drone strikes have killed "unidentified" individuals (Currier and Elliot 2013).
13 Not surprisingly, there is a 17 percentage-point gender gap, with drone support rates at 50% for women and 67% for men; there is also a racial gap in drone support with whites (66%) much more supportive than non-whites (blacks, 46%; Hispanics 39%) (Pew Research Center 2015).
14 However, 48% of Americans are very concerned about killing innocent civilians (Pew Research Center 2015).
15 Freedman (2010) provides a similar argument.
16 See also Fisk and Ramos (2014).
17 In the Christian tradition, St. Augustine and Thomas Aquinas are often thought of as the forefathers of the just war tradition. However, and though the criteria may vary, the idea that a war can be fought in a just and honorable manner exists in virtually all faith and non-faith traditions.
18 For an alternative view, see Davis et al. 2014.

BIBLIOGRAPHY

Abizaid, General John P., and Rosa Brooks. 2014. "Recommendations and Report of the Task Force on U.S. Drone Policy." Washington, DC: Stimson Center, June. http://www.stimson.org/images/uploads/ task_force_report_final_web_062414.pdf.

Bakircioglu, Onder. 2009. "The Future of the Preventive Use of Force: The Case of Iraq." *Third World Quarterly* 30 (7): 1297–1316.

Barnes, Jonathan. 1982. "The Just War." In *The Cambridge History of Later Medieval Philosophy*, edited by Norman Kretzmann, Anthony Kenny, and Jan Pinborg. Cambridge, UK: Cambridge University Press, 771–784.

Becker, Jo, and Scott Shane. 2012. "Secret 'Kill List' Proves a Test of Obama's Principles and Will." *New York Times*, May 29. http://www.nytimes.com/2012/05/29/world/obamas-leadership-in-war-on-al-qaeda.html?pagewanted=all.

Bergen, Peter. 2011. "Increased U.S. Drone Strikes in Pakistan Killing Few High-Value Militants." *Washington Post*, February 21. http://www.washingtonpost.com/wp-dyn/content/article/2011/02/20/AR2011022002975.html.

Bergen, Peter. 2013. "Drone Wars: The Constitutional and Counterterrorism Implications of Targeted Killing." Testimony presented before Congress, April 23. http://newamerica.net/sites/newamerica.net/files/articles/TESTIMONY_BERGEN_DRONES.pdf.

Betts, Richard. 1982. *Surprise Attack*. Washington, DC: Brookings.

Boyle, Michael J. 2013. "The Costs and Consequences of Drone Warfare." *International Affairs* 89 (1): 1–29.

Brennan, John O. 2011. "Remarks of John O. Brennan, 'Strengthening Our Security by Adhering to Our Values and Laws.'" Washington, DC: White House Office of the Press Secretary, September 16. https://www.whitehouse.gov/the-press-office/2011/09/16/remarks-john-o-brennan-strengthening-our-security-adhering-our-values-an.

Brooks, Rosa Ehrenreich. 2004. "War Everywhere: Rights, National Security Law, and the Law of Armed Conflict in the Age of Terror." *University of Pennsylvania Law Review* 153 (2): 675–761.

Brunstetter, Daniel, and Megan Braun. 2011. "The Implications of Drones on the Just War Tradition." *Ethics & International Affairs* 25 (3): 337–358.

Brunstetter, Daniel, and Megan Braun. 2013. "From *Jus ad Bellum* to *Jus ad Vim*: Recalibrating Our Understanding of the Moral Use of Force." *Ethics & International Affairs* 27 (1): 87–106.

Buchanan, Allen, and Robert O. Keohane. 2004. "The Preventive Use of Force: A Cosmopolitan Institutional Proposal." *Ethics & International Affairs* 18 (1): 1–22.

Bureau of Investigative Journalism. 2015. "Somalia: Reported U.S. Covert Actions, 2001–2015." http://www.thebureauinvestigates.com/2012/02/22/get-the-data-somalias-hidden-war/.

Bush Administration. 2002. "The National Security Strategy of the United States of America." Washington, DC: The White House. September. http://nssarchive.us/NSSR/2002.pdf.

Bush, George W. 2003. "Text of President Bush's 2003 State of the Union Address." *Washington Post*, January 28. http://www.washingtonpost.com/wp-srv/onpolitics/transcripts/bushtext_012803.html.

Christopher, Russell. 2012. "Imminence in Justified Targeted Killing." In *Targeted Killings: Law and Morality in an Asymmetrical World*, edited by Claire Finkelstein, Jens David Ohlin, and Andrew Altman, 253–284. Oxford, UK: Oxford University Press.

Coady, C. A. J. 2008. *Morality and Political Violence*. New York: Cambridge University Press.

Colonomos, Ariel. 2013. *The Gamble of War: Is It Possible to Justify Preventive War?* New York: Palgrave Macmillan.

Copeland, Dale C. 2000. *The Origins of Major War*. Ithaca, NY: Cornell University Press.

Currier, Cora, and Justin Elliot. 2013. "The Drone War Doctrine We Still Know Nothing About." *ProPublica*, February 26. http://www.propublica.org/article/drone-war-doctrine-we-know-nothing-about.

Daalder, Ivo H., and James M. Lindsay. 2003. *America Unbound: The Bush Revolution in Foreign Policy*. Hoboken, NJ: Wiley.

Davis, Lynn E., Michael J. McNerney, James S. Chow, Thomas Hamilton, Sarah Harting, and Daniel Byman. 2014. "Armed and Dangerous? UAVs and U.S. Security." Santa Monica, CA: RAND Corporation. http://www.rand.org/pubs/research_reports/RR449.html.

Dinstein, Yoram. 2010. *The Conduct of Hostilities under the Law of International Armed Conflict*. New York: Cambridge University Press.

Doyle, Michael W. 2008. *Striking First: Preemption and Prevention in International Conflict*. Edited, with an introduction by Stephen Macedo. Princeton, NJ: Princeton University Press.

Enemark, Christian. 2013. *Armed Drones and the Ethics of War: Military Virtue in a Post-Heroic Age*. New York: Routledge.

Fearon, James D. 1995. "Rationalist Explanations for War." *International Organization* 49 (3): 379–414.

Finkelstein, Claire, Jens David Ohlin, and Andrew Altman, eds. 2012. *Targeted Killings: Law and Morality in an Asymmetrical World*. Oxford, UK: Oxford University Press.

Fisk, Kerstin, and Jennifer M. Ramos. 2014. "Actions Speak Louder than Words: Preventive Self-Defense as a Cascading Norm." *International Studies Perspectives* 15 (2): 163–185.

Ford, Brandt S., 2013. "*Jus Ad Vim* and the Just Use of Lethal Force-Short-of-War." In *Routledge Handbook of Ethics and War: Just War Theory in the 21st Century*, edited by Fritz Allhoff, Nicholas Evans, and Adam Henscke, 63–75. New York: Routledge.

Gaddis, John Lewis. 2005. "Grand Strategy in the Second Term." *Foreign Affairs* 84 (1): 2–15.

Galliott, Jai. 2015. *Military Robots: Mapping the Moral Landscape*. Burlington, VT: Ashgate.

Gentili, Alberico. (1612) 2003. *De Jure Belli Libri Tres*. Translated by A. C. Campbell. Boston: Elibron Classics.

Gilpin, Robert. 1981. *War and Change in World Politics*. Cambridge, UK: Cambridge University Press.

Gray, Colin S. 2007. "The Implications of Preemptive and Preventive War Doctrines: A Reconsideration." Carlisle, PA: Strategic Studies Institute, July. http://www.strategicstudiesinstitute.army.mil/pdffiles/PUB789.pdf.

Grotius, Hugo. (1625) 2010. *On the Law of War and Peace*. Whitefish, MT: Kessenger Publishing.

Guiora, Amos. 2013. *Legitimate Target: A Criteria-Based Approach to Targeted Killing.* Oxford, UK: Oxford University Press.

Hudson, Leila, Colin S. Owens, and David J. Callen. 2012. "Drone Warfare in Yemen: Fostering Emirates through Counterterrorism?" *Middle East Policy* 19 (3): 142–156.

Johnston, Patrick B., and Anoop K. Sarbahi. 2015. "The Impact of U.S. Drone Strikes on Terrorism in Pakistan." RAND Corporation and University of Minnesota. Typescript. http://patrickjohnston.info/materials/drones.pdf.

Kaag, John, and Sarah Kreps. 2012. "The Moral Hazard of Drones." *New York Times*, July 22. http://opinionator.blogs.nytimes.com/2012/07/22/the-moral-hazard-of-drones/.

Kaag, John, and Sarah Kreps. 2014. *Drone Warfare.* Cambridge, UK: Polity Press.

Kaufman, Whitley. 2005. "What's Wrong with Preventive War? The Moral and Legal Basis for the Preventive Use of Force." *Ethics & International Affairs* 19 (3): 23–38.

Kegley, Charles W., and Gregory A. Raymond. 2003. "Preventive War and Permissive Normative Order." *International Studies Perspectives* 4 (4): 385–394.

Kennedy, Edward M. 2006. *America Back on Track.* New York: Viking.

Koh, Harold Hongju. 2014. "Ending the Forever War: One Year after President Obama's NDU Speech." *Just Security*, May 23. http://justsecurity.org/10768/harold-koh-forever-war-president-obama-ndu-speech/.

Kreps, Sarah. 2014. "Flying under the Radar: A Study of Public Attitudes towards Unmanned Aerial Vehicles." *Research and Politics* 1 (1): 1–8.

Kreps, Sarah, and John Kaag. 2012. "The Use of Unmanned Aerial Vehicles in Contemporary Conflict: A Legal and Ethical Analysis." *Polity* 44:260–285.

Kreimer, Seth F. 2003. "Too Close to the Rack and the Screw: Constitutional Constraints on Torture in the War on Terror." *University of Pennsylvania Journal of Constitutional Law* 6 (2): 278–325.

Lake, David A. 2010. "Two Cheers for Bargaining Theory: Assessing Rationalist Explanations for the Iraq War." *International Security* 35 (3): 7–52.

Lee, Steven. 2005. "A Moral Critique of the Cosmopolitan Institutional Proposal." *Ethics & International Affairs* 19 (2): 99–107.

Levy, Jack. 1987. "Declining Power and the Preventive Motivation for War." *World Politics* 40 (1): 82–107.

Levy, Jack. 2008. "Preventive War and Democratic Politics." *International Studies Quarterly* 52 (1): 1–24.

Lewis, Michael. 2013. "Drones: Actually the Most Humane Form of Warfare." *Atlantic*, August 21. http://www.theatlantic.com/international/archive/2013/08/drones-actually-the-most-humane-form-of-warfare-ever/278746/.

Lin, Patrick. 2010. "Ethical Blowback from Emerging Technologies." *Journal of Military Ethics* 9 (4): 313–331.

Luban, David. 2004. "Preventive War." *Philosophy and Public Affairs* 32 (3): 207–248.

Lubell, Noam. 2010. *Extraterritorial Use of Force against Non-State Actors.* New York: Oxford University Press.

McMahan, Jeff. 1996. "Realism, Morality and War." In *The Ethics of War and Peace*, edited by Terry Nardin, 78–92. Princeton, NJ: Princeton University Press.
Moore, James, and Wayne Slater. 2004. *Bush's Brain: How Karl Rove Made George W. Bush Presidential*. Hoboken, NJ: Wiley.
Morgenthau, Hans J. 1948. *Politics among Nations: The Struggle for Power and Peace*. New York: Alfred A. Knopf.
New America Foundation. 2015a. "World of Drones." Washington, DC. http://securitydata.newamerica.net/world-drones.html.
New America Foundation. 2015b. "Drone Wars in Pakistan: Analysis." Washington, DC. http://securitydata.newamerica.net/drones/pakistan-analysis.html.
New America Foundation. 2015c. "Drone Wars Yemen: Analysis." Washington, DC. http://securitydata.newamerica.net/drones/yemen/analysis.
O'Connell, Mary Ellen. 2012. "Unlawful Killing with Combat Drones: A Case Study of Pakistan, 2004–2009." In *Shooting to Kill: Socio-Legal Perspectives on the Use of Lethal Force*, edited by Simon Bronitt, Miriam Gani, and Saskia Hufnagel, 263–292. Oxford, UK: Hart Publishing.
Otto, Roland. 2011. *Targeted Killings and International Law*. Heidelberg, Germany: Springer.
Peterson, Scott. 2002. "Can Hussein Be Deterred?" *Christian Science Monitor*, September 10. http://www.csmonitor.com/2002/0910/p01s03-wosc.html.
Pew Research Center. 2015. "Public Continues to Back U.S. Drone Attacks." Washington, DC, May 28. http://www.people-press.org/2015/05/28/public-continues-to-back-u-s-drone-attacks/.
Plaw, Avery. 2008. *Targeting Terrorists: A License to Kill?* Burlington, VT: Ashgate.
Powell, Robert. 2006. "War as a Commitment Problem." *International Organization* 60 (10): 169–203.
Renshon, Jonathan. 2006. *Why Leaders Choose War: The Psychology of Prevention*. New York: Praeger.
Rengger, Nicholas. 2008. "The Greatest Treason? On the Subtle Temptations of the Preventive Use of Force." *International Affairs* 84 (5): 949–961.
"Rep. Schiff Offers Amendment Sunsetting Authorization for Use of Military Force—AUMF Never Intended to Authorize a War Without End." 2013. Washington, DC: Press Release, July 24. http://schiff.house.gov/press-releases/rep-schiff-offers-amendment-sunsetting-authorization-for-use-of-military-force-aumf-never-intended-to-authorize-a-war-without-end/.
Ripsman, Norrin, and Jack Levy. 2008. "Wishful Thinking or Buying Time? The Logic of British Appeasement in the 1930s." *International Security* 33 (2): 148–181.
Ryan, Missy. 2015. "Obama Administration to Allow Sales of Armed Drones to Allies." *Washington Post*, February 17. http://www.washingtonpost.com/world/national-security/us-cracks-open-door-to-the-export-of-armed-drones-to-allied-nations/2015/02/17/c5595988-b6b2-11e4-9423-f3d0a1ec335c_story.html.
Schmidt, Brian C., and Michael C. Williams. 2008. "The Bush Doctrine and the Iraq War: Neoconservatives versus Realists." *Security Studies* 17 (2): 191–220.

Schmitt, Eric. 2013. "Embassies Open, but Yemen Stays on Terror Watch." *New York Times*, August 11. http://www.nytimes.com/2013/08/12/world/embassies-open-but-yemen-stays-on-terror-watch.html?pagewanted=1.

Schweller, Randall L. 1992. "Domestic Structure and Preventive War: Are Democracies More Pacific? *World Politics* 44 (2): 235-269.

Shane, Scott. 2015. "Drone Strikes Reveal Uncomfortable Truth: U.S. Is Often Unsure about Who Will Die." *New York Times*, April 23. http://www.nytimes.com/2015/04/24/world/asia/drone-strikes-reveal-uncomfortable-truth-us-is-often-unsure-about-who-will-die.html?_r=0.

Silverstone, Scott. 2007. *Preventive War and American Democracy*. New York: Routledge.

Singer, Peter W. 2009. *Wired for War: The Robotics Revolution and Conflict in the 21st Century*. New York: Penguin Press.

Smith, Megan, and James Igoe Walsh. 2013. "Do Drone Strikes Degrade al Qaeda? Evidence from Propaganda Output." *Terrorism and Political Violence* 25 (2): 311–327.

Sofaer, Abraham D. 2005. "Preventive Force: Issues for Discussion." Stanford, CA: Stanford University Group on Preventive Force. http://www.princeton.edu/~ppns/conferences/reports/pf_paper.pdf.

Solis, Gary D. 2010. *The Law of Armed Conflict*. New York: Cambridge University Press.

Stern, Jessica. 2003. "How America Created a Terrorist Haven." *New York Times*, August 20. http://www.nytimes.com/2003/08/20/opinion/how-america-created-a-terrorist-haven.html.

Strawser, Bradley. 2013. *Killing by Remote Control*. Oxford, UK: Oxford University Press.

Thucydides. 1972. *History of the Peloponnesian War*. Edited by M. I. Finley. Translated by Rex Warner. Harmondsworth, UK: Penguin.

Totten, Mark. 2010. *First Strike: America, Terrorism, and Moral Tradition*. New Haven, CT: Yale University Press.

Trachtenberg, Marc. 2007. "Preventive War and U.S. Foreign Policy." *Security Studies* 16 (1): 1–31.

Truman, Harry S. 1950. "Radio and Television Report to the American People on the Situation in Korea." Public Papers of the Presidents, Harry S. Truman Library and Museum, September 1. http://trumanlibrary.org/publicpapers/index.php?pid=861&st=&st1=.

Tuck, Richard. 1999. *The Rights of War and Peace*. New York: Oxford University Press.

Van Evera, Stephen. 1999. *Causes of War: Power and the Roots of Conflict*. Ithaca, NY: Cornell University Press.

Van Schaack, Beth. 2014. "The United States' Position on the Extraterritorial Application of Human Rights Obligations: Now Is the Time for Change." *International Law Studies, U.S. Naval War College* 90: 20–65.

Walzer, Michael, 2006. *Just and Unjust Wars: A Moral Argument with Historical Illustrations*, 4th ed. New York: Basic Books.

Weisiger, Alex. 2013. *Logics of War: Explanations for Limited and Unlimited Conflicts.* Ithaca, NY: Cornell University Press.

Whetham, David. 2013. "Drones and Targeted Killing: Angels or Assassins?" In *Killing by Remote Control*, edited by Bradley Strawser, 69–83. Oxford, UK: Oxford University Press.

Wills, Garry. 2004. "What is a Just War?" *New York Review of Books*, November 18, 32–35.

Yoo, John. 2004. "Using Force." *University of Chicago Law Review* 71:729–32.

Zenko, Micah, and Sarah Kreps. 2014. "Limiting Armed Drone Proliferation." Council Special Report No. 69 (June). New York: Council on Foreign Relations. www.cfr.org/drones/limiting-armed-drone-proliferation/p33127.

PART I

Understanding Preventive War

2

Preventive Force

The Logic of Costs and Benefits

JENNIFER TAW

The preventive use of force entails the discretionary, anticipatory use of armed force to foil the emergence or development of a prospective threat.[1] As discussed in the previous chapter, preventive force can be conceptualized on a continuum encompassing everything from a single missile strike on a high-value target to the launching of a large-scale conventional war. It is inherently contradictory—an offensive action intended to have a defensive effect against a threat that has not yet developed. As such, it does not benefit from the efficient inaction of deterrence, since it requires actual engagement and not just the threat of retaliation. It does not have the legitimacy of defense, unlike preemptive force, since it is offensive in the absence of an imminent threat. It presumably has the efficiency advantage, however, of removing the possibility of an anticipated threat entirely, cauterizing it, and thus stopping a potentially costly pathway of extended conflict.[2]

Kenneth Waltz wrote in *Man, the State, and War* that "though a state may want to remain at peace, it may have to consider undertaking the preventive use of force; for if it does not strike when the moment is favorable, it may be struck later when the advantage has shifted to the other side" (1959, 7). Elihu Root similarly cautioned that a state should resort to the preventive use of force to avoid a "condition of affairs in which it will be too late to protect itself" (1916, 109–11). It was ostensibly this logic that led the United States into Iraq and Israel to bomb Osirak, and underpins Israel's current desire to attack Iran's nuclear development facilities.

There is an enormous literature on the preventive use of force; the topic has been addressed in theoretical and philosophical considerations of war, peace, and politics for centuries (see Chapters 1, 10, and 11 in this volume). Writing on the subject spiked, however, after President George W. Bush justified the 2003 U.S. invasion of Iraq in terms of the preventive use of force (though he erroneously referred to it as preemptive war). Most of the literature falls into one of two categories, which are also related to two of this volume's foci: assessments of the ethics of the preventive use of force and analyses of both precipitating and permissive causes. Indeed, many of the chapters in this volume deal with precisely these issues.[3]

However, this chapter addresses another one of this volume's concerns: In terms of practical considerations, what are the costs and benefits of employing preventive force? What are the advantages and disadvantages relative to alternative responses and nonresponse? Answering these questions requires defining (1) the inherent characteristics of the preventive use of force, (2) the types of threats amenable to such actions, and (3) the conditions under which these threats might be better addressed using preventive force, compared to other responses. Such calculations do not take place in a vacuum, however, and are susceptible to political considerations. For example, Miroslav Nincic (Chapter 3) illustrates how the American public's responses to the use of preventive military force can affect decision-makers' calculations when considering this option relative to others. Further, because preventive military action is undertaken in response to an anticipated threat, rather than one that has materialized, measuring the action's relative costs and benefits depends in part on accurate threat assessment. This chapter therefore also includes a discussion of tendencies toward threat exaggeration and cognitive limitations on measuring risk.

By looking at the fundamental features of the preventive use of force and the conditions under which the strategy may be employed, this chapter offers a pragmatic perspective on preventive military action thus far absent from the literature. Ultimately, this chapter concludes that the preventive use of force is only rarely, and under very specific conditions, the *best* option in the face of a perceived, emergent threat. It may nonetheless be the *preferred* option in some circumstances (though note that Nichols, Chapter 4, argues that preventive force is becoming the preferred option in many circumstances). If the benefits appear high (decision-makers perceive the incipient threat to be extreme and/or believe there would be a political boost from taking preventive action) and the potential costs—initial expenses, retaliation, and escalation—are low relative to a state's ability to absorb them, a government may opt for preventive military action. In other words, the threats that could arise must be perceived as severe and as having predictable enough consequences to justify the costs of precipitant action and/or the government undertaking such action must be wealthy and secure enough that the costs and risks of such action are effectively diminished relative to the perceived benefits. Keeping these points in mind is useful when considering actions on the preventive force continuum, whether these be preventive war or limited uses of force, such as deploying drones to achieve preventive ends.[4]

Costs and Benefits

The goals of preventive military action are to retain an advantage, to cauterize or defer a developing threat, to jam the mechanism before the tables turn. A state must be willing and able to undertake what is essentially an act of war to prevent a future greater war in which it may not have an advantage. It must be willing to assume the risks of retaliation, escalation, and disapprobation in order to achieve its end. The threat must therefore be perceived as sufficient and certain enough to justify the preventive action. Moreover, the state must be convinced that the costs it may incur through such action will still be fewer than the potential costs it would suffer if it failed to use preventive force.

As with any cost-benefit analysis, one must begin by assessing the inherent costs and benefits of the preventive use of force. The costs and benefits of the preventive use of force are best measured by determining whether preventive force is not only better than inaction, but also better than alternative interim responses to the perceived threat. Such alternatives can include diplomacy, deterrence, or economic means to address the concern in its early stages. The initial anticipated costs of such action conform to those for any use of force, especially an offensive military action. They include the economic and human costs of the attack, any domestic political fallout, any international disapprobation, the political and opportunity costs in terms of the relationship with the entity that was targeted, costs to other national interests (economic ties, soft power, and so forth), and the possibility for both retaliation and escalation once hostilities are initiated. These are standard concerns for any offensive military action.

The anticipated benefits of preventive attacks are different in several ways from those of other forms of military force, however. Unlike the material and political benefits yielded by conquest or the defense of key interests and territories, the primary benefit of the preventive use of force is ideally the elimination of the possibility of the emergence or development of a threat. There is no victory or possibility of victory. There is, at best, the elimination or deferral of a threat. This is far harder to measure, prove, and tout as a success compared to the more concrete outcomes of other forms of warfare. The other potential benefits of the preventive use of force are likewise amorphous. For instance, a show of force could have a deterrent effect on potential enemies, but it could also serve to strengthen their resolve. The Iranian government, having seen the effects of Israel's attack on the Osirak reactor, did not decide against nuclearization, but instead ensured that its own nuclear development facilities would be far less vulnerable than Iraq's had been. Further, a government or policymaker might experience a rally-round-the-flag or diversionary (wag-the-dog) domestic political boost with the deployment of military forces for preventive ends, but in the absence of an extant threat, this would be harder to achieve (see Nincic, Chapter 3).

Finally, the preventive use of force could yield positive international responses—but again, the government or policymakers responsible would have to be able to demonstrate persuasively that a serious threat would have emerged absent their action and, moreover, that its preventive behavior was not more costly to the system than the potential threat might have been. The United States' difficulties rallying international support for the 2003 invasion of Iraq and the disapprobation the United States subsequently suffered illustrates this challenge. Ultimately, with the preventive use of force, there is no ground gained; there are no foreign forces visibly and dramatically repelled. The need for preventive force—a military attack into a sovereign territory targeting either a resident non-state actor or the state itself—therefore becomes a far harder case to make to domestic and international audiences, both before and after it is used.

These nebulous benefits, in turn, also impose unique costs. If the incipient threat is hard to prove, may never have emerged, may be considered a misreading or misunderstanding of trends, or may be otherwise brought into doubt, the preventive military action becomes politically and ethically unjustifiable, as do the costs associated with it. This is true even if the preventive military action is very carefully tailored to be as precise and limited as possible.

Normative proscriptions against the preventive use of force abound, both in the United States and abroad. Hans J. Morgenthau wrote in his seminal 1948 work *Politics among Nations* that attitudes toward war had changed, so that one could not "consider seriously the possibility of preventive war, regardless of its expediency in view of the national interest" (1993, 232). He further explained that

> the moral condemnation of war as such has manifested itself in recent times in the Western world. When war comes, it must come as a natural catastrophe or as the evil deed of another nation, not as a foreseen and planned culmination of one's own foreign policy. Only thus might the moral scruples, rising from the violated moral norm that there ought to be no war at all, be stilled, if they can be stilled at all. (1993, 232)

Harry S. Truman similarly observed:

> It goes without saying that the idea of "preventive war"—in the sense of a military attack not provoked by an attack on us or allies—is generally unacceptable to Americans. A surprise attack upon the Soviet Union, despite the provocations of recent Soviet behavior, would be repugnant to many Americans. Although the American people would probably rally in support of the war effort, the shock of responsibility for a surprise attack would be morally corrosive. (Quoted in Etzold and Gaddis 1978, 430–432)

And as George Kennan and Bernard Brodie cautioned, respectively, "a democratic society cannot plan a preventive war," and "a policy of preventive war has always been 'unrealistic' in the American democracy" (quoted in Schweller 1992, 242–244). Expectations of domestic and international disapprobation in the event that a preventive use of force is undertaken therefore must be included in any cost-benefit analysis of such action. In practical terms, this means that the preventive use of force is an enormous political risk; the anticipated threat must be great enough to justify it. Accordingly, Nincic (Chapter 3) carries out an in-depth analysis of polling data to effectively demonstrate how public opinion can factor into a state's cost-benefit calculation for preventive war.

The potential costs of preventive military action are high, but it would make rational sense to use such force when two conditions hold: *the threat a country's policymakers anticipate is bigger than the risks inherent to preventive action* and *the threat cannot be equally or better mitigated by any other means.* It is important to note at the outset that this calculation will be different for different countries, depending on their circumstances. The strongest countries, which can bear the burden of the costs of the use of force, retaliation, and criticism, have more freedom to resort to preventive military action than those with less ability to absorb such costs. In effect, powerful countries assume less risk with such actions, which lowers the cost of action relative to inaction. However, lower cost options to both preventive military action and inaction may be available and superior, even for great powers.

Threats

Cost-benefit analyses must be undertaken with a goal in mind. In this case, states at a minimum want to ensure their own survival and freedom, and often their values, too; they therefore must defend against threats to their sovereignty, sources of power, status, and principles. The extent to which each of these is vulnerable will depend on individual states' circumstances, with a whole range of variables coming into play: their relative military power, their geographic location, their international relationships (good and bad), their economic strength, their internal political stability and demographics, and so forth.

Taking each essential interest in turn, we can begin with sovereignty. Threats to sovereignty—that is, efforts to reduce a country to an ungoverned territory or a territory under new management—may include efforts to conquer some or all of a state's territory or to overthrow or destabilize a state's government and institutions. Russia's recent assertion of control over Crimea and continued threat to Ukrainian territory and even other states in eastern Europe is a stark reminder that although conquest has become rare, it has not been eliminated. American troops remain in Afghanistan, where the United States and its allies toppled the Taliban, and Iraq is not the same country it was before external powers forced their way into the country and overthrew its government. Taiwan remains concerned about China's intentions towards it; Sudan and South Sudan are poised at the edge of a fractious and tenuous peace; South Korea and North Korea each continue to fear the other's possible attempt to unify the peninsula under its own rule; and Israel remains on constant alert for fear of its neighbors attempting to wipe it off the map.

Meanwhile, many states are concerned about defending their sovereignty from within. Internal wars, insurgencies, secessionist movements, terrorist activities, and organized crime are threatening the standing governments in countries as varied as Spain, Somalia, Mexico, Thailand, and Syria. In some cases, those who are challenging their states from within would be satisfied with political concessions that would largely leave their governing structures and borders in place; others, however, want nothing short of the overthrow of existing regimes so that they can restructure their countries into entirely new entities.

Nonmilitary threats like pandemics that could annihilate people or devastate the institutions of the state would likewise be considered existential. Because preventive military strikes would make no sense as a response to these problems, we can leave them off the table.[5]

Lesser challenges to sovereignty involve incursions or attacks on a state's people or property by external terrorist groups, foreign drone strikes, and any other activities that effectively reduce a country's territorial integrity and control, though they may not threaten its survival.[6]

Another essential interest of the state is its power. Power can be measured in myriad ways, but is usefully understood as the accumulation of those resources and abilities that give the state durability and potential for growth domestically and the ability to defend and assert itself internationally.[7] The conventionally recognized elements of national power are economies, militaries, and political institutions. Threats to a state's power thus can include weakening or undermining a state's economy or political institutions or reducing, eliminating, or making obsolete its military capabilities. International sanctions like those in place against Iran, North Korea, and Cuba are one representation of this kind of threat; economic sanctions are intended to have negative effects on a country's material wellbeing, weaken its political institutions by demonstrating their failures, flaws, and/or corruption, and often impede its ability to modernize or maintain or develop its military.

Denial of access to key strategic areas or resources is a more direct attack on a state's power and independence of action. While sanctions work indirectly, if at all, denying a state mobility for either trade or defense purposes or cutting it off from necessary resources like energy or industrial inputs is a more direct, immediate threat. These dynamics are in play worldwide. China is in the process of slowly circumscribing American access to Asian waters and building up its own military capabilities specifically to challenge the U.S. Navy in the South China Sea. Meanwhile, Iran developed fast attack boats to challenge U.S. vessels in the Strait of Hormuz, should tensions between the two states ever escalate to kinetic conflict. As for cutting off resources, in 2010 China blocked Japanese access to rare earths for some time in retaliation for Japanese interdiction of a Chinese fishing boat. In 2014, Russia removed the price supports it had proffered to Ukraine, thereby raising the cost of Russian natural gas in Kiev by 80 percent. The implicit threat, however, was that Russia could turn off the tap altogether. These kinds of actions can seriously circumscribe a state's power.

Intertwined with threats to states' power are threats to their international status. This is because power is important not just absolutely, but relatively, since states are concerned about economic and military competition. Efforts to reduce states' status can include ejecting them from international organizations or alliances, reducing trade ties, removing or refusing them diplomatic recognition, or otherwise shoving them to the sidelines of international interactions. The 2014 decision to expel Russia from the G-8, for example, was an attack on Russia's status. India and the United States, and the United States and Venezuela, each undertook tit-for-tat expulsions of each other's diplomats in the spring of the same year.

The aforementioned threats to status are symbolic, but threats to status can also be more concrete as competing powers arise in the military, economic, diplomatic, or sociocultural realms. In fact, Douglas Lemke (2003) characterized most scholarship on preventive war as focusing on states' defense not of their territory and sovereignty, per se, but of their position in the international system; that is, states use preventive force when they are beginning to decline relative to other potentially hostile powers. China's current economic growth and military modernization, combined with its more assertive foreign policy in Asia, is thus perceived as a challenge to America's status in Asia and even globally. Russia's reemergence as a regional military power, a feat made possible by enormous material gains buttressed by its natural gas exports, threatens the United States and the European Union. These kinds of threats to status have less to do with imagery and more to do with the emerging potential for real world challenges.

Finally, organized groups of people, including states, often perceive threats to their values, whether religious, political, or socioeconomic. The entire Cold War is frequently cast as a conflict over competing ideologies. The recent report (Flood 2014) that the CIA disseminated *Doctor Zhivago* in the Soviet Union in the late 1950s, and that the Soviets subsequently banned it, speaks to the extent to which each side tried to control the narrative. U.S. actions in Vietnam are often explained by their intention to prevent the spread of communism (the falling of more dominoes), rather than as a local war against the North Vietnamese government and southern Vietnamese guerrillas. In this case, the preventive use of force was intended to draw a line in the sand in a country that otherwise had little strategic significance. The war was not to defeat communism or even specific Communist states, but to cut off at the pass the spread of the ideology before it began to tilt the global advantage towards the Communists.

These kinds of concerns arise in different forms in different times. Today, countries in the West worry about a rising China because the state does not share Western values and is increasingly well placed to undermine Western efforts at international influence, as it has done by subverting U.S. and European sanctions against Sudan, Syria, and Iran, among other things. Violent challenges by al-Qaeda offshoots like al-Shabaab in Somalia, al-Qaeda in the Arabian Peninsula (AQAP) in Yemen, and Boko Haram in Nigeria are also undertaken in the name of religion and societal values. Indeed, one could almost depict the early al-Qaeda attacks against the United States as that transnational actor's stab at preventive action: globalization and Westernization were perceived as creeping threats against Islamic values and the eventual creation of an independent caliphate (Musharbash 2005).

This conceptualization of threat as being related to values, ideology, and accepted practices is consistent with Samuel Huntington's warnings in *The Clash of Civilizations*, in which he posited that "the greatest divisions among humankind and the dominating source of conflict will be cultural. Nation states will remain the most powerful actors in world affairs, but the principal conflicts of global politics will occur between nations and groups of different civilizations" (1993, 22).

Conditions for Preventive Action

Lag and Predictability

While the above cost/benefit and threat categories begin to help identify the critical interests states might seek to protect with preventive force, threats to these interests must share some characteristics in order for preventive responses to be possible. First, the threats must be lagged and have predictable consequences. Emphasis on lag derives from the definition of preventive military action, which requires that the threat being addressed is immature rather than imminent. This means any threat that might be ameliorated with the preventive use of force must be observable in its early stages (before it develops into a full-blown concern), and it should have predictable intent or repercussions (that is, be identifiable as an incipient threat). This last point is significant. In the absence of reasonable certainty that an emerging threat not only will fully materialize but also will impose serious and perhaps existential costs on a country, it becomes difficult to justify the preventive use of force, especially in light of less costly and potentially more constructive alternatives to preventive military action.

Climate change is an ideal example of a predictable lagged existential threat: it is itself observable, as are its effects, which are intensifying over time and foreseeable. If we wish to avoid the direst predictions, or, at this point, at least prepare for them, climate change needs to be addressed proactively. Yet, while climate change meets the stipulations for lag and predictability, military action clearly would not be an appropriate response. We need an additional identifier.

Targetable Political Perpetrators

A threat will be amenable to preventive force only if it emanates from a targetable political entity. This can be anything from an individual political leader, to a government, to a non-state actor, to a coalition, but the potential threat must come from a person or people. And that person or those people must be targetable with military force; if they are wholly dispersed in a population, or unidentifiable, or in some other way made inaccessible, preventive military action will not be an option.

Putting It All Together

Table 2.1 illustrates the extent to which the various types of threats mentioned thus far meet the initial, pragmatic criteria for preventive response (threats are to key national interests, are lagged, have predictable intent or consequences, and are undertaken by targetable political actors).[8]

TABLE 2.1 Types of Threat and Criteria for Preventive Response

	Observable prior to maturity	Predictable in their intent or consequences	From identifiable political entities	From targetable political entities	Examples
Foreign efforts to conquer some or all of a state's territory	sometimes	yes	yes	yes	Russia's invasion of Crimea
Foreign efforts to overthrow or destabilize a state's government and institutions	sometimes	yes	yes	yes	Russia's support for rebels in Ukraine
Internal threats	yes	yes	yes	no	Rebels in Ukraine
Incursions or attacks on a state's people or property by external non-state actors	sometimes	no	sometimes	sometimes	Hezbollah rocket attacks on Israel

TABLE 2.1 *(cont.)*

	Observable prior to maturity	Predictable in their intent or consequences	From identifiable political entities	From targetable political entities	Examples
Attacks on a state's power bases					
Sanctions	no	no	yes	yes	UN sanctions on Syria
Denial of access to strategic areas	sometimes	yes	yes	yes	Iran's buildup of fast-boat capabilities in the Strait of Hormuz
Denial of access to strategic resources	sometimes	yes	yes	yes	China's short-lived denial of rare earths to Japan
Symbolic attacks on status					
Ejection from international organizations or alliances	no	yes	yes	yes	Russia's ejection from the G8
Reduced trade ties	no	yes	yes	yes	Brazil's cancellation of defense contracts with U.S. following NSA scandal
Removal or refusal of diplomatic recognition	no	yes	yes	yes	Brazil canceling U.S. visit after NSA scandal
Rise of competing powers	yes	no	yes	yes	China's rise to challenge the U.S. in Asia
Threats to values/ideology/religion	yes	no	sometimes	sometimes	Mutual perception by the West and Islamic extremists

None of the threats listed conforms entirely to the criteria for considering the preventive use of force. Internal threats come close, but, at least within the United States, the perpetrators would not be targetable with military force; such emerging concerns would be met with political and police responses. In more autocratic states, or in unstable states (such as Ukraine in 2014), however, the rise of internal threats has indeed led to domestic military responses. Yet such actions have tended to take place after the threat has emerged, making them defensive rather than preventive. If these groups were perceived as emerging threats not only to their own governments, but to other states as well, then there is a possibility for unilateral or even multilateral preventive military action. One could argue that this was the case in Yemen, where prior to the fall of the government to Houthi rebels, the U.S. and Yemeni forces were collaborating to eliminate al-Qaeda extremists. However, for the Yemeni government, that threat was extant, not potential; arguably, only the United States was undertaking preventive military action, and its efforts would fall into the incursions category in table 2.1. The same could be said of U.S. drone strikes in places like Pakistan and Somalia.

The rise of competing powers conforms in every way except predictability, yet this variable is crucial. What if rising powers create opportunities rather than risks? If Britain in the late 1800s had attempted to check American power with war, for example, the outcome in terms of global stability and Britain's own national interests would arguably have been much worse than what occurred with the peaceful hegemonic shift that eventually took place. The same debate about threat versus opportunity has taken place in the United States with respect to China since the late 1980s.

Other categories of threats on the list may allow for the effective, preventive use of force even though they do not clearly meet all of the criteria outlined above. Incursions or attacks on a state's people or property by terrorist groups, for example, might be anticipated, could be escalatory, could be assumed to be threatening (though to what extent is debatable), and would emanate from identifiable and targetable political entities (though external terrorist organizations based in other states would pose a dilemma in terms of targeting). The aforementioned U.S. drone strikes in Yemen, Somalia, and Pakistan would be a response to perceived threats in this category. Likewise, attacks on a state's values or ideology, at home or abroad, could definitely be observable early on, escalatory, and from identifiable political entities, though it might be hard to predict their effect (for instance, how many people would be persuaded or indoctrinated) and those entities might be difficult or impossible to target. This is in part because such perceived threats are often understood to be insidious, rather than blatant. The Red Scare in the United States in the 1950s played upon fears of Communist infiltration just as the 2013 French decision to outlaw the *niqab*, or full face veil, reflects fear of growing fundamentalist Muslim influence in France. In both cases, perceptions of a menace against the national way of life did play out in the preventive use of force abroad: the United States took its war against the Communists to Asia, Latin America, and Africa, and today France is expanding its counterterrorism efforts in North and West Africa.

Some of the threat types mentioned previously can be entirely removed from consideration of the practicability of preventive military action. Illegal immigration would be taken off the list (no identifiable or targetable political entity responsible). Likewise, symbolic efforts to undermine a state's status are not observable prior to maturity nor are they, in any serious sense, escalatory. Sanctions can be removed from the list, since they cannot be surreptitiously imposed. A response to drone strikes would be defensive rather than preventive. The very fact of a potential enemy having drones might raise concern, but that would fall more into the category of the rise of competing actors.

Significantly, most external efforts to conquer a state's territory or overthrow its government should be removed from the list, since they cannot really be undertaken incrementally. The kinds of massing of troops or strategic movement of resources and material necessary to conquer a country or overthrow its regime would be more consistent with imminent warfare, and therefore would be met with preemptive, rather than preventive, military force. An exception to this might be a state that begins by conquering a neighbor but is perceived as being intent on continuing in the direction of the state considering preventive action (a juggernaut of sorts). This kind of World War II scenario of expansionism could be considered a lagged threat for the states not yet in immediate danger.

We now have six types of threats that meet the initial criteria thus far laid out for preventive responses. Having identified the kinds of potential threats that might justify the use of preventive force does not mean, however, that there is a clear dichotomous choice between preventive military action and total inaction; there are always other options. Table 2.2 lays out both the costs and benefits of preventive action and various alternative responses to these potential threats.

TABLE 2.2 Preventive Action: Costs, Benefits, and Alternative Actions

	Possible kinetic preventive responses	Likely benefits	Likely costs	Possible alternative responses prior to threat fully materializing	Possible alternative responses if threat fully materializes
Juggernaut or foreign efforts to overthrow a state's government	Missile or air strikes on strategic targets; Conventional war	Surprise; Bring the war to the opponent; Slow the advance; Signal determination	Guarantees engagement; Deploying troops abroad leaves fewer for defense; Retaliation	Deterrence; Diplomacy; Collective security	Defense; Coordination with friends and allies; Economic sanctions; Diplomacy
Rise of competing power or change in capabilities (i.e., nuclearization)	Missile or air strikes on strategic targets; Conventional war	Slow or reverse opponent's rise; Destabilize opponent; Undercut opponent's ability to wage war; Delay possibility of future full-blown warfare	Guarantees adversarial relationship; Forego future economic and cooperative opportunities; Spark retaliation; Spark war; Domestic and international political backlash	Diplomacy; Trade; Institutional ties; Deterrence; Collective security; Strengthened defense	Balancing; Defense; Deterrence; Negotiation
Threats to values, etc.	Conventional war; Proxy war; Attacks on leaders; Terrorism	Create fear; Create economic costs; Weaken or eliminate leadership	Backlash in target state or group; International disapproval; Spark retaliation; Spark war	Counter-proselytize; Reliance on soft power; Aid for education and development; International institutional responses	Containment; Soft power; Economic incentives; Negotiation

TABLE 2.2 *(cont.)*

	Possible kinetic preventive responses	Likely benefits	Likely costs	Possible alternative responses prior to threat fully materializing	Possible alternative responses if threat fully materializes
Non-state actors' attacks on people or property	Attacks on leaders; Attacks on meeting places	Decapitate organizations; Create fear and prevent association; Weaken or eliminate group	Rally support to the group; Enhance group's recruitment; Push group underground or to diversify; Spark more overt conflict; Spark tensions or conflict with host nation	Strengthened defense; Soft power; Diplomacy; Cooperation with foreign governments and/or IOs to police and/or monitor	Attacks on leaders; Attacks on meeting places; Negotiation; Economic sanctions
Denial of access to strategic resources	Attacks on means of denial; Attacks to weaken opponent overall	Create pressure on decision-makers to allow access	Expedite decision to limit access; Expedite limits to access; Reduce cooperation and likelihood of future cooperation; Undermine relations; Spark war	Diplomacy; Economic incentives; Diversification of demand or suppliers	Attacks on means of denial; Attacks to weaken opponents overall; Negotiation; Economic sanctions
Denial of access to strategic areas	Attacks on means of denial; Attacks to weaken opponent overall	Reduce opponent's ability to deny access; Maintain access to key strategic areas	Expedite decision to limit access; Expedite limits to access; Reduce cooperation and likelihood of future cooperation; Undermine relations; Spark war	New defense technology and tactics, techniques, and procedures; Negotiation; Economic incentives	Attacks on means of denial; Attacks to weaken opponents overall; Negotiation; Economic sanctions

Table 2.2 provides a useful tool. Consider the light it sheds on two different circumstances involving the rise of a competing power. On the one hand, the mitigating responses could be enough to prevent the threats from fully emerging or, better yet, to turn them into opportunities. As China has undertaken its "peaceful rise" over the past few decades, for example, every one of the mitigating strategies listed has been implemented, as the United States and its friends and allies have attempted to tame the growing power with trade agreements, international organizations, and international laws even as they have also strengthened their own military capabilities and alliances. Of course, war could break out over Taiwan or events in the South China Sea, but the years of peace did not leave the United States in a particularly weakened or precarious position relative to China. Not only are mitigating strategies still being implemented, but the deterrent effect of each state's military development over the past decades remains strong.

On the other hand, Israel chose to strike both Osirak and al-Kibar for fear that Iraq's and Syria's abilities to develop nuclear weapons would be fundamental game-changers creating unacceptable levels of threat to the Israeli state and populace. Neither preventive attack resulted in hot war; indeed, Israel understood from the outset that such a risk was mitigated by the support of the United States and raising the costs of retaliation. Moreover, Israel had nothing to lose politically or economically; it had antagonistic relations with both states and was already regularly and roundly criticized internationally. The constellation of relations, in turn, precluded any hope for effectively negotiating either state's cessation of nuclear development: Israel had no leverage, and the international community was hesitant to respond to either state's nuclear developments. Neither the French nor the Americans believed Osirak signaled any serious threat of imminent nuclear weapons development, and when North Korean complicity in building al-Kibar in Syria was revealed, key American policymakers still preferred to respond diplomatically (Abrams 2013). In Israel's cost-benefit analysis, therefore, the potential threat was bigger than the risk assumed with the preventive strike, and there was no better means for responding to it. In fact, intense American and allied economic and political pressure on Iran to open its facilities to inspection and halt any nascent weapons production has been partly a function of understanding Israel's cost-benefit analysis on the use of preventive force. While Iran would present a much riskier target for Israel than either Osirak or al-Kibar did—not least because of the lessons Iran learned from Israel's attack on Iraq's facility—Israel's calculation remains essentially the same.

Table 2.2 also captures U.S. actions against al-Qaeda offshoots in Somalia and Yemen, among other places. The benefits are tangible and immediate; the actual costs of the operations are minimal; the projected costs are amorphous and hard to prove; and the alternatives are limited, complex, and unlikely to yield any short-term gains. The threats the groups pose to the United States are certainly not existential, but given the United States' ability to absorb the costs of preventive strikes, eliminating the potential for these groups to organize attacks on the U.S. homeland is not unreasonable.

The table suggests that preventive strikes against potential denial of access efforts might be acceptable if the risk is low (great power versus a much weaker state, for example) and relations are already lousy; even then, mitigating efforts would likely be preferable even for powerful states. Preventive military action to avert denial of resources is too self-fulfilling a prophecy to make sense unless the resources are important enough and the denier is weak and insignificant enough that its hand can be forced, its retaliation minimized, and the relationship with it corrupted without concern. If these conditions pertain, however, then the denier is likely weak enough to be amenable to economic threats or promises or other forms of negotiation. Whether the preventive use of force makes sense in the face of threats to values seems to depend on how vulnerable values are perceived to be. Al-Qaeda's attacks on the United States and the West suggests fears that Westernization poses an existential threat, meriting the assumption of great risks. That the United States retaliated, however, should not have been surprising.

Finally, the argument in favor of preventive action in the face of a juggernaut seems strong, but even that is debatable. As a case in point, in early 2014, there were some perceptions that Russia was poised to become a juggernaut; its actions in Crimea and then in support of pro-Russian forces in eastern Ukraine created ripples of concern throughout eastern Europe, prompting NATO to move material and manpower closer to its members' borders with Russia. Months later, a full-blown war had not materialized; yet preventive military action at the outset of the hostilities would have been a self-fulfilling prophecy guaranteeing war. Moreover, a little doubt as to the juggernaut's intentions can be enough to cause a state to hesitate, at which point it will have to rely on a strong defense rather than the advantages of moving the fight to the potential aggressor. And this may not be such a terrible thing, if its defense is all the more solid for having all of its forces at home and all the resources at hand to sustain them, especially if the juggernaut might lose, rather than gain, momentum as it approaches.

Political Rationality and Threat Exaggeration

Thus far the assumption has been that the decision to use preventive force is a rational calculation that involves weighing the costs and benefits of preventive military action relative to either inaction or to alternative responses to a potential threat. Yet it is necessary to point out that limited information imposes limitations on rationality, a particularly relevant concern when potential, rather than materialized, threats are being identified and assessed.

To the first point, any foreign policy calculation will include not only the kinds of criteria outlined above, but also—as is discussed further in Chapter 3—decision-makers' personal equations, with an eye to their own images, agendas, and supporters. In their respective studies of the domestic variables associated with foreign policy decision-making, Alex Mintz and Karl DeRouen (2010) and Bruce Bueno de Mesquita and David Lalman (1992) identified the key roles of politics and public opinion in decision-makers' analyses. In their—and others'—work on poliheuristic theory (including Mintz and Geva 1997; Redd 2005; Sathasivam 2003), they examined specific cases and found that foreign policy decisions were deeply affected by decision-makers' assumptions about public support and the political viability of any option. This means that if politicians anticipate negative fallout from failing to anticipate or prevent a crisis, for example, they are likely to be more supportive of preventive action. A president has reason to expect serious political ramifications if terrorists succeed in attacking the U.S. homeland during his or her tenure and therefore becomes more likely to take actions to avoid that risk.

Even assuming apolitical rationality, however, the biggest challenge in trying to develop a persuasive cost-benefit analysis for preventive military force is that, insofar as it hinges on the assessment of how serious a potential threat will be, such assessments are inherently subjective. Studies of political psychology consistently demonstrate tendencies to exaggerate threats. Though he was focusing on terrorism, John Mueller's observations about this tendency are generalizable. Noting that it is "common to exaggerate and to overreact to foreign threats," he added that "alarmism and overreaction can be harmful" and can "help create the damaging consequences" opponents seek (2005, 208). Janice Gross Stein (2013) details how emotional and cognitive processes can affect threat perception. And Robert Jervis's work on prospect theory is particularly illuminating: "more than the hope of gains, the specter of losses activates, energizes, and drives actors, producing great (and often misguided) efforts that risk—and frequently lead to—greater losses" (1992, 187). He continues, "People will choose the risky alternative when the choice is framed in terms of avoiding losses when, in the exact same case, they would take the less risky course of action if the frame of reference is the possibility of improving the situation" (188). These observations illuminate George W. Bush's now famous call to preventive war: "The greater the threat, the greater the risk of inaction—and the more compelling the case for taking anticipatory action to defend ourselves, even if uncertainty remains as to the time and place of the enemy's attack" (Bush Administration 2002, 15).[9]

Were the Israelis correct that a nuclear-armed Saddam Hussein or Bashar al-Assad would have threatened their existence? Is the United States correct that the threats posed by al-Shabaab and their ilk merit the costs involved in hunting them down preventively? Should preventive military force be launched against Putin to foil his apparent territorial aspirations beyond Crimea? Was the U.S. war in Vietnam necessary to prevent the dominoes from falling in Southeast Asia? Did the war prevent the empowerment of the Communist bloc or the spread of the communist ideology? Ultimately, the cost-benefit analysis can only be as good as the risk assessment upon which it is premised.

Conclusion

The unique nature of the preventive use of force—an offensive action for defensive ends in the face of tremendous uncertainty and near universal international disapproval—already demands that the strategy be undertaken only in extraordinary circumstances (for a contrasting view, see Nichols, Chapter 4). The very nature of the action requires that the perceived threat be lagged with predictable intentions or effects and imposed by targetable political actors. In weighing the costs and benefits of preventive military action, it is evident that using preventive force will be the best course of action only under two conditions: the potential threat must be bigger than the risks inherent in preventive action and the threat cannot be equally or better mitigated by any other means.

To the first condition, considerations include the anticipated scale and scope of the potential threat, the costs integral to the scale and scope of the preventive response (on a continuum from a deterrent show of force to declared war), and the potential for retaliation and escalation. Given that any form of preventive military action bears at least some material and political costs, the anticipated threats must be to essential interests (sovereignty, sources of power, status, and/or principles) in order to justify undertaking what is essentially a self-fulfilling prophecy of at least some violence at some price.

Of course, powerful states (or those with powerful backers) can undertake preventive military action with relative impunity because they can absorb the costs of the preventive use of force and may be able to deter or quell retaliation or escalation. The war in Iraq is a case in point; at a price of over $2 trillion and counting, over 4,000 coalition lives, and an estimated half-million Iraqi lives, the United States prosecuted a war justified on the basis of the possibility of Saddam Hussein's government developing and distributing weapons of mass destruction. The U.S. National Security Advisor at the time, Condoleezza Rice, warned with respect to Hussein: "The problem here is that there will always be some uncertainty about how quickly he can acquire nuclear weapons. But we don't want the smoking gun to be a mushroom cloud." Then-President Bush similarly said, "There was a risk, a real risk, that Saddam Hussein would pass weapons or materials or information to terrorist networks, and in the world after September the 11th, that was a risk we could not afford to take" (CNN 2004). It turned out that the only WMD in the desert were overlooked remnants of the 1991 Gulf War. But, despite the tremendous costs, the years of warfare, and the fact that the potential threat upon which the entire tragedy was predicated never existed, the United States emerged relatively unscathed. For a powerful country, the preventive use of force against a weaker state or non-state actor can be an affordable option. Nichols (Chapter 4) provides further evidence of powerful states unrestrained with the cases of the U.S. intervention in Libya and Russian involvement in the Ukraine.

That does not mean, however, that it will be the best option. A cost and benefit analysis is intended to help identify best options. When strategies for tempering, mitigating, or even capturing and turning the anticipated threat—or to effectively respond to it if it develops fully—exist and will be less expensive in terms of immediate material outlays, opportunity costs, and political responses (on public opinion and political calculations, see Nincic, Chapter 3), then they will be a better option. Indeed, this is how foreign policy is weighted, with diplomacy and development frontloaded to help reduce threats, create transparency, mitigate security dilemmas, foment cooperation and communication, and promote constructive engagement.

As norms of international behavior have evolved, the use of force is increasingly restricted to defense and deterrence. The preventive use of force, as an offensive action, undermines this trend. And relaxing the norm against preventive action is problematic precisely because of the inherent uncertainty involved. The human tendency to exaggerate threats, combined with people's propensity to accept risks in order to protect against losses, could lead to some very bad decision-making. Furthermore, even if a nascent threat is correctly identified, left alone, or mitigated by other means, it may never emerge; if preventive action is taken, however, it becomes a self-fulfilling prophecy.

That said, even though preventive action will not be the best option for almost every hypothetical and real-world case offered in this chapter, an exception, in a purely cost-benefit sense, is the U.S. use of drones against terrorist targets. In that circumstance, the strikes represent a relatively low-cost option for the United States, with a limited chance of retaliation or escalation, and no clear, viable alternatives. The argument could be made that any U.S. action is unnecessary, since the individuals being targeted pose no real threat to America in any of the aforementioned essential interests; one could further argue that the strikes could incur a long-term cost by helping the extremists recruit and by crystallizing anti-U.S. sentiment among them (see Chapter 1 in this volume for further discussion). In the end, though, the groups already exist and are growing. They have stated their aggressive intent and they have carried out attacks against U.S. targets. The drone strikes are affordable and, for the United States, fairly risk-free. Until something changes to affect that calculation—greater international or domestic outcry, an increased capacity on the part of the terrorist organizations to retaliate against the United States, local governments' withdrawal of overt or tacit approval of such strikes, or a political opportunity to engage with the terrorists and/or their supporters in a more constructive way—American drone strikes against terrorist targets arguably remain a rare form of preventive military action that passes the cost-benefit analysis test.[10]

NOTES

1 For the purposes of this chapter, I am not making a distinction between the preventive use of force, preventive military action, or preventive strikes. Rather, because of the nature of the inquiry, I am bundling those terms in reference to any action intended to stop or defer the emergence or development of a threat. The preventive use of force has a very specific meaning in international politics, and I am deliberately avoiding a discussion at this point of declaring versus not declaring war. However, I am also assuming that any overt preventive military action could readily be perceived as an act of war, declared or not. This distinguishes the preventive use of military force from covert preventive actions like cyber-attacks and clandestine kinetic activities.
2 Short of eliminating a future threat, preventive force could be used to defer or delay the emergence of a prospective menace in the hopes that in the future, were it to develop, it could be better dealt with.
3 However, this chapter engages only the ongoing debates about the ethics, origins, and implications of the preventive use of force insofar as how they affect and are affected by more practical considerations.
4 The term "drone" is used here in reference to unmanned aerial vehicles (UAVs), also known as remotely piloted vehicles (RPVs). "Drone" has become common parlance, while RPV and UAV tend to be used more commonly in military circles.
5 The fact that the preventive use of force is seriously considered in light of possible emergent military threats but that intensive preventive action on the same scale to ameliorate nascent but observable threats like global warming and its effects is a topic worthy of a conference of its own.
6 Interestingly, illegal immigration could also be classified this way.
7 This is consistent with how Morgenthau describes power in his seminal *Politics among Nations* ([1948] 1993).
8 This list is not exhaustive.

9   It is worth noting that President Bush referred to this concept as preemptive, not preventive war, but his statements here and in his 2006 speech at the National Defense University were understood to be—and more consistent with—the latter. President Bush may have chosen his words deliberately, either to give the impression that the threat was imminent or to avoid the legal and political problems inherent in advocating preventive action.

10 The author would like to express her thanks to remarkably helpful research assistants Hee Yoon (Chloe) Lee and Dante Toppo as well as to her 2014 Freshman Honors seminar students.

BIBLIOGRAPHY

Abrams, Elliot. 2013. "Bombing the Syrian Reactor: The Untold Story." *Commentary*, February 1. http://www.commentarymagazine.com/article/bombing-the-syrian-reactor-the-untold-story/.

Bakircioglu, Onder. 2009. "The Future of Preventive Wars: The Case of Iraq." *Third World Quarterly* 30 (7): 1297–1316.

Bueno de Mesquita, Bruce, and David Lalman. 1992. *War and Reason*. New Haven, CT: Yale University Press.

Bush Administration. 2002. "The National Security Strategy of the United States of America." Washington, DC: The White House. September. http://nssarchive.us/NSSR/2002.pdf.

Caraley, D.J., ed. 2004. *American Hegemony: Preventive War, Iraq, and Imposing Democracy*. New York: Academy of Political Science.

CNN.com. 2004. "Report: No WMD Stockpiles in Iraq." October 7. http://www.cnn.com/2004/WORLD/meast/10/06/iraq.wmd.report/.

Crawford, Neta C. 2003. "The Slippery Slope to Preventive War." *Ethics & International Affairs* 17 (1): 30–36.

Dolan, Chris J. 2005. *In War We Trust: The Bush Doctrine and the Pursuit of Just War*. Burlington, VT: Aldershot.

Doyle, Michael W. 2008. *Striking First: Preemption and Prevention in International Conflict*. Edited, with an introduction by Stephen Macedo. Princeton, NJ: Princeton University Press.

Etzold, Thomas H., and John Lewis Gaddis, eds. 1978. *Containment: Documents on American Policy and Strategy, 1945–1950*. New York: Columbia University Press.

Flood, Alison. 2014. "CIA Used Doctor Zhivago as a Literary Weapon during the Cold War." *Guardian*, April 9. http://www.theguardian.com/books/2014/apr/09/cia-published-doctor-zhivago-in-cold-war-declassified-documents-reveal.

Huntington, Samuel P. 1993. "The Clash of Civilizations?" *Foreign Affairs* 72 (3): 22–49.

Jervis, Robert. 1992. "Political Implications of Loss Aversion." *Political Psychology* 13 (2): 187–204.

Kaufman, Whitley. 2005. "What's Wrong with Preventive War? The Moral and Legal Basis for the Preventive Use of Force." *Ethics & International Affairs* 19 (3): 25–38.

Lemke, Douglas. 2003. "Investigating the Preventive Motive for War." *International Interactions* 29 (4): 273-92.

Levy, Jack S. 2011. "Preventive War: Concept and Propositions," *International Interactions* 37 (1): 87-96.

Luban, David. 2004. "Preventive War." *Philosophy & Public Affairs* 32 (3): 207-48.

Mintz, Alex, and Nehemia Geva. 1997. "The Poliheuristic Theory of Foreign Policy Decisionmaking." In *Decisionmaking on War and Peace: The Cognitive-Rational Debate*, edited by Nehemia Geva and Alex Mintz, 81-101. Boulder, CO: Lynne Rienner.

Mintz, Alex, and Karl DeRouen, Jr. 2010. *Understanding Foreign Policy Decision-Making*. New York: Cambridge University Press.

Morgenthau, Hans J. 1993. *Politics among Nations: The Struggle for Power and Peace*, brief ed., revised by Kenneth W. Thompson. New York: McGraw Hill.

Mueller, John. 2005. "Simplicity and Spook: Terrorism and the Dynamics of Threat Exaggeration." *International Studies Perspectives* 6 (2): 208-234.

Musharbash, Yassin. 2005. "The Future of Terrorism: What al-Qaeda Really Wants." *Spiegel*, August 12. http://www.spiegel.de/international/the-future-of-terrorism-what-al-Qaeda-really-wants-a-369448.html.

Redd, Steven B. 2005. "The Influence of Advisers and Decision Strategies on Foreign Policy Choices: President Clinton's Decision to Use Force in Kosovo." *International Studies Perspectives* 6:129-150.

Rengger, Nicholas. 2008. "The Greatest Treason? On the Subtle Temptations of the Preventive Use of Force." *International Affairs* 84 (5): 949-61.

Renshon, Jonathan. 2006. *Why Leaders Choose War: The Psychology of Prevention*. Westport, CT: Praeger Security International.

Root, Elihu. 1916. "'The Real Monroe Doctrine,' Presidential Address at the Eighth Annual Meeting of the American Society of International Law, Washington, April 22, 1914." In *Addresses on International Subjects, by Elihu Root*, edited by Robert Bacon and James Brown Scott, 109-111. Cambridge, MA: Harvard University Press.

Sathasivam, Kanishkan. 2003. "'No Other Choice': Pakistan's Decision to Test the Bomb." In *Integrating Cognitive and Rational Theories of Foreign Policy Decisionmaking*, edited by Alex Mintz, 55-76. New York: Palgrave MacMillan.

Schweller, Randall. 1992. "Domestic Structure and Preventive War: Are Democracies More Pacific?" *World Politics* 44 (2): 235-269.

Silverstone, Scott A. 2007. *Preventive War and American Democracy*. New York: Routledge.

Stein, Janice Gross. 2013. "Threat Perception in International Relations." In *The Oxford Handbook of Political Psychology*, edited by Leonie Huddy, David O. Sears, and Jack S. Levy, 365-394. Oxford, UK: Oxford University Press.

Waltz, Kenneth. 1959. *Man, the State, and War.* New York: Columbia University Press.

3

Preventive War and Its Domestic Politics

MIROSLAV NINCIC

The continuing prospect of state-sponsored terror attacks and the possibility that weapons of mass destruction might come into threatening hands presage preventive U.S. military operations, with some rising to the level of wars (Nichols 2008). Decisions in this context reflect leadership views of what the U.S. national interest and its identity requires. They are also guided by domestic political considerations, especially by public attitudes that set the boundaries of permissible U.S. policy. In this regard, government calculations concerning preventive war, like those concerning all wars, involve expectations regarding national support for the operation and its implications for the political standing of major incumbents,[1] yet these calculations, and the expectations on which they rest, have not been seriously studied where preventive war is concerned.

At the planning stage, the value of preventive or preemptive war is discussed mainly at the elite level—among military strategists, political leaders, public intellectuals, and so forth. But when it comes to implementing such policies, especially on the battlefield, a much wider national debate must be expected, as the political consequences of military operations direct debate toward the issues most relevant to the public, a public that cannot be excluded from foreign policy debates (Nincic 1992a; Nincic 1992b). Accordingly, this chapter addresses one of this volume's main concerns: how the preventive use of force, when it rises to the level of war, is encouraged or constrained by the nation's democratic context, including the preferences of the nation's public.

I will begin by outlining the basic logic shaping popular evaluation of the desirability of war, after which I will describe the manner in which preventive wars, as a subset of defensive wars, differ from other members of that subset (reactive and preventive operations). From there, I will attempt to show how the public responded to the main instance of a preventive U.S. war (Operation Iraqi Freedom) both at the point of the war's initiation and subsequently, as estimates of the operation's costs and effectiveness could be formed. I will conclude with some observations on the political implications of shifts in war support and on the manner—admittedly modest—in which it could be affected by drone warfare.

The Nature of Preventive War

Preventive wars are a subset of defensive wars—that is, of those intended to protect something that has already been attained, something that lies at the heart of the national interest. Where the war's purpose is to acquire something new, it is not defensive; a war seeking territorial expansion, the spread of a system of beliefs, or even an advantage in a broader geopolitical struggle can rarely be justified as defensive (for example, U.S. motives in the Spanish-American War or the Vietnam War could scarcely be considered as such). Plainly, not every defensive use of force amounts to a war—which we define as a sustained and multi-unit military operation, generally waged against an adversary state amid some expectation of casualties on both sides. This is a level of engagement of which some uses of force fall short. Obviously, too, the definition does not encompass cases where "war" is used in a metaphorical sense, as in the global war on terror.

Depending on the clarity and immediacy of the threat, defensive wars can be reactive, preemptive, or preventive. If reactive, a crucial interest has been attacked, the threat has fully materialized, and the war's evident purpose is to thwart the hostile action. Preemptive wars differ from reactive wars only because the assault has not yet actually occurred. There is, however, no ambiguity about the threat and its implications: it is just around the corner, only preemption can stop it from being actualized.

Preventive war has a different purpose. Aiming to protect the nation from a compelling threat, it shares with reactive and preemptive wars a defensive intent.[2] While a reasonable calculus would normally guide the resort to preventive war, as Taw (Chapter 2) points out, the danger addressed is inevitably hypothetical and projected into the future (no actual steps have been taken by an adversary to build nuclear weapons, no military mobilization has yet occurred, and so on). Not surprisingly, then, the contours of a conjectural threat may be hard to pin down; it is subject to some time-discount, and the associated probabilities often are interpreted through a subjective and political prism. With preventive war, the usual goal is to eliminate *conditions* creating an as-yet partially unformed and not-yet-imminent danger. In Nichols's view, preventive war is directed "against the foundations of future threats, rather than against an imminent danger of attack" (Chapter 4).

Sometimes a grey zone separates preemption and prevention, but the difference is generally apparent. For instance, a preemptive U.S. operation might target a rogue state possessing fully operational nuclear weapons and delivery systems if that state appears to be taking steps to launch an attack against the United States or one of its allies. Preventive military action could be advocated when the adversary is acquiring significant stockpiles of enriched uranium or plutonium, is developing appropriate delivery vehicles, and exhibits interests hostile to those of the United States. Here, the purpose of the nuclear fuel and of the missiles may remain debatable (they may be designed purely for deterrent purposes), the danger is not imminent, and worries focus on the circumstances that may produce an actualized threat. Thus, different interpretations of the apparent danger are possible and subject to political debate. Further, the government could intentionally conflate preventive and preemptive purposes, intending the former but deeming it politically more expedient to invoke the latter. This appears to have been the case with the Bush Doctrine which, as Ivo Daalder observed at the time, "suffers from considerable conceptual confusion, most importantly by conflating the notion of prevention with that of preemption" (2002, 23).

The Limited Empirical Record

Empirical research comparing the domestic environment of preventive wars with that of non-preventive, yet defensive, wars could lead to a firmer grasp of the issue, but the historical record provides limited possibilities for such investigation. In recent decades, the United States has engaged in a number of military actions abroad, but not all rose to the level of war, and, of those that did, only two could have been preventively framed: the war on terror waged in Afghanistan since 2001 (Operation Enduring Freedom) and the military intervention in Iraq launched in 2003 (Operation Iraqi Freedom), but the former at least is a very imperfect case.

With the Afghanistan-centered war, the theme of preventing further terrorist attacks against the United States was present from the outset, justifying the need to destroy the Taliban infrastructure in that country. The problem is that the desire to prevent future terrorist activity launched from Afghanistan and to decapitate its leadership could not be disentangled from a desire for retributive justice after the 9/11 attacks. The president's articulated determination to "smoke them out" and to bring Osama bin Laden to justice "dead or alive" reflected a desire to avenge (Knowlton 2001), a commitment to *lex talionis* that many Americans almost certainly identified with. Since a wish to punish those guilty of past abominations coexisted with a desire to prevent their future crimes, it is hard to consider this a pure case of preventive war.

Operation Iraqi Freedom, on the other hand, was framed in unambiguously preventive terms: a concrete menace was identified and the need to quash it dominated arguments in support of military intervention. The fact that the threat soon proved groundless does not mean that it was not initially taken seriously by the public, possibly even by the Bush administration. But conjectural threats, by their nature, may be poorly understood and can be mischaracterized, with implications for domestic support.

The best way to assess how a preventive war fares in this regard is to compare it to one that is non-preventive. In order to remain in the domain of military operations involving a substantial commitment of U.S. forces and meaningful combat activity in a post–Cold War context, we might compare Operation Iraqi Freedom to the military intervention intended to repel Saddam Hussein's invasion and occupation of Kuwait (Operation Desert Storm, preceded by Operation Desert Shield). That clearly was a reactive war, meant to defeat an ongoing aggressor, not to deal with a future and hypothetical danger. In both cases, the adversary was Saddam Hussein's regime. The similarities and differences between the two provide a credible basis for comparative assessments, and the role of the domestic political setting should be examined at two stages— the initiation of the operation and its subsequent course—since the circumstances shaping public attitudes may differ somewhat between the two.

How the Public Assesses War's Merits

The public is not usually assumed to form its views on military involvements via its own independent examination of the evidence. Often, it resorts to simple cognitive shortcuts that may have little to do with the merits of a war: for example, Republicans are more likely to endorse a war fought by a Republican president; the same applies to Democrats (the role of partisanship will be further addressed in a later section). People also form their opinions on the basis of cues and information initially received from the media by policy elites, who then convey their interpretation of that information to the general public ("opinion followers"), as originally described in the two-stage communications flow theory (Lazarsfeld, Berelson, and Gaudet 1948). John Zaller (1992) has extended our understanding of how most people respond to elite messages (pointing out that their receptivity depends on prior beliefs and on the considerations currently foremost in the person's mind).

Even accepting that public opinion reflects, in part, elite interpretations of the evidence, and that these interpretations are often manifold, it remains the case that peoples' assessments of wars generally reflect a simple "dynamic citizen cost/benefit calculation" (Eichenberg 2005, see also Larson 1996) revolving around three considerations. The first bears on the war's stakes or objectives: How substantial and clear are they, and do they justify the liabilities of armed conflict? The second bears on costs: How great are these costs, especially in terms of U.S. casualties, and are they warranted by the importance of the stakes? The third assessment involves judgments on the effectiveness of force at attaining its objectives—judgments that can be cast in absolute terms (how successfully are military operations attaining their goals?) or in relative terms (could other tools of policy be more effective in this regard?). While scholars differ on the weights of the various factors, they tend to agree that all three matter and that they are, to some extent, mutually conditioned.

Objectives

Extensive studies by the Chicago Council on Foreign Relations regarding the foreign policy objectives most important to the U.S. public indicate how they are ranked and thus suggest, by extension, which goals Americans would most willingly fight for.[3] Figures 3.1 and 3.2 list the ten objectives considered "very important" by the largest number of respondents at the time the United States was engaged in its two most recent wars. Promoting the country's economic interests—mainly protecting American jobs—was considered vital by many. While the 2002 survey, following on the heels of the 9/11 attacks, predictably found the aim of combating terrorism to be especially pressing, economic ends retained a high rank, while promoting human rights and democracy abroad were not considered vital priorities. Protecting weaker nations against aggression was close to the bottom of the list in 1991; nor did it even make the top ten in 2002. Curbing nuclear proliferation was either near the middle or the top of the rankings.

Beyond the importance of the objectives, it probably also matters whether the war's goal is to protect something one has already acquired or to secure something yet to be obtained. As prospect theory demonstrates, the general tendency is to place a greater value on the former than the latter, even if the two are of objectively equal worth, and to be willing to incur greater risks for that which one is defending (Kahneman and Tversky 1979; Thaler 1980; Nincic 1997; McDermott 2007). Thus, greater risk-tolerance would likely accompany the objective of protecting an ally than that of gaining one, though both might be of comparable geopolitical or economic importance.

Costs

It is generally recognized that the costs most salient to the public are U.S. casualties,[4] and that these costs, once experienced, can sway national commitment to war. In his pioneering study of domestic support for the wars in Korea and Vietnam, John Mueller (1973) concluded that this support declines in proportion to the logarithm of U.S. casualties, implying that (1) the public is indeed casualty sensitive and (2) those whose initial support is weakest stand to withdraw it most quickly as casualties mount, meaning that remaining support is apt to be increasingly stable, as residual supporters are less and less responsive to arguments against the war. As Mueller reiterated more recently, "The public gave substantial support to the military ventures in Korea, Vietnam, and Iraq

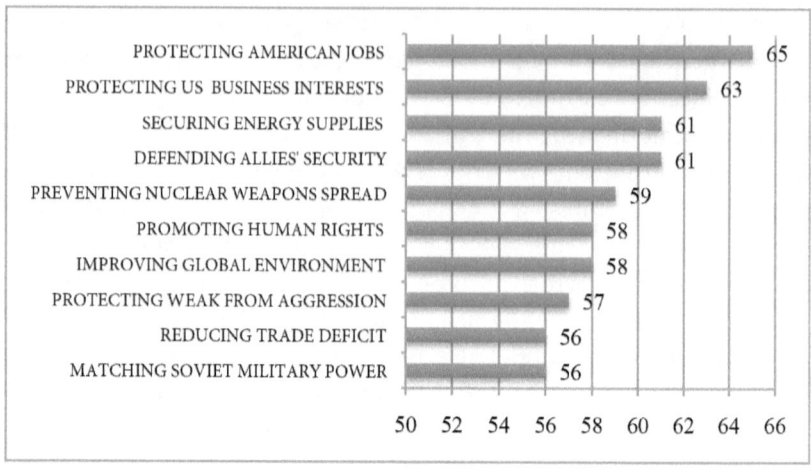

Figure 3.1: "Very Important" US Foreign Policy Goals, 1991 (%)
Source: Chicago Council on Foreign Relations.

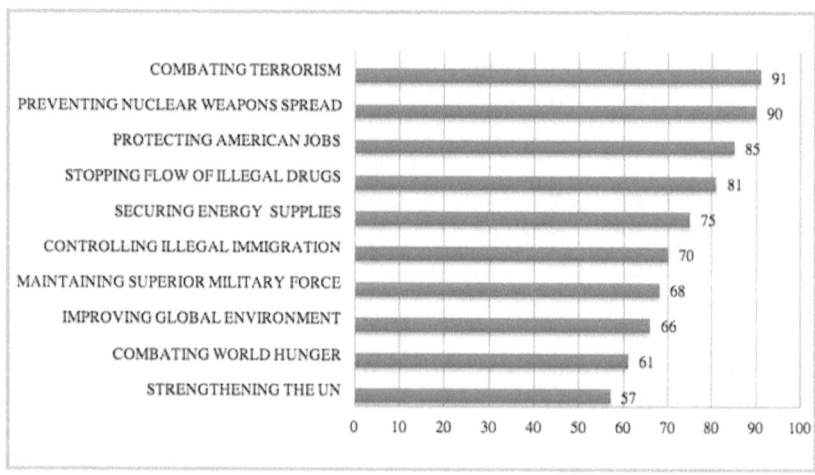

Figure 3.2: "Very Important" Foreign Policy Goals, 2002 (%)
Source: Chicago Council on Foreign Relations.

as the troops were sent in. In all cases, support decreased as casualties—whether of draftees, volunteers, or reservists—mounted" (2005, 788). Some have concluded that the prospect of casualties and their implications for domestic politics reflect not merely casualty sensitivity but "casualty phobia" (Record 2000)—a significant constraint on the pursuit of foreign policy goals by military means.

Effectiveness

Christopher Gelpi (a critic of Mueller's views on casualties) characterizes the public not as "casualty phobic" but as "defeat sensitive," observing that "the public's willingness to bear the human cost of war has varied substantially during different phases of the wars in Korea, Vietnam, and Iraq. The key variable is the perceived likelihood that the mission will succeed" (2006). Eric Larson and Bogdan Savych agree that "those who are more confident in a successful outcome are more likely to support the operation than those who are less confident" (2005, 213). According to an extensive study by Christopher Gelpi, Peter Feaver, and Jason Reifler, "When the public believes that the mission will succeed, it continues to support the mission, even as costs mount. When the public thinks victory is unlikely, even small costs will cause support to plummet" (2005/6, 16).

One may ask whether estimates of the effectiveness of U.S. force may depend, as does casualty tolerance, on the character of the objectives? It should, not only because certain ends are easier to attain by armed force than others, but also because the nature of these ends determines one's ability to gauge military success, while murkiness in this area stands to reflect on war support. It is difficult to judge how well goals are being advanced if one is uncertain what they are. As Jennifer Taw (Chapter 2) observes, with preventive war, "There is no victory or possibility of victory. There is, at best, the elimination or deferral of a threat. This is far harder to measure, prove, and tout as success than the more concrete outcomes of other forms of warfare."

The issue here of whether or not the manner in which costs, benefits, and effectiveness are evaluated depends on the character of the war. A first question is how action directed against a hypothetical threat affects the domestic political environment, especially popular attitudes, to which war-related decisions are sensitive. The various possibilities are not necessarily compatible. A defensive purpose could increase initial public support, but doubts about the threat's credibility could also reduce support and restrain rally effects. Anticipated in the future, its outlines are harder to circumscribe than those of a peril that has already materialized, with two consequences: (1) when contemplating military action, it is easier to disagree on the reality of the menace, and (2) once action is undertaken, it is harder to gauge how far an ill-defined danger has been suppressed, and thus how effective the military operations are. The latter problem might be reduced if the threat was associated *exclusively* with one individual or small set of individuals, or if it was unarguably linked to a specific aspect of the adversary's capacity. Where that is the case, eliminating the person(s) or capacity could indicate success. But although it is expedient to demonize the adversary's leader in order to fuel anger against the regime that person represents, meaningful threats can rarely be so narrowly circumscribed or their elimination so clearly ascertained. The war's effectiveness is, then, debatable, and support may suffer. In this regard, preventive war differs from the other two types of defensive wars, although not necessarily from a number of non-defensive wars.

On the basis of the above, two hypotheses suggest themselves. The first is that, other things being equal, a threat that appears certain (or virtually so) is likely to produce greater support for military action than is a hypothetical threat to which debatable probabilities are assigned (the usual case with preventive war). This should be especially apparent at the time the war is initiated, when objectives dominate national debates and before the matter of costs and effectiveness can be reliably judged. The second hypothesis proposes that preventive wars also are less likely to enjoy continuing support than are other defensive wars—but, here, costs and estimates of likely success play a decisive role. Thus, I will consider national responses to preventive war both at war's outset and as the operation progresses, the two being linked in important ways: on the one hand, a war that is enthusiastically backed at the point of initiation is one whose support has further ability to drop before declines become a serious problem for the administration in charge; on the other hand, a war whose support soon collapses in spite of its initial popularity may deter leadership from initiating preventive operations in the future.

War Initiation and Domestic Support

At the point of initiation, public feelings about a military operation hinge mostly on its goals, since costs and the probability of success often are murky at the outset (Eichenberg 2005). Given the early dominance of objectives (that is, of the threat[s] that military action is supposed to avert), comparing the 1991 Gulf War with the 2003 Iraq War must begin by juxtaposing their proclaimed stakes, for only if the threats were of comparable weight would it be feasible to determine whether, of the two, the preventive purpose produced stronger support.

Operation Desert Storm was justified mainly by the need to reverse a powerful state's invasion of a weaker one. Operation Iraqi Freedom was meant to prevent Saddam Hussein's acquisition of weapons of mass destruction, but justifications for military action went somewhat beyond that. It was unclear, in 1990 or 1991, whether many people would have endorsed the idea of fighting Iraq merely to protect Kuwait. As figure 3.1 indicates, protecting weak nations from aggression ranked quite low as a foreign policy priority (and did not even make the figure 3.2 list). At the same time, the defense of U.S. economic interests, including securing energy supplies, topped the hierarchy of foreign policy goals. Accordingly, the president sought to make the operation more palatable to Americans by declaring, in the run-up to the war, that "'our jobs, our way of life, our own freedom and the freedom of friendly countries around the world would all suffer if control of the world's great oil reserves fell into the hands of Saddam Hussein'" (Apple 1990). The overlapping objectives notwithstanding, most Americans claimed to have a lucid view of the war's purpose: a few weeks before U.S. troops in Saudi Arabia took the battle to Kuwait, 74 percent of Gallup's respondents declared they had "a clear idea of what the United States military involvement in the Iraqi situation is all about" (Gallup 1990).

With regard to the 2003 Iraq War, Saddam's alleged nuclear program was the main threat touted by the administration. This was coupled with warnings about the regime's other weapons of mass destruction. President Bush claimed that the United States or its allies could be attacked "on any given day" with chemical or biological weapons, likening the threat to the 1962 Cuban Missile Crisis (Sanger 2002). The Iraqi dictator was also charged with abetting terrorists, the implication being that this might enable another 9/11 style attack with Iraqi-supplied weapons of mass destruction (Gordon 2003).

Since combatting terrorism and preventing the spread of nuclear weapons were at the top of the 2002 list of top foreign policy goals (figure 3.2), it could be inferred that although the stakes of both wars seemed important to the public, Operation Iraqi Freedom should have garnered even more support. If not (despite the higher-ranked objectives), the value of a preventive justification would be brought into doubt.

Table 3.1 compares, for these two wars, the public's initial response to U.S. war initiation. Two aspects of that response are examined: (1) the extent of the presidential rally effect accompanying the decisions to engage in war and (2) public approval of the decision as expressed in national opinion surveys. For the first, I calculated the increase in the president's approval rating as measured by the two Gallup surveys bracketing the initiation of military activity. For the second, two questions on war approval (asked at the approximate juncture at which the intervention began) were used.

TABLE 3.1 Domestic Responses to Initiation of Two Iraq Wars

Military Operation	Rally Effect (peak magnitude and duration)	Percentage Supporting War
Iraq 1991 (Operation Desert Storm)	19% (Jan. 11–13 to Oct. 17–20, 1991)[a]	75% (Jan. 19, 1991)[c] 79% (Jan. 16, 1991)[d]
Iraq 2003 (Operation Iraqi Freedom)	13% (Mar. 14–15 to Mar. 22–23, 2003)[b]	64% (Mar. 20, 2003)[e] 71% (Mar. 21, 2003)[f]

Question wordings:
a. "Do you approve or disapprove of the way George W. Bush is doing his job as President?" Gallup.com.
b. Same as above.
c. "Do you agree/disagree that the U.S. should take all action, including military force, to make sure Iraq withdraws its forces from Kuwait?" (reporting "agree"). Gallup.com.
d. "Do you think the United States did the right thing in starting military action against Iraq, or should the United States have waited longer to see if the trade embargo and other economic sanctions worked?" (reporting "right thing"). Gallup.com.
e. "Do you think the U.S. made a mistake in sending ground troops to Iraq, or not?" (reporting "did not"). Gallup.com.
f. "Do you think the U.S. made the right decision or the wrong decision in using military force in Iraq?" (reporting "right decision"). pewresearch.org.

The results do not favor the preventive war. The more substantial rally, both in terms of magnitude and duration, accompanied Operation Desert Storm, which was plainly reactive and also did better with regard to direct evidence of approval. Regarding support for the wars at the point of their initiation, then, the preventive justification did not trump the reactive purpose—despite the fact that the direct importance of the objectives to U.S. interests might have predicted the opposite.

Although it was too early for the costs of the war to be witnessed or for judgments regarding the effectiveness of U.S. force to develop, *expectations* of casualties and eventual success could have explained the different levels of enthusiasm for the two operations. There is, however, no evidence that the Gulf War initially looked better than the Iraq War in these respects. Table 3.2 displays the percentage of people who anticipated, in the run-up to each war, whether it would result in (1) a quick war with few U.S. casualties and a quick victory, (2) a long war with many casualties and ultimate U.S. victory, or (3) withdrawal without victory. Slightly more people expected a favorable outcome from the Iraq War; however, the preventive war did less well in terms of initial support despite these perceptions.

TABLE 3.2 Expectations Regarding War Outcomes

Expected Outcome	Gulf War[a] (Nov. 1990)	Iraq War[b] (Feb. 2003)
Short war, few casualties, victory	38%	46%
Long war, many casualties, victory	41%	42%
Withdrawal, no victory	12%	8%
Don't know/Not sure	9%	4%

a. Time/CNN; b. Time/CNN/Harris Interactive.

This (admittedly limited) body of evidence suggests that a preventive justification may not boost initial support, possibly because of the hypothetical nature of the apparent threat. Additional evidence points in the same direction.

In 1964, as America's involvement in Vietnam was gaining momentum and when it was feared that China might soon develop a nuclear capacity, people were asked whether one way out for the United States would be to conduct a preventive war against Communist China.[5] The idea was opposed by a 62 percent to 17 percent margin. Similarly, very few Americans think it acceptable to overthrow foreign governments that pose a weapons of mass destruction (WMD) threat to the United States, although, as table 3.3 indicates, the plausibility and imminence of the threat can make a difference in support for preventive measures.

TABLE 3.3 When the United States Is Justified in Overthrowing Foreign Governments

When it has strong evidence that they are acquiring WMD that might be used to attack the U.S.	31%
When it has strong evidence that the U.S. is in imminent danger of being attacked with WMD by the other country.	41%
Only of the other country attacks first.	9%
It can use force to stop another country from invading, but that does not give it the right to overthrow the government.	15%

Question wording: "Do you think that using military force against countries that may seriously threaten our country, but have not attacked us, can often be justified, rarely be justified, or never be justified [... condition stated, e.g., acquiring WMDs]."
Source: PIPA 2003.

Continuing War Support

Once war is underway and as immediate surges of patriotic zeal are spent, continued backing for the effort depends, again, on the salience and clarity of the threat(s) (that is, on its [their] objectives) and also now on observed costs (especially in U.S. lives) and on indications of how successfully the threat is being dealt with.

A Quantitative Snapshot

What impact did costs and estimates of effectiveness have on public support for the Iraq War? A glance at relevant trends provides a rough answer (see figure 3.3). General (dis)approval for the war is here measured by the percentage of those affirmatively answering the question:

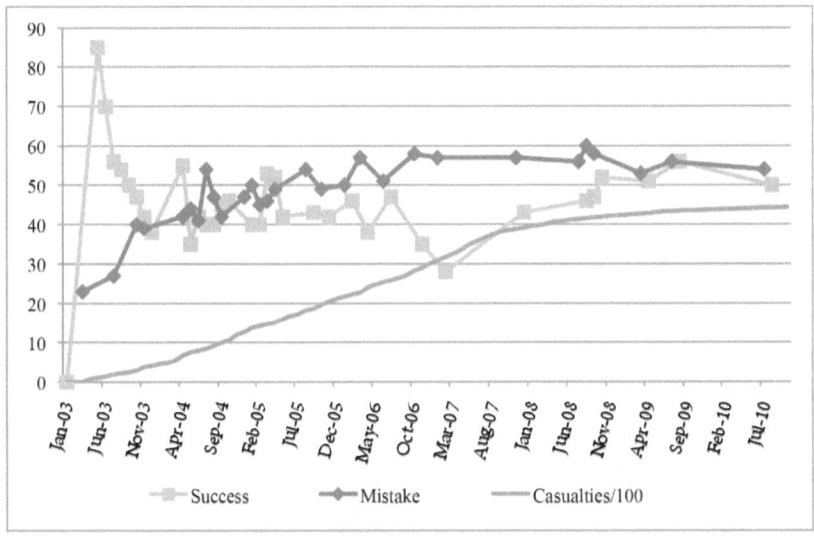

Figure 3.3: Attitudinal Correlates of the Iraq War, 2003-2010 (%)

"Do you think the U.S. made a mistake in sending ground troops to Iraq, or not?"[6] Casualties are as reported in the online Icasualties database.[7] Perception of the success of U.S. military operations are given by those responding "very well" to the question (routinely asked by Gallup): "In general, how would you say things are going for the U.S. in Iraq: very well, moderately well, moderately badly, very badly?

Figure 3.3 indicates that both estimates of success and cumulative casualties had an impact and that this impact was most sharply felt during the war's first two years. By 2006, however, all three curves had flattened out. As Mueller (2005) has pointed out, disapproval of war is linked to casualties by a logarithmic relation, and a comparable assumption (for a similar reason) could be made about perceptions regarding success. A more precise notion of the situation can be gleaned from a simple statistical analysis. Table 3.4 displays the result of a regression of the "mistake" variable on the natural logarithms of "success" and "casualties." Serial correlation of the error term (as reflected in the Durban-Watson statistic) appeared problematic, so estimation was carried out using Newey-West standard errors.

TABLE 3.4 Iraq War as "Mistake"

Variable	Estimated Coefficient	Newey-West Standard Error	T statistic
log(success)	-12.03	3.59	3.35
log(casualties)	9.25	0.81	11.40

$F_{(2,24)} = 66.79$ N=27 Intercept=25.23

Both "success" and "casualties" display a high level of statistical significance. The impact of casualties is particularly unambiguous, but the coefficient for "success" is significant at any of the conventionally accepted levels. These results provide support both for the importance of costs (as reflected in U.S. casualties) and perceived success, and a closer look at the impact of both during the Iraq War furnishes a more direct sense of their respective impact in a preventive war.

The Impact of Costs

The costs of the wars considered here were more modest than for America's other two post–World War II military entanglements (Vietnam and Korea). With a seven-year duration, Operation Iraqi Freedom led to 4,486 American battle deaths, as compared to Korea's 36,516, absorbed over three years. At the same time, however, the casualty-sensitivity of domestic war support has increased quite significantly. With America's major military involvements, popular support drops off most rapidly in the early phases of the involvement, after the initial rally effect has worn off and once patriotic fervor confronts the reality of military embroilment. Such was the case with Operation Iraqi Freedom, but as Mueller notes, "casualty for casualty, support has declined far more quickly than it did during either the Korean War or the Vietnam War" (Mueller 2005, 115). By early 2005, for example, when U.S. combat deaths were around 1,500, public war support had dropped to 50 percent, a decline not registered during the Vietnam War until after the 1968 Tet offensive, when nearly 20,000 Americans had died.

Eventually, the impact of casualties on further declines in support became less pronounced—probably because those with a weaker initial commitment to the war shed their support early on, while remaining supporters were increasingly hard-core and less likely to be swayed by additional casualties.

War's Effectiveness and Purpose

During the initial period, declining enthusiasm for the Iraq War also coincided with deteriorating appraisals of the war's purpose and effectiveness. The initial weeks of the U.S.-led invasion appeared undeniably successful: Baghdad fell to coalition forces on April 9, symbolizing the end of Saddam's rule. By the end of the month, the active combat phase of the operation ended, and on May 1, President Bush delivered his ill-considered Mission Accomplished speech. But the flush of victory soon dissipated with the growing Sunni insurgency (and the activities of the Shia-led Mahdi Army), which marked the emergence of sectarian conflict within Iraq and bloody attacks on coalition forces.

At that time, more and more Americans were reevaluating their thinking on the war. Whereas at the beginning of the invasion, according to Gallup (figure 3.3), only 23 percent of Americans considered the war a mistake, by the end of the year, that number hovered around 40 percent. In January, and before the rally-related surge, 53 percent of those asked felt that the current situation on Iraq was worth going to war over.[8]

After 2005, assessments of the war's success fluctuated in a fairly narrow range, centering on about 50 percent. Doubts that goals were being realized may have reflected, in part, uncertainty about what these were supposed to be, as the conjectural nature of the threats invoked often meant limited clarity about the war's precise aims. As a leading analyst of the determinants of public support for military intervention has explained:

> A failure to articulate clear objectives or a disconnection between the declaratory (or original) and perceived objectives (as in Somalia) can result in high levels of criticism from political leaders and in confusion and declining support from the public. Thus, uncertainty about the stakes, interests, *may* lead to a lack of clarity about the benefits of an intervention; to the extent that this uncertainty results in a discounting or undervaluation of the expected benefits, support will usually fall. (Larson 1996, 11 n. 8)

Even during the war's early phases, many Americans doubted that there was a peril requiring military action. Despite the administration's vigorous initial claims, the Iraqi Survey group—a 1,400-member fact-finding team set up by the Pentagon and CIA to hunt for the alleged WMDs—reported in October that none had been found. At the same time, no evidence emerged of Saddam's links to al-Qaeda. The Bush administration scrambled to redefine the threat, claiming that although no WMDs were present, the Iraqi dictator had wanted to build them and that "Saddam was systematically gaming the system, using the U.N. oil-for-food program to try to influence countries and companies in an effort to undermine sanctions. He was doing so with the intent of restarting his weapons program once the world looked away" (El Deeb 2004). This was a far cry from the President's earlier claim that Iraq could attack the United States with WMDs "on any given day." The public grew more and more confused about the purpose and direction of the war. By December 2003, only 41 percent thought that Iraq's threat required action, while an equal percentage believed that it could have been contained without resort to force.[9] In December, 2003, only 35 percent of the respondents felt that President Bush had clearly explained his "plans for bringing the situation in Iraq to a successful conclusion," a number that dropped to 28 percent by July 2005.[10] The following year, only 34 percent felt that the U.S. goals in Iraq had been clearly explained.[11]

Amid declining enthusiasm for the war, support levels nevertheless fluctuated, the shifts coinciding with major junctures in the war's progress. Declines followed the growth of the Sunni insurgency; improved support tracked the progress of the U.S. surge (an increase of 20,000 combat troops) and the Awakening Movement in Iraq. Still, less than half of Americans tended to consider the intervention a success or, relatedly, to have a clear idea of what it was about. As claims about WMDs and terrorists lost their credibility, the main residual goal appeared to be Saddam Hussein's elimination; indeed, in December 2003, 62 percent of respondents felt that the war could not be won unless Saddam were captured or killed.[12] Yet, just after his capture, 72 percent thought that insurgent attacks against U.S. troops would increase or stay the same.[13] Right after his death, 59 percent thought it would make no difference to Iraq's stability,[14] while 90 percent said that attacks on U.S. troops would either increase or stay the same.[15] The fact that so few were thrilled by

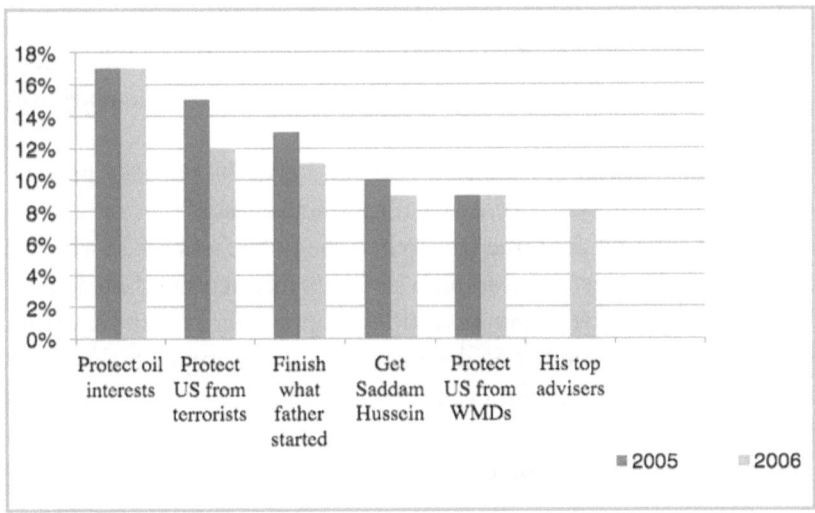

Figure 3.4 Why Bush Went to War, 2005 and 2006
Source: CBS News/New York Times.

an achievement previously considered so necessary suggests widespread confusion about the purpose of military involvement in Iraq.

When respondents were asked in late 2005 why the president had gone to war, not even 20 percent identified a single purpose, and only 9 percent identified Iraqi WMDs as the objective (figure 3.4); in March 2006, the purpose enjoying largest agreement (protecting the United States from terrorists), was chosen by merely 12 percent of the sampled public (figure 3.4). The administration had failed to muster a broad national consensus for *any* given objective.[16]

With preventive wars, initial support may suffer because the threat appears insufficiently compelling, while subsequent backing can be undermined if the threat is not quickly confirmed. In any case, the dangers justifying a preventive war stand to impress Americans less than the clear and immediate ones associated with reactive or preemptive wars. In this regard, preventive wars may resemble *non*-defensive operations more than the other two categories (reactive and preemptive) of defensive wars. Thus, many may have wondered in the fall of 1950 and for the next three years (that is, after North Korean forces had been repelled across the thirty-eighth parallel) what the purpose of continuing U.S.

military operations was. Similarly, and in addition to its toll in U.S. casualties, support for the Vietnam War suffered from murkiness about what, precisely, the United States was seeking to achieve amid objectives several times redefined.

As public backing for the Iraq War (and also the partially preventive war in Afghanistan) decreased, the attractiveness of preventive wars in general suffered the same fate. When the public was asked whether military force against dictators who seriously threaten the United States but have not attacked it can be justified, support for preventive action over the years 2003 and 2006 declined. This change suggests that commitment to the notion of preventive war dropped in tandem with enthusiasm for this particular instance of a war thus justified when threats rather than bellicose action is involved (figure 3.5).[17]

Partisanship and Elections

At this stage, two observations about public support for war are necessary: the first is that attitudes are, to a significant extent, correlated with partisan leanings; the second is that popular preferences are most likely to affect national policy when the political consequences of ignoring them are greatest—namely, at election time.

The Impact of Partisanship

Partisan affiliation seems to be the main filter through which most people interpret and evaluate their nation's wars, and, in the absence of a clear basis for evaluating the desirability and outcome of a military entanglement, a simple cognitive shortcut for many is to follow the lead of politicians they identify with, rather than to try to shape their opinions around an independent analysis of available information. As in the case of preventive wars, where the objectives are ambiguous and the meaning of success is debatable, partisan cues may play an especially powerful role in shaping peoples' attitudes.

Partisan cleavages were manifest with regard to Operation Iraqi Freedom. Figure 3.6 displays the different reactions of Republicans and Democrats to various aspects of the war. Although most Americans were aware that no WMD had been found, some clung to the opposite belief;

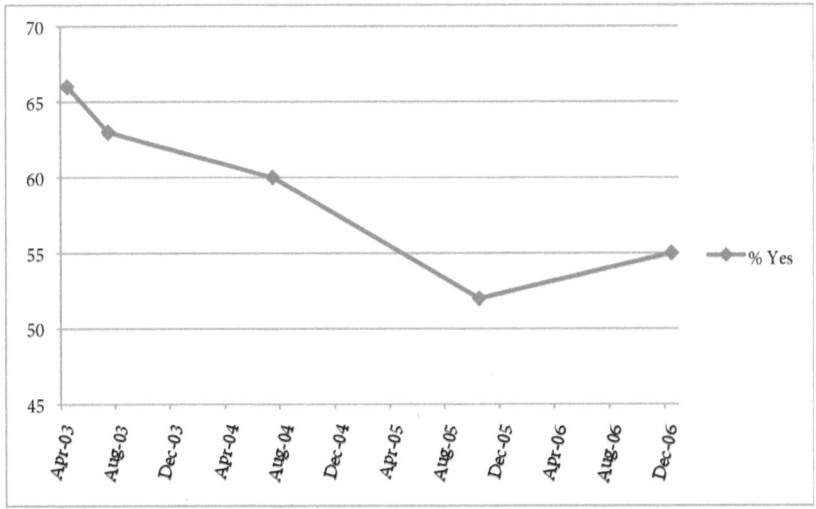

Figure 3.5 Public Support for Preventive War, 2003-2006 (%)
Source: Pew Research Center for the People and the Press and the Press Iraq Poll.

within that group, in 2006, Republicans were more than four times as likely as Democrats (and more likely than Independents) to maintain that conviction.[18] Republicans were significantly more likely, in 2007, to believe that the war was worth its costs.[19] They tended more strongly than Democrats to think that Saddam's death helped the United States achieve its goals in Iraq.[20] Significantly, they were also more inclined to state that they had a clear idea of what the war was all about.[21]

Further light is shed on partisan preferences when it comes to the conditions under which another government could be removed by force (table 3.5). The preventive (and least restrictive) condition *a*, and the one by which the Bush administration sought to justify Operation Iraqi Freedom, is endorsed by twice as many Republicans as Democrats (independents being closer to Democrats, here). Conditions *b* and *c* are chosen by more Democrats than Republicans; yet *b* appears to involve preemption rather than prevention, while *c* implies a reactive use of force. Option *d* precludes the forceful removal of a government (though not necessarily the use of force) under any of these conditions. It is least often chosen, though Democrats are more likely to choose it than Republicans.

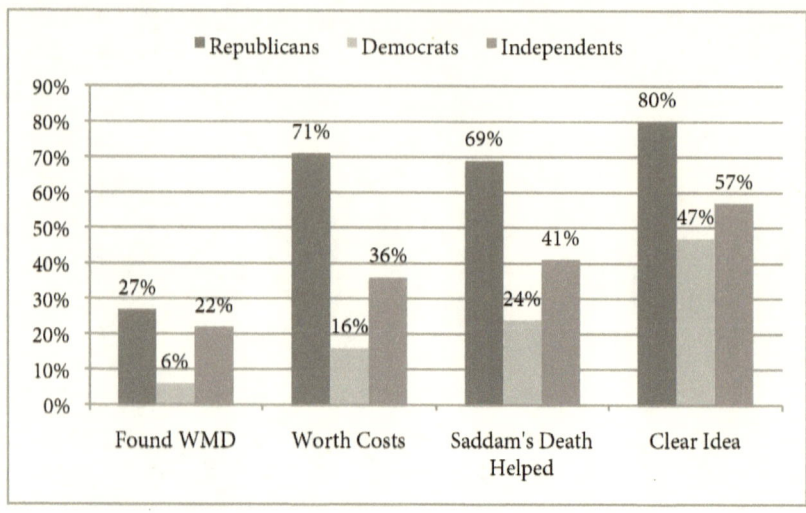

Figure 3.6 Partisanship and the Iraq War
Source: Pipa/Knowledge Networks Poll 2006; ABC News/Washington Post Poll 2007; Gallup/USA Today Poll 2007; CNN Poll 2006.

TABLE 3.5 Partisanship and Preventive War

Conditions justifying military action	Republicans	Democrats	Independents
a. When they have strong evidence that it is acquiring WMD that might be used to attack them.	44	22	28
b. When they have strong evidence that they are in imminent danger of being attacked with WMD by the other country.	42	46	40
c. Only if the other country attacks first.	4	12	13
d. They can use force to stop another country from invading, but that doesn't give them the right to overthrow the government.	8	16	18

Question wording: "Do you think that using military force against countries that may seriously threaten our country, but have not attacked us, can often be justified, rarely be justified, or never be justified?" Source: PIPA 2003.

The Electoral Connection

Though politicians must consider the impact of their policies on electoral outcomes, the extent to which these outcomes are shaped by U.S. war involvement is unclear. While there is some evidence that foreign wars can influence the results of presidential elections (e.g., Nincic 2008), congressional elections rarely revolve around external affairs: few voters know how their congressional representatives (or even senators) stand on those issues, while the high number of "safe" congressional constituencies produced by artful redistricting further diminishes the likelihood that serious challenges would stem from a politician's foreign policy stances. At the margins, however, wars can affect electoral outcomes, as tables 3.6 and 3.7 show. Table 3.6 displays the percentage of people whose 2006 and 2010 congressional votes were determined mainly by the Iraq War. These numbers were capable of making some, though hardly decisive, differences to electoral outcomes.

TABLE 3.6 The Iraq War and Congressional Election Issues

2006			2010	
	McLaughlin and Associates[a] (November)	Fox News/ Opinion Dynamics[b] (November)	CNN/ORC[c] (March)	
War in Iraq	30%	21%	Economy	43%
Jobs/Economy	28%	28%	Health Care	23%
Terrorism	18%	17%	Education	11%
Corruption in Congress	18%		Federal Deficit	8%
All equal		30%	Iraq/Afghanistan Wars	7%
Other/Don't Know/Refused	7%	4%	Energy	2%
			Other/ Don't Know/ Undecided	1%

Question wordings:
a. "Which issues from this following list were most important in deciding your vote for Congress and U.S. Senate?" (November)
b. "Which one of the following issues will be most important to your [2006] vote for Congress?"
c. "Which of the following will be the most important issue when you decide to vote for Congress this year [2010]?"

The situation is not very different where presidential elections were concerned.

TABLE 3.7 The Iraq War and Presidential Election Issues

2004 (Time/SRBI, August)[a]		2008 (CNN/ORC, October)[b]	
Economy	26%	Economy	57%
War on Terrorism	24%	War in Iraq	13%
Iraq	17%	Health Care	13%
Moral Issues (Gay Marriage/Abortion)	16%	Terrorism	11%
Health Care	11%	Illegal Immigration	5%
Other	2%	Other	2%
Don't know/refused		Don't Know/Undecided/Refused	0%

Question wordings:
a. "Which of the following issues is most important to you in deciding how you vote for President in November [2004]?"
b. "Which of the following issues will be most important to you when you decide how to vote for President [in 2008]?"

Even in 2004, when the war had proved more difficult than had earlier been hoped, and when the U.S. economy was doing reasonably well, Iraq dominated the voting intentions of less than a quarter of respondents. By 2008, as the recession began to be felt, Iraq was dwarfed by the economy, tying its ranking with health care. A PEW exit poll indicated that the Iraq War was most important in determining the vote of only 10 percent of the electorate. The difference between presidential and most congressional elections is that, with the former, the outcome is often decided by rather slim margins (considerably lesser, in any case, than 10 percent), so that even percentages such as displayed in table 3.7 must be taken seriously, especially since success or failure in war may produce an aura effect on other dimensions of the public's evaluation of the president. Thus, it has been argued that the 2008 election was largely determined by the Iraq War. According to Gary C. Jacobson,

> Although the economic crisis certainly played a part, the Obama presidency is first and foremost a legacy of George Bush's decision to invade Iraq. Gradual disenchantment with the war was the main reason the

public ended up with decisively negative views of Bush's performance as president. As opinion turned against the war and the president, the Republican Party absorbed collateral damage to its image and appeal as an object of individual identification, costly in this area of strong party-line voting. (2010, 131)

Addressing Costs

To summarize: domestic support for U.S. armed involvement rests on (1) national assessments of the objectives for which the war is fought—that is, on the value placed on these objectives and the clarity with which they present themselves to the public, (2) the war's costs, especially as weighed against the importance of its stakes, and (3) the perceived effectiveness of military operations at advancing these goals. When seeking to understand how these judgments play out in preventive war, we are limited by the scant U.S. experience with this sort of military involvement. Focusing, however, on Operation Iraqi Freedom—a war justified in preventive terms—and comparing it to Operation Desert Storm, an unambiguously reactive war, we are led to two observations.

The first is that the goal of prevention evokes a guarded public reaction—as reflected in the weaker initial backing for the Iraq War than for the Gulf War. The contingent and conjectural nature of the alleged threat may account for much of this reticence. Limited clarity with regard to objectives, and the fact that the threats initially invoked may not be borne out, makes it hard for many people to assess what, exactly, force is meant to achieve and, consequently, how effectively it is being applied. Thus, the case both for the objectives (neutralizing a threat) and the ability of force to achieve them is more problematic with preventive wars than when the operations are reactive or preemptive. Secondly, because it is hard to rely on arguments involving stakes and effectiveness, it is especially important that an apparently preventive war should focus on the third pillar of domestic support—that is, on controlling its costs with regard to U.S. casualties. It is inconceivable that the Iraq War could have maintained domestic support with casualties comparable to those sustained in Korea or Vietnam.[22]

Circumstances that helped minimize Iraq War casualties included the extensive use of airpower (especially in the war's initial stages), the improved equipment of U.S. soldiers (such as better body armor, blood-clotting bandages, and so forth), and the increased reliance on precision-guided weapons and unmanned weapons systems. The latter are critical to limiting exposure to enemy fire; they may also reduce the number of boots on the ground needed for many operations.

In addition, advanced technology such as land robots—which came into their own during Operation Iraqi Freedom—reduced casualties. The Talon robot, for instance, proved effective at finding and removing improvised explosive devices (IEDs), a major and initially intractable source of casualties in Iraq and Afghanistan (Wilson 2007). Subsequently, the Talon was armed with machine guns, yielding the Special Weapons Observation Reconnaissance system (SWORD), capable of rapid gunfire under the remote control of an operator located up to half a mile away (*NBC News* 2005). Development is proceeding on advanced robots capable of missions such as reconnaissance and logistics (Magnuson 2012).

Unmanned aerial vehicles ("drones"), including the armed Predator and Reaper systems, played a less prominent role in the Iraq War, but have been used extensively in Afghanistan as well as against terrorist targets in Pakistan and Yemen.[23] In addition to their effectiveness at target destruction, both long- and short-range drones should contribute to U.S. casualty reduction, involving no risk of loss of pilot life (being remotely operated) or of lives within the search and rescue operations that usually accompany strike aircraft.[24]

If the need to minimize American combat deaths is crucial to maintaining public support for a war claimed to be preventive, then unmanned military systems, which carry many other implications, could be one of the effort's central components. This could mean a more sustained commitment to the war and, perhaps, an improved likelihood that its goals would be achieved (especially if all that stands between success and failure is something that can be destroyed militarily). At the same time, two other consequences of low- or no-cost preventive operations must be considered. First, if the public's main concern with preventive operations is U.S. casualties, then an ability to avoid combat deaths could imply military action without democratic dialogue, or action reflecting an exclusively elite consensus on what the United States should fight for. Second, an ability to target the enemy with unmanned weaponry may encourage the resort to force at a variety of levels, some short of actual warfare, while minimal early casualties might encourage the escalation of involvement to the point where its aggregate goals come to require troop commitments, with the quagmires facing extrication once this happens.

The most likely future area within which U.S.-led preventive war could occur is Iraq and Syria (should a full ISIS sweep of the latter occur). Limited U.S. air strikes against ISIS have taken place in both countries. Military action meeting our definition of preventive war would almost surely be justified by the danger of expanding terrorism and by broader threats to the Middle East and its oil supplies. The results of a CNN poll conducted in early June 2015 indicated that the public would oppose sending U.S. ground troops to fight the Islamic State in Iraq or Syria by a 55 to 43 percent margin.[25] The margin is significant but not enormous, and a U.S. president might believe that intervention would be greeted with a sufficient rally effect to vindicate its initial phase. At that point, the challenges encountered in sustaining public support would have to be addressed—the research presented here having indicated what some of these challenges might be.

NOTES

1 As Stephan Sonnenberg observes in his contribution to this volume (Chapter 5), "The brief history of efforts to constrict countries' right to wage war has shown that legal prohibitions on nations' right to wage war have typically failed. Instead, *domestic* political considerations have tended to be the most effective barriers to war."
2 In this sense, although every war presumably is designed to prevent something undesirable from occurring, not every war is a preventive one, as the term is used here; that is, it does not involve a threat to an interest located at (or near) the pinnacle of national interests.
3 Known since 2006 as the Chicago Council on Global Affairs.
4 The term "casualties" is here used as a synonym for deaths.
5 The question's full wording is: "One way out suggested in Vietnam and in Asia for the United States is to conduct a preventive war against Red China, especially before the Chinese have been able to develop atomic bombs to start a nuclear war on their own. Would you favor or oppose such a preventive war against China?" Harris Survey, December 1964, www.harrispollonline.com.
6 Gallup, January 2003–July 2010, http://www.gallup.com/poll/1633/iraq.aspx.
7 Icasualties.org, http://icasualties.org/Iraq/Index.aspx.
8 Gallup/CNN/USA Today, January 3–5, 2003.

9   Question wording: "Which comes closer to your opinion? . . . Iraq was a threat to the United States that required immediate military action. Iraq was a threat that could have been contained. Iraq was not a threat to the United States." CBS News/New York Times Poll, December 2003.
10 Question wording: "Do you think George W. Bush has explained clearly his plans for bringing the situation in Iraq to a successful conclusion, or do you think he has not explained his plans clearly enough?" Pew, December 2003 and July 2005.
11 Question wording: "Do you think the Bush Administration has clearly explained what the US's (United States) goals are in Iraq, or haven't they done that yet?" CBS News Poll, April 2006.
12 Question wording: "If Saddam Hussein is not captured or killed, then do you think the United States will have won the war against Iraq or not?" CBS News/New York Times Poll, December 10–13, 2003.
13 Question wording: "Now that Saddam Hussein has been captured, do you think the attacks against US (United States) troops in Iraq will increase, decrease, or stay about the same?" CBS News Poll, December 21–22, 2003
14 Question wording: "Now that Saddam Hussein has been executed, do you think Iraq will become more stable, less stable, or won't it make any difference to Iraq's stability?" CBS News Poll, January 1–3, 2007.
15 Question wording: "As you may know, Saddam Hussein has been executed by Iraqi authorities. Now that he has been executed, do you think attacks against the US (United States) troops in Iraq will increase, decrease, or stay about the same?" CBS News Poll, January 1–3, 2007.
16 Question wording: "Why do think the Bush administration decided to go to war against Iraq?" CBS News/New York Times Poll, December 2005 and March 2006.
17 Question wording: "Do you think using military force to remove dictators of countries that may threaten the United States, but have not attacked us is . . . usually the right thing to do, sometimes the right thing to do, rarely the right thing to do, never the right thing to do?" Pew Research Center for the People and the Press and the Press Iraq Poll. Various dates. Reporting "usually" or "sometimes the right thing to do."
18 Question wording: "Since the war in Iraq ended, is it your impression that the US (United States) has or has not found Iraqi weapons of mass destruction?" Pipa/Knowledge Networks Poll, March 2006.
19 Question wording: "All in all, considering the costs to the United States, do you think the war in Iraq was worth fighting, or not?" ABC News/Washington Post Poll, July 2007.
20 Question wording: "Do you think the execution of Saddam Hussein last week will help or hurt the US in its efforts to achieve its goals in Iraq?" Gallup/USA Today Poll, January 2007.
21 Question wording: "Do you feel that you have a clear idea of what the war in Iraq is all about—that is what we are fighting for?" CNN Poll, September 2006.

22 One national newspaper observed, at an early point in the Iraq War, that "Operation Iraqi Freedom illustrates how the United States has forged an era that is closer to bloodless war—for U.S. combat troops, at least—than anyone could have imagined after the carnage in Vietnam" (Cauchon 2003).
23 Several terms are used to refer to these weapons (UAV, RPA); however, as "drones" is the word most often used, it is the one employed in this chapter.
24 For a description of the various drone categories, see Davis et al. 2014.
25 The full question wording is: "Now here are a few questions about the militant group often referred to as ISIS that controls some areas of Iraq and Syria. The group is also sometimes referred to as ISIL or the Islamic State. As you may know, the US has conducted air strikes against ISIS forces in Iraq and in Syria, but there are no US combat troops fighting ISIS forces in either country.) . . . Do you favor or oppose the United States sending ground troops into combat operations against ISIS forces in Iraq or Syria?" (CNN, June 2015).

BIBLIOGRAPHY

Apple, R.W., Jr. 1990. "Bush Says Iraqi Aggression Threatens 'Our Way of Life.'" *New York Times*, August 16. http://www.nytimes.com/1990/08/16/world/confrontation-in-the-gulf-bush-says-iraqi-aggression-threatens-our-way-of-life.html.

Cauchon, Dennis. 2003. "Why U.S. Casualties Were Low." *USA Today*, April 20. http://usatoday30.usatoday.com/news/world/iraq/2003-04-20-cover-usat_x.htm.

Daalder, Ivo F. 2002. "Policy Implications of the Bush Doctrine on Preemption." New York: Council on Foreign Relations, November 16. http://www.cfr.org/international-law/policy-implications-bush-doctrine-preemption/p5251.

Davis, Lynne E., Michael J. McNemey, James Chow, Thomas Hamilton, Sarah Harting, and Daniel Byman. 2014. "Armed and Dangerous? UAVs and U.S. Security." Santa Monica, CA: Rand Corporation. http://www.rand.org/pubs/research_reports/RR449.html.

De Tocqueville, Alexis. (1835) 2004. *Democracy in America*, volume 1. New York: Alfred A. Knopf.

Eichenberg, Richard. 2005. "Victory Has Many Friends: The American Public and the Use of Military Force: 1981–2004." *International Security* 30 (1): 140–177.

El Deeb, Sarah. 2004. "Bush, Cheney Admit Iraq Had No WMD, Take New Tack." *San Diego Union-Tribune*, October 8. http://www.utsandiego.com/uniontrib/20041008/.

Gallup. 1990. "Iraq: Poll on Persian Gulf War, Dec. 13–16." http://www.gallup.com/poll/1633/iraq.aspx.

Gallup/CNN/USA Today. 2003. "Poll Results: President Bush, Postwar Iraq, and the California Recall." January 3-5. http://usatoday30.usatoday.com/news/polls/tables/live/0827.htm

Gelpi, Christopher. 2006. "The Cost of War: How Many Casualties will Americans Tolerate." Foreign Affairs. https://www.foreignaffairs.com/articles/north-korea/2005-09-01/cost-war.

Gelpi, Christopher, Peter D. Feaver, and Jason Reifler. 2005/6. "Success Matters: Casualty Sensitivity and the War in Iraq." *International Security* 30 (3): 7–46.

Gordon, Michael R. 2003. "Bush Enlarges Case for War by Linking Iraq with Terrorists." *New York Times*, January 29. http://www.nytimes.com/2003/01/29/world/state-union-iraq-issue-bush-enlarges-case-for-war-linking-iraq-with-terrorists.html.

Jacobson, Gary C. 2010. "George W. Bush, the Iraq War, and the Election of Barack Obama." *Presidential Studies Quarterly* 40 (2): 207–224.

Kahneman, Daniel, and Amos Tversky. 1979. "Prospect Theory: An Analysis of Decision under Risk." *Econometrica* 47 (2): 263–292.

Knowlton, Brian. 2001. "Terror in America: 'We're Going to Smoke Them Out,' President Airs His Anger." *New York Times*, September 19. http://www.nytimes.com/2001/09/19/news/19iht-t4_30.html.

Larson, Eric. 1996. *Casualties and Consensus: The Historical Role of Casualties in Domestic Support for U.S. Military Operations*. Santa Monica, CA: Rand Corporation.

Larson, Eric, and Bogdan Savych. 2005. *American Public Support for Military Operations from Mogadishu to Baghdad*. Santa Monica, CA: Rand Corporation.

Lazarsfeld, Paul F., Bernard Berelson, and Hazel Gaudet. 1948. *The People's Choice*. New York: Columbia University Press.

McDermott, Rose. 2001. *Risk-Taking in International Politics*. Ann Arbor: University of Michigan Press.

Magnuson, Stew. 2012. "Robot-Makers Ponder Next Moves as Wars Wind Down." *National Defense Magazine*, January 9.

Mueller, John. 1973. *War, Presidents, and Public Opinion*. New York: John Wiley and Sons.

Mueller, John. 2005. "The Iraq Syndrome." *Foreign Affairs*, November/December: 44–54.

NBC News. 2005. "Army Readies Robot Soldier for Iraq." January 24. http://www.nbcnews.com/id/6852832/ns/technology_and_science-tech_and_gadgets/t/army-readies-robot-soldier-iraq/.

Nichols, Thomas N. 2008. *Eve of Destruction: The Coming Age of War*. Philadelphia: University of Pennsylvania Press.

Nincic, Miroslav. 1992a. *Democracy and Foreign Policy: The Fallacy of Political Realism*. New York: Columbia University Press.

Nincic, Miroslav. 1992b. "A Sensible Public: New Perspectives on Popular Opinion and Foreign Policy." *Journal of Conflict Resolution* 36 (4): 772–789.

Nincic, Miroslav. 1997. "Loss Aversion and the Domestic Context of Military Intervention." *Political Research Quarterly* 50 (1): 97–120.

Nincic, Miroslav 2008. "External Affairs and the Electoral Connection." In *The Domestic Sources of American Foreign Policy*, edited by Eugene R. Wittkopf and James M. McCormick, 139–156 . Lanham, MD: Rowman and Littlefield.

Page, Benjamin I., and Robert Y. Shapiro. 1992. *The Rational Public: Fifty Years of Trends in Americans' Policy Preferences*. Chicago: University of Chicago Press.

Record, Jeffrey. 2000. "Failed States and Casualty Phobia." Occasional Paper #18, September. Montgomery, AL: Center for Strategy and Technology, Air War College, Maxwell Air Force Base.

Sanger, David. 2002. "Bush Sees 'Urgent Duty' to Pre-Empt Attack." *New York Times*, October 8. http://www.nytimes.com/2002/10/08/politics/08BUSH.html.

Thaler, Richard. 1980. "Toward a Positive Theory of Consumer Behavior." *Journal of Economic Behavior and Organization* 1:39–60.

Wilson, Clay. 2007. "Improvised Explosive Devices (IEDs) in Iraq and Afghanistan: Effects and Countermeasures." Washington, DC: Congressional Research Service, August 28.

Zaller, John. 1992. *The Nature and Origins of Mass Opinion*. New York: Cambridge University Press.

4

Sovereignty and Preventive War in the Twenty-First Century

A Retrospective on Eve of Destruction: The Coming Age of Preventive War

THOMAS M. NICHOLS

Nearly a decade ago, I wrote a book called *Eve of Destruction: The Coming Age of Preventive War* (2008).[1] In it, I argued that the Westphalian norms of state sovereignty that governed the international system for over 300 years were in the process of rapid collapse and that the major powers were now more prone to resort to military action over threats—or worse, over *perceived* or *asserted* threats. As Stephen Krasner (1999) and others have argued, such "Westphalian norms" are mostly a convenient fiction, and may never have really existed, since nations obviously breached them when they felt it necessary to do so. But states in the nineteenth and twentieth centuries at least pretended they cared about sovereignty, and this provided at least some brake on their use of force. By the end of the twentieth century, however, the major powers seemed to be through with even the pretense of Westphalian sovereignty.

In this chapter, I reconsider these claims and reflect on their application in light of recent international events: What did *Eve of Destruction* get wrong or right, and what has changed in the ensuing decade? I begin by outlining my previous arguments and predictions from *Eve of Destruction* in more detail. Next, I investigate several cases as demonstrations of the continued erosion of state sovereignty and the rise of the preventive war norm: U.S. counterterrorism efforts, the Libya intervention, and the recent Russian interventions in Georgia and the Ukraine. I then discuss the case of Syria, where the U.S. did not intervene in the face of compelling evidence of human rights abuses and violations of international law. I conclude with thoughts on the future of intervention and preventive war.

In Brief: *Eve of Destruction*

Rejecting short-term explanations that were overly focused on single events or countries, *Eve of Destruction* identified three larger, global, and interrelated phenomena, all of them related to the end of the Cold War, which weakened the sovereignty norm and consequently lowered the threshold for the use of military force. First, the multiplicity of humanitarian disasters in the 1990s (themselves a result of internecine conflicts spawned by the post–Cold War power vacuum) challenged the international community, but especially the more militarily capable Western powers, to act without the previous constraints—or the convenient excuses—of Cold War tensions. Second, relatively weak rogue regimes (or to use Anthony Lake's [1994] phrase, "outlaw" or "backlash" states) posed threats to stronger powers in a way they had never been able to do before.[2] Freed from the constraints of Cold War alliances and client relationships, these states became more unpredictable and hence more dangerous. Some of these regimes intensified efforts to gain weapons of mass destruction, including nuclear arms, and at least one, North Korea, succeeded. Third, terrorist organizations operating with the support of these rogue regimes (or from bases in "failed" states) demonstrated the ability to inflict major damage on large nations far out of proportion to their own size. The terror attacks against Washington and New York in 2001, and the subsequent mass casualty attacks in London, Moscow, Madrid, and elsewhere, underscored a new vulnerability among the major powers that was no longer notional or hypothetical.

The book made three general predictions: (1) the United States and other nations would continue to conduct interventions and military strikes practically at will in regions they deemed threats to their security; (2) these actions would be more often *preventive* than *preemptive*—that is, against the foundations of future threats rather than against an imminent danger of attack—a distinction that had already been drained of any difference in the public mind and would soon disappear; and (3) the collapse of the sovereignty norm would lead to more military activity in general, as opportunism would overcome principle and stronger powers would, in effect, do as they pleased in the absence of any international prohibitions against the discretionary use of military force.

Since all of the increased military action described and predicted in *Eve of Destruction* depends on the weakening of norms regarding sovereignty, the first question should be whether those norms have continued to dissolve, or if instead the 1990s and early 2000s were merely an anomalous period in international affairs. To paraphrase noted political scientist Yogi Berra, the future of the sovereignty norm is not what it used to be, even after only a decade. The book was completed before the 2008 U.S. presidential election had even begun, without any knowledge of who the candidates would be or what their positions were. What actually emerged during that contest is important, because the outcome confirmed, at least in the American case, some of the central contentions of the book.

Counterterrorism Policy from Bush to Obama

During his two years in the U.S. Senate and then during the 2008 campaign, presidential candidate Barack Obama consistently opposed President Bush's foreign policies, and repeatedly noted in his primary fight against New York Senator Hillary Clinton that he would not have voted (as she did) to support the invasion of Iraq (Pindell and Klein 2007). He issued searing criticisms of many Bush administration policies, such as the detainee facility at Guantánamo, which he promised to close immediately if elected. He also scored Bush for abuses of executive power in the execution of foreign policy. Interestingly, however, Obama avoided specific criticism of Bush's overseas antiterrorism activities, including drone strikes. Obama was later criticized, particularly by American progressives, for having adopted, or at least refused to rescind, too many of Bush's policies. At the same time, conservative critics wrongly claimed that Obama had flip-flopped on issues like drones (see, for example, Baker 2013).[3]

Whatever his rhetoric in his competition with Clinton, Obama's behavior on the campaign trail foreshadowed predictions by other scholars—with which I agreed and reported in the book—that "neither Bush nor his successors, whatever their party," in the words of John Gaddis in 2005, would be able to ignore the lessons of 9/11 (Gaddis 2005, 2). Or as Robert Lieber wrote in 2006, the "next president, whether a Democrat or Republican," will have something close to a "Bush doctrine" of striking at will but "will call that doctrine by another name" (Lieber 2007, 229).

In 2007, candidate Obama said that he would have no hesitation about striking terrorists inside Pakistan, even if the Pakistanis did not approve. "If we have actionable intelligence about high-value terrorist targets," Obama said during a foreign policy speech, "and [Pakistani] President Musharraf won't act, we will" (Balz 2007). Four years later, President Obama ordered a raid inside Pakistan that killed terror leader Osama Bin Laden. The furious Pakistanis set up a commission to investigate the event; it later called Obama's order a "contemptuous disregard of Pakistan's sovereignty, independence and territorial integrity in the arrogant certainty of . . . unmatched military might," a charge that could easily have substituted Bush's name a decade earlier (Strange 2013).

Leaks from the Pakistani commission's report went virtually unnoticed in the United States, a testimony itself to how little Americans of either party or of any political persuasion cared about objections related to sovereignty, especially with regard to anything connected to post-9/11 operations. Bin Laden was dead. For the American people—who overwhelmingly and unsurprisingly approved of Obama's action—that was good enough.[4]

While few tears were shed anywhere for Osama Bin Laden, critics of the Obama administration were stunned to see the new president engage in preventive killing in other cases as enthusiastically as his predecessor, and perhaps even more so. "Mr. Obama," a *New York Times* profile noted in 2012,

> is the liberal law professor who campaigned against the Iraq war and torture, and then insisted on approving every new name on an expanding "kill list," poring over terrorist suspects' biographies on what one official calls the macabre "baseball cards" of an unconventional war. . . . Nothing else in Mr. Obama's first term has baffled liberal supporters and confounded conservative critics alike as his aggressive counterterrorism record. (Becker and Shane 2012)

But this behavior is baffling only if such actions are viewed purely through a partisan spectrum, or if they are categorized as the result of one presidential administration as opposed to another.

Eve of Destruction argued that preventive attacks, small or large, were soon to be the new norm. While there are obvious differences between Bush and Obama in style and tone, there is a fundamental similarity in their approaches to counterterrorism activities that reflects the changed environment surrounding foreign policy described in the book. Obama's critics have since pointed to the administration's continuation of many of Bush's counterterrorism policies as hypocrisy, but it is exactly the behavior that would have been expected from any American administration, of either party, once in office.[5] Every president learns the hard lesson that the promises and soaring rhetoric made in parking lots and town halls across America get pushed aside during the first national security brief of the first morning in the Oval Office.

In early 2014, the Obama administration backed away from its earlier drone campaigns, with the number of strikes in Pakistan, especially, dropping precipitously over the previous twelve months (Wong 2014). Whether this was due to public pressure, international objections, or a change in the nature of the threat is difficult to determine. Micah Zenko, one of the scholars at the Council on Foreign Relations who created an unofficial tally of 2013 drone deaths, later said that "attempts to correlate the decline in strikes to a decline in specific threats are blocked by secrecy, diplomatic contingency and political convenience" (quoted in Ackerman 2013). It is noteworthy, however, that among many reasons given, including "war weariness" on the part of the American people, objections to operations over the sovereign territory of another nation were conspicuously absent.

If there was something that could be called the "Bush Doctrine," some of its most troubling aspects clearly survived into the succeeding Obama administration. But targeted killings, while technically acts of preventive force, could be considered part of the overall American strategy against terrorism, thus were unlikely to be changed in the near term by any president. In 2013, Obama tried to signal a rhetorical end of the first phase of the "global war on terror," but his policies remained essentially unchanged (Shinkman 2013; Eviatar 2013).

What about other actions, however, such as humanitarian intervention involving the preventive use of more significant military force? Was the NATO attack on Serbia in 1999, for example, led by U.S. President Bill Clinton and U.K. Prime Minister Tony Blair—the so-called "Clinton Doctrine" of preventive action against humanitarian disaster—an exception or a harbinger? (Conservative pundit and Clinton critic Charles Krauthammer [1999] scathingly but memorably referred to the Clinton Doctrine as a generic "anti–son of a bitch policy.") Have these policies remained embedded in American or Western thinking in any recognizable form? In other words, a major test of the central thesis in *Eve of Destruction* rests not only on whether the Bush Doctrine would survive into the twenty-first century, but perhaps even more on whether the twentieth-century Clinton Doctrine would return.

Here, the record of the past decade is mixed, not least because the issue of preventive force has been unwisely alloyed to the broader debate over the "responsibility to protect," or the notion that powerful nations have a positive duty to alleviate large-scale human death and suffering if it is within their power. Now known by its own acronym, "the R2P," it is a subject too large to cover here. As I noted in the book, however, it is important to be mindful of the difference in language between preventive war and the R2P. A duty to *help* is not the same as a duty to *prevent*, as Lee Feinstein and Anne-Marie Slaughter noted in 2004.

Libya and Operation "Odyssey Dawn," 2011

The Obama administration in 2011 had a core of interventionists on its national security team. Anne-Marie Slaughter, an advocate of a more activist U.S. foreign policy, joined the Obama administration as the chief advisor to Secretary of State Hillary Clinton. Samantha Power, the author of a Pulitzer Prize–winning book on Western inaction in the face of genocide, was brought into the National Security Council (Power 2002). Susan Rice, Obama's UN ambassador, had served in the Clinton administration and was, by her own admission, scarred by U.S. inaction during the Rwandan genocide in 1994.[6] When disaster loomed in Libya, these past experiences of President Obama's closest advisors played an important role in the decision to intervene.

The Libya situation was not actually a genocide or a mass killing in progress. Instead, the eventual attack on Libya was a classic case of preventive war, conducted because of a threat to do great harm. When the shock waves of the 2011 Arab Spring reached Libya and protests broke out in the eastern regions of the country, dictator Moammar Qaddafi reacted with characteristic brutality. As the fighting progressed and the body count grew, other Arab nations approached the United Nations with requests for intervention, including a no-fly-zone. Enraged, Qaddafi made clear that he intended to show "no mercy," warning the rebellious residents of Benghazi in March 2011 that "we are coming for you tonight" and "we will find you in your closets" (Kirkpatrick and Fahim 2011).

Whatever the interventionist leanings of his advisors, President Obama was initially hesitant to commit to the use of force to stop Qaddafi from committing a massacre in Benghazi.[7] Nonetheless, in explaining his actions to the American people, Obama specifically referenced his fear of an imminent and massive act of violence: "We knew that if we waited one more day, Benghazi, a city nearly the size of Charlotte, could suffer a massacre that would have reverberated across the region and stained the conscience of the world" (White House 2011).

The White House, however, was slow to act, and it is here that one of the central propositions of *Eve of Destruction* was tested. If the United States demurred, would other nations take advantage of the lowered bar to military action, or was intervention solely an American (or Anglo-American) impulse? Thus, it is important to note that when disaster appeared imminent in Libya, the most aggressive response came from France. This should not be surprising; Paris, like Washington, had a long history with Qaddafi and had been looking for ways to kill him—literally—for decades (Nichols 2008, 84). French President Nicolas Sarkozy and British Prime Minister David Cameron moved with a speed and violence that surprised many Americans who had internalized the stereotype of a pacifist Europe. Sarkozy and Cameron called for ousting Qaddafi weeks before the threat to Benghazi, which was followed by a nonbinding U.S. Senate resolution asking the UN Security Council to impose a no-fly zone. On March 17, 2011, the Security Council approved Resolution 1973, marking only the third time (after Korea in 1950 and Iraq in 1991) that the United Nations approved military action against a member state.

Two days later, a French warplane fired the first shots against Libyan regime forces (BBC 2011). Although France and Britain led the way for intervention, they were not capable of a sustained campaign against Libya without deeper U.S. and NATO participation. Accordingly, the United States launched "Operation Odyssey Dawn" (later NATO "Operation Unified Protector"), albeit reluctantly, in a manner described clumsily by an unnamed Obama administration official as "leading from behind."[8]

Although some observers struggled to find an "Obama Doctrine" in the Libyan operation, there is no evidence that the White House saw itself as establishing any such thing.[9] Instead, as in earlier interventions in Somalia and Kosovo, the ghastly realities of the post–Cold War era muscled their way onto the agenda of Western powers whose leaders clearly preferred dealing with domestic issues rather than with the moral complexities of collapsing dictatorships and civil wars.

The Russian Invasions

The move to embrace more permissive norms regarding the use of force is not merely a U.S. phenomenon. The recent Russian invasions are yet another example of a global shift in norms (for other examples, see Plaw and Reis, Chapter 9). The first intervention occurred in 2008 in Georgia. The George W. Bush administration, caught flat-footed in part because of the actions of the Georgians (who actually shot first), condemned the Russian invasion, but (mistakenly) avoided levying any serious consequences on Moscow (Tapper 2014). In the end, the Americans accepted the status quo ante—that is, a return of Russian forces to areas of Georgia they were already occupying. If there was a concern for Georgian sovereignty, it was not evident in any U.S. or UN response.

In 2014, a more serious conflict erupted between Ukraine and Russia. Unlike the Georgian situation, Russian intervention in Ukraine's internal politics threatened to disrupt not only European stability but potentially to bring the United States and Russia into a direct confrontation of the kind not seen since the Cold War. The general political deterioration of Ukraine is too complicated to recount here, but a few aspects are worth summarizing. Ukraine's borders, like those of other former Soviet states—including Russia—were artifacts of Soviet antinationalist engineering in the 1930s. When the Soviet Union fell, large numbers of Russians found themselves in what was now eastern Ukraine, including the Crimean peninsula, a traditional part of Russia given by Soviet leader Nikita Khrushchev as a "gift" to Ukraine in 1954. In 2013, when faced with the choice of closer association with the European Union or with the Russian Federation, Ukraine's pro-Russian president bowed to Moscow's pressure, a deeply unpopular move that resulted in street protests in Kiev and other areas of Ukraine and eventually to the former president fleeing to Russia.

Pro-Russian forces, encouraged and armed by Russian President Vladimir Putin, seized Crimea, which Putin promptly annexed as part of Russia. Once again, the claim was the protection of the rights of ethnic Russians. Once again, the U.S. administration, this time under President Obama, was caught off guard. Violence broke out in major Ukrainian cities in the east, and the Russians sent tens of thousands of troops to the Ukraine border, poised to invade and raising the specter of the forcible partition of a major European nation by an outside power.

The Russians cited NATO's actions in Kosovo as a precedent, although this was more of a rhetorical feint than any kind of legal precedent (Stojanovic 2014; Somin 2014). The UN Security Council condemned the Russian action by a vote of 13-1-1, with Russia exercising its veto and China, as usual (and cleverly), abstaining. The United Nations vote, of course, meant nothing, and Crimea was absorbed into the Russian Federation, while a significant Russian force remained in close proximity to Ukraine's borders. In May 2014, a pro-Western candidate was elected president of Ukraine, but the country's future remains in peril.

Again, the issue of sovereignty evaporated almost as soon as it was raised. This time, however, the country involved a European nation of 50 million people, rather than a disintegrating or failed smaller state like Yugoslavia or Somalia. The Ukrainians, in vain, protested Russian actions; they cited the Budapest Memorandum of 1994, an agreement among the United States, the United Kingdom, the Russian Federation, and Ukraine in which the three other powers agreed to respect Ukraine's territorial integrity in return for Ukraine relinquishing thousands of Soviet nuclear arms left on its territory (Pifer 2014).[10]

The Budapest Memorandum was not a treaty, and it required almost nothing of the signatories (except, obviously, of Ukraine). And "nothing" is exactly what happened. The agreement mandated that any issues be taken to the UN Security Council, and the United States duly executed that requirement. Many Ukrainians and other observers lamented that the removal of the Soviet arsenal had left Ukraine open to Russian attack, but this was a misunderstanding not only of the situation in 2014 but of the circumstances in which the original memorandum had been negotiated. Indeed, had Ukraine retained a nuclear arsenal, it is highly likely that Russia would have used the pretext of instability, perhaps during the 2004 "Orange Revolution," to intervene earlier on the openly preventive pretext of securing those weapons (Nichols 2014). Nonetheless, Russian actions in 2014 showed not only that Moscow cared little for its previous commitments to Ukrainian sovereignty, but that Washington and London were of no mind to make them an issue either.

Traditional realists would argue that cases such as Syria and Ukraine simply underscore the ancient truth that anarchy and power, rather than law or ideas, rule international life. As is so often the case with realist arguments, however, this is a truism that does not explain very much. During the Cold War, aggression across borders, forced redrawing of maps, and the violation of territorial integrity were the stuff of major crises, often leading to war. In the twenty-first century, these have become daily events, to the point where the international community, whatever its expressions of organized outrage in Turtle Bay or Geneva, has become inured to them, and thus accepting of them by default. As of mid-2015, the international community has accepted what is in effect the de facto partition of Ukraine. The only remaining question is whether Russia will enlarge the conflict.

Syria: The Disconfirming Case?

Given the evidence presented thus far, it would seem that preventive war is here to stay. Not only has the concept survived through multiple changes of leadership in major powers, but also some of the most prominent advocates of forceful intervention, such as Samantha Power, occupy senior positions in the U.S. government. Although Power and others were not supporters of the Bush Doctrine of preventive attack against military threats, they are vocal supporters of the use of Western power to prevent humanitarian disasters—an important difference in some respects, but almost indistinguishable in terms of the use of preventive force and norms regarding sovereignty.

What, then, explains the lack of Western action in the Syrian civil war? Obviously, the situation in Syria is much different than in other recent cases such as Libya. As journalist David Remnick wrote in the *New Yorker*, "Part of Obama's anti-doctrinal doctrine is that it insists on the recognition of differences in a way that Bush's fixed ideas did not. Complex as Libya was, and remains, Syria is infinitely more so" (2011). Yet the sheer scale of killing, including the use of chemical weapons against civilians, would seem to be a perfect test case, if not of prevention, then at least of the violability of the sovereignty norm in the face of barbarous killing. Of course, deaths are not an automatic trigger for action; the international community's response—or lack of it—to the genocide in Darfur, Sudan, showed starkly that there was no magic threshold above which Western saviors would parachute into mass bloodshed (see, for example, Straus 2005). The eventual answer in Sudan would be to partition the country, a short-term solution that may yet prove unviable.

Still, Syria is a different case. Unlike central African nations whose problems might be viewed, fairly or not, as intractable, Syria sits on the shores and borders of the U.S. European Command and NATO itself. Through 2012, the Syrian death toll had passed 50,000 and was rising with no end in sight. In August of that year, President Obama reiterated American unwillingness to get involved in yet another chaotic Middle Eastern mess, but he also emphasized, in a statement that would later haunt him, that there could be actions which would force Washington's hand: "We have been very clear to the [Syrian President Bashar] Assad regime, but also to other players on the ground, that a red line for us is we start seeing a whole bunch of chemical weapons moving around or being utilized. That would change my calculus. That would change my equation" (Obama, 2012).

The equation changed a year later. In August 2013, chemical attacks at Ghouta, an area near Damascus, killed thousands of civilians. A UN report a month later confirmed the agent was Sarin gas, delivered by rocket. Although the United Nations declined to assign blame for the attack, the United States and its allies determined that the Syrian regime, rather than the rebels, had launched the weapons. Attempts to undermine that finding have since been debunked, but the key point here is that Washington and its allies were at that moment convinced of the Syrian regime's guilt.[11] Other nations and international organizations, such as the Arab League, also believed that Damascus had ordered the attack, but widespread outrage did not translate into widespread calls for intervention.

Nonetheless, the 2013 chemical attacks forced a simmering fight within Washington to a public boil. The U.S. military, or at least a significant portion of it including Joint Chiefs Chairman Martin Dempsey, were opposed to involvement in Syria of almost any kind. Two months before the Ghouta attacks, Secretary of State John Kerry pressed for a military response in the face of evidence of previous Syrian chemical attacks. The chairman, according to journalist Jeffrey Goldberg, fired back at the secretary, and "loudly."

Dempsey threw a series of brushback pitches at Kerry, demanding to know just exactly what the post-strike plan would be and pointing out that the State Department did not fully grasp the complexity of such an operation. Dempsey informed Kerry that the Air Force could not simply drop a few bombs, or fire a few missiles, at targets inside Syria. To be safe, the United States would have to neutralize Syria's integrated air-defense system, an operation that would require 700 or more sorties. At a time when the U.S. military was exhausted, and when sequestration was ripping into the Pentagon budget, Dempsey is said to have argued that a demand by the State Department for precipitous military action in a murky civil war was not welcome. As former Reagan administration official Elliott Abrams later complained, this was not military advice but a political preference cloaked in military terms; he also noted that, in 2007, when George W. Bush asked Dempsey's predecessor, General Peter Pace, if the United States could strike Syria's nuclear program, "Pace told him we could do it, period" (Abrams 2013).

As in previous interventions (particularly the Bosnian situation, in which then-Secretary of State Madeleine Albright chided Colin Powell for a reluctance to use force), civilian and military policymakers were at odds over what violence could achieve. Indeed, retired Major General Robert Scales claimed to speak for a wide spectrum of senior officers when he said, in effect, that the military had no faith in the administration's judgment on such matters, a startling breach of U.S. civil-military traditions even from a retired general officer. This led to an unusual and public rebuke of Gates from another, and more senior, retired general, David Barno (2013).

Note, however, that in these debates, neither Dempsey nor others made the argument that the United States *should not* intervene in Syria; the argument, rather, was that it *could not* intervene. There was no serious case made at senior levels in the White House (at least publicly) about trespassing with force upon Syrian sovereignty. If anything, the debate about Syria both within the United States and between the United States and other nations revolved around Syria's transgressions against the norms and moral values of the international community rather than the other way around, especially if judged by Secretary Kerry's impassioned speech on August 25, 2013, in which he referred to the chemical attacks as "a moral obscenity" (Fisher 2013).

Kerry's speech, as well as U.S. military preparations, signaled that the Obama administration had settled on striking Syrian regime targets. Naval and air assets were moved into place, and plans were apparently drawn up. Dempsey's criticisms vanished into a quiet admission that if the president wanted to strike Syria, it could be done. "I'm confident," Dempsey told the Senate in September 2013, "in the capabilities we can bring to bear to deter and degrade. And it won't surprise you to know that we will have not only an initial target set but subsequent target sets should they become necessary" (*Washington Post* 2013a). This was hardly a ringing endorsement of force, but was a far cry from the dressing down he delivered to Kerry only a few months earlier. For a time, it seemed as though the predictions in *Eve of Destruction* would be fulfilled again in Syria.

So what happened? Why did the Americans at the last minute pull back from military action, dissolve into recriminations between the legislative and executive branches, and then finally just throw the whole matter into the hands of the Russians, effectively outsourcing U.S. security policy in the Middle East to the Kremlin for the foreseeable future (Nichols and Schindler 2013)?

Single cases, of course, are replete with unique circumstances, including accidents and other contingent events that complicate drawing overarching conclusions. In the case of Syria, it is difficult to know how the August crisis might have turned out differently had, for example, British Prime Minster David Cameron more adeptly handled a vote on action in Syria in Parliament, where he was handed a stinging defeat, or if John Kerry had not, in a moment of frustration, thrown out a casual comment that the Russians then seized as the cornerstone of an agreement.

Any U.S. plan to use force disappeared somewhere along the stretch of Pennsylvania Avenue between Capitol Hill and the White House. Congress, as is its tradition, claimed to want involvement in the decision, but in truth wanted no part of the responsibility for committing American military power in Syria (or anywhere else, for that matter).[12] Perhaps more important, the balance of political forces inside the White House tilted against the interventionists—or, more accurately, the White House chief of staff achieved a last minute change in the president's policy. As the *New York Times* reported in a detailed October 2013 story, there was already friction between Chief of Staff Denis McDonough and Samantha Power, especially over Syria. Power had come to believe that America's offers of support to the Syrian rebels were empty.

> "Denis, if you had met the rebels as frequently as I have, you would be as passionate as I am," Ms. Power told Mr. McDonough at one meeting, according to two people who attended.
> "Samantha, we'll just have to agree to disagree," Mr. McDonough responded crisply. (Mazzetti, Worth, and Gordon 2013)

And yet, after the Syrian gas strikes, Obama apparently decided to move toward military action—"within hours," according to the *Times* report—at least until a last-minute discussion with McDonough.

On August 30, a week after the Ghouta attack, Obama and McDonough took a walk on the White House grounds. Afterward, the president called National Security Adviser Susan Rice and others into the Oval Office to inform them he was delaying military action and sending the matter to Congress. This led to "a vigorous debate that lasted two hours," according to unnamed "senior administration officials" (Rampton and Mason 2013). NBC political reporter Chuck Todd put it a bit more bluntly: "The plan was immediately met with robust resistance from a whiplashed Obama [national security] team," who were gearing up to support the use of force after Kerry had just laid out "the administration's strongest case yet for action against Assad" (2013).

To ask partisans in Washington what happened in the summer of 2013 was to get a full spread of conflicting explanations. Obama's supporters initially argued that the president had outfoxed the Syrians and the Russians, forcing them to agree to destroy Syria's chemical stocks.

Critics, in contrast, argued that the Russian deal was a face-saving exit that required very little of Syria other than to dispose of the weapons left now that it was through with them. By mid-2014, both sides claimed they were right. Syria had signed on to U.N. chemical weapons conventions and shed most of its arsenal. On the other hand, a year after Ghouta, Kerry charged Syria with new chemical attacks, this time with chlorine dropped from helicopters, but said the administration would refuse to "pin ourselves down" to any particular response (Wroughton and Osborn 2014).

An administration that had campaigned on ending, rather than starting, any further wars in the Middle East did in fact smother the early impulse, as French President Francois Hollande put it at the time, to "punish those who took the heinous decision to gas innocents" (Corbe 2013). Again, however, the failure to act as a matter of *expedience* rather than as a matter of *principle* is an important distinction. President Obama's eventual address to the American nation on Syria had all the marks of being cobbled together from two different speeches: the one he planned to give at the outset of military action, and the one that was hastily put together when McDonough convinced him to throw the question of military force to Congress and thus, effectively, to bury it as an option. At the outset, Obama said:

> Now, after careful deliberation, I have decided that the United States should take military action against Syrian regime targets. This would not be an open-ended intervention. We would not put boots on the ground. Instead, our action would be designed to be limited in duration and scope. But I'm confident we can hold the Assad regime accountable for their use of chemical weapons, deter this kind of behavior, and degrade their capacity to carry it out.

In the same speech, he then pivots away from this threat: "This morning, I spoke with all four congressional leaders, and they've agreed to schedule a debate and then a vote as soon as Congress comes back into session" (*Washington Post* 2013b).

Adding to the fractured nature of the speech was the president's assertion that he did not really need Congress to act, but would ask anyway. "I believe I have the authority to carry out this military action without specific congressional authorization. I know that the country will be stronger if we take this course, and our actions will be even more effective. We should have this debate." This passage betrays the incoherence of the policy behind it: to say a debate is needed is one thing, but to say a debate is needed after announcing that a decision to use force has *already been made* is quite another.

Where the future of preventive war is concerned, however, the important point is that nowhere in the president's remarks, nor in the fiery charge led by Secretary Kerry a few days earlier, are there any traditional concerns for sovereignty or about levying a punitive attack on another nation for internal actions. Echoing the Clinton Doctrine, all of these statements treat Syria as a transnational problem, a kind of cancer whose diagnosis was certain but whose treatment was yet to be determined. The United Nations Security Council, a central player in the Libyan events, was once again relegated to the status of observer, and by late September of 2013, the only negotiations that mattered over Syria were the discussions between the United States and Russia.

The French and British reactions are likewise interesting in this regard. Hollande, perhaps looking to replicate French success in spurring the Libyan intervention, took a fundamentally moralist position, eschewing any talk of long-term outcomes and seeking instead to inflict punishment directly on Assad for his sins. Cameron took much the same position, and one that would have been nearly indistinguishable from his predecessor, Blair, two decades earlier (United Kingdom 2013). It probably says more about Cameron's clumsiness as a political leader than it does about any retreat from preventive war that his own party handed him a defeat in Parliament.[13]

In the end, however, the fact of the matter is that for over two years the United States chose not to act in a situation that should have provided the clearest case for the use of force in a new age of preventive war and virtually unrestrained humanitarian intervention. It is possible that a future administration will take action against the Assad regime should it survive in power and the carnage in Syria continues, but for the foreseeable future the United States and NATO have turned away from that path. Whether this was due to war weariness, failed alliance management, contingent facts on the ground, or a simple failure of nerve among U.S. and Western leaders, it is nonetheless undeniable that the

Syrian case should raise serious questions about the central contentions in *Eve of Destruction*.

The Future of Intervention and Preventive War

In May 2014, President Obama gave a speech on U.S. foreign policy at West Point, symbolically returning to the same institution where President George W. Bush laid out a call to arms for the war on terror in 2002 (*New York Times* 2002). In it, Obama seemed to step back from the muscular foreign policy activism embodied in the Clinton and Bush Doctrines of the previous two decades:

> Today, according to self-described realists, conflicts in Syria or Ukraine or the Central African Republic are not ours to solve. . . . A different view, from interventionists from the left and right, says that we ignore these conflicts at our own peril, that America's willingness to apply force around the world is the ultimate safeguard against chaos, and America's failure to act in the face of Syrian brutality or Russian provocations not only violates our conscience, but invites escalating aggression in the future.
>
> And each side can point to history to support its claims, but I believe neither view fully speaks to the demands of this moment. . . . To say that we have an interest in pursuing peace and freedom beyond our borders is not to say that every problem has a military solution. (*Washington Post* 2014)

The White House marketed the speech as a major address on foreign affairs, and reactions revealed the usual partisan divide: supporters saw in the President's remarks a new call for prudence and measure after too many years of war, while critics saw little more than the demolition of straw men as a means of covering a string of foreign policy failures.

Once again, however, analysts searching for an "Obama Doctrine" in the speech were to be disappointed. In the end, the president promised to oppose an approach to foreign policy that no one advocated:

> Here's my bottom line: America must always lead on the world stage. If we don't, no one else will. The military that you have joined is, and always

will be, the backbone of that leadership. But U.S. military action cannot be the only—or even primary—component of our leadership in every instance. Just because we have the best hammer does not mean that every problem is a nail.

It would have been hard to find serious political figures on the American right or left, isolationist or interventionist, who represented the extremes between which the president claimed to be navigating. Likewise, the speech contained no radical alteration of policy, and no initiatives followed in its wake.

And so American foreign policy is set to continue along the fitful interventionist route of the previous twenty years. Just days before the West Point address, for example, two former Bush and Obama administration officials, along with an Iran expert from the Council on Foreign Relations, suggested that Congress in effect vote to "pre-authorize" war with Iran if talks with the Islamic Republic over its nuclear weapons program failed. Such an idea might have been radical two decades ago, but now is standard fare in America's newspapers of record and top foreign policy journals.

Especially in the wake of the Syrian debacle, it is difficult to imagine the United States going to war with Iran. Indeed, one of the arguments in *Eve of Destruction* was that "Iraq-like" actions were unlikely in any case (see also Taw, Chapter 2; Nincic, Chapter 3). But there are other options short of major war, especially with new technologies—like drones—that allow for long-distance, low-risk military operations. There are also unforeseeable contingencies, such as another Libyan situation in which the United States is drawn into conflict by the more activist policies of other nations, with Israel the obvious candidate in Iran. Much can change along the road to intervention or prevention.

One thing definitely has not changed: the words "preventive" and "preemptive" are still used as imprecisely as they were when the book was written, and perhaps even more so today. "Preventive war" remains a term of opprobrium, and so even as the major powers continue to consider—or even to undertake—preventive military actions, they are usually described as "preemptive." The label of "prevention" is used only in retrospect and only with disapproval, unless it is attached to words like "diplomacy."

The other reality, of course, is that the United Nations Security Council remains almost completely useless. This was a prediction and part of a plea for change in the book, but it was hardly a moment of clairvoyance on my part. A moment of successful cooperation on the Libya resolution turned to ashes during the Russian invasion of Ukraine, proving yet again that the institution of the veto has long outlived its usefulness. While the Libyan resolution was a success, the lesson of Libya might only be that the trigger for UN action is an openly barbaric threat by a universally despised and obviously insane dictator in a small and weak country. That is not much of a bar to clear, and unlikely to happen very often, if ever again.

In the decade since *Eve of Destruction* was undertaken, the world has slid deeper toward the preventive use of force, but the worst scenarios have not yet come to pass. Whether this is because the book's claims were too large or because it is too early to know if darker days are ahead is a matter for the reader to decide. There is evidence for both interpretations: the United States, at least for a while, has lost any interest in large-scale military actions. It has not, however, lost its appetite for the preventive use of force on a smaller scale, as those on President Obama's "kill list" can attest (see, for example, Sonnenberg, Chapter 5; Fair, Chapter 8). Russia, once the vocal proponent of preventive action anywhere in the world, has instead settled for the complete abandonment of any pretenses about military force on its own borders. The Europeans, inconstant and bickering among themselves as always, nonetheless demonstrated a willingness to use preventive force against Libya, while faltering over the very same issue a year later in Syria.

If the greatest fears expressed in *Eve of Destruction* have not been realized, the record of the past decade is nonetheless ground for continued concern. The Ukraine crisis in particular should raise a significant alarm: the complete Russian disregard for any concept of sovereignty in a state whose borders Moscow had once vowed to respect brought the world closer to a major power confrontation than at any time since the worst Cold War days of the early 1980s. This kind of collision of interests among the great powers, stomping about like elephants in a room full of mice, was one of the concerns that led to the writing of *Eve of Destruction*, and I am no more optimistic that we will avoid that future now than I was a decade ago.

NOTES

1. Although published in late 2008, *Eve of Destruction* was begun in 2004, with the final draft completed in late 2006 and early 2007.
2. Lake listed five major outlaw states in 1994: North Korea, Iran, Iraq, Cuba, and Libya.
3. The idea that Obama flip-flopped on the issue of drones in particular became such an article of faith among Republicans after Obama took office that the writers at Politifact.com finally waded into the controversy, declaring the charge false. "Should President Barack Obama Apologize for Flip-Flop on Drone Strikes? Louisiana Lawmaker Says Yes," Politifact.com, October 5, 2011.
4. Obama's overall approval rating spiked by at least nine points after the raid, with even higher ratings for his handling of terrorism. See Cohen and Balz 2011.
5. Writing in *Salon*, Glenn Greenwald (who would later become the Obama administration's nemesis in the Edward Snowden spy scandal) said in 2012 that "the Democratic Party owes a sincere apology to George Bush, Dick Cheney and company for enthusiastically embracing many of the very Terrorism [sic] policies which caused them to hurl such vehement invective at the GOP for all those years."
6. See, for example Crowley 2011.
7. Whether that slaughter would have taken place is now unknowable, as is often the case with evidence for a preventive strike.
8. The provenance of the quotation was debated as soon as it appeared. Reporter Ryan Lizza stands behind his assertion it came from inside the White House. See Rogin 2011.
9. Fareed Zakaria (2011), for one, made a more coherent "model" out of the Libyan operation than the White House itself ever claimed or tried to establish.
10. Ambassador Steven Pifer, who helped negotiate the agreement, noted that it was the best that could be done under the circumstances of the time. See Pifer 2014.
11. Shortly after the attacks, Seymour Hersh tried to make this case in the pages of the *London Review of Books*; his argument was dismantled in the pages of *Foreign Policy* shortly after. See Higgins 2013.
12. As scholar Gordon Silverstein (1996) has noted, there is a long tradition of Congressional "blame avoidance" in foreign policy.
13. Hollande is said to have referred to Cameron's defeat in Parliament as "a schoolboy error" of basic politics. See Telegraph 2013.

BIBLIOGRAPHY

Abrams, Elliot. 2013. "Syria and the 700 Sorties." *Pressure Points*, CFR.org, June 19. http://blogs.cfr.org/abrams/2013/06/19/syria-and-the-700-sorties/.

Ackerman, Spencer. 2013. "Fewer Deaths from Drone Strikes in 2013 after Obama Policy Change." *Guardian*, December 31. http://www.theguardian.com/world/2013/dec/31/deaths-drone-strikes-obama-policy-change.

Baker, Peter. 2013. "Obama's Turn in Bush's Bind." *New York Times*, February 9. http://www.nytimes.com/2013/02/10/world/obamas-turn-in-bushs-bind-with-defense-policies.html?_r=0.

Balz, Dan. 2007. "Obama Would Take Fight to Pakistan." *Washington Post*, August 2. http://www.washingtonpost.com/wp-dyn/content/article/2007/08/01/AR2007080101233.html.

Barno, David. 2013. "U.S. War Decisions Rightfully Belong to Elected Civilian Leaders, Not the Military." *Washington Post*, September 12. https://www.washingtonpost.com/opinions/us-war-decisions-rightfully-belong-to-elected-civilian-leaders-not-the-military/2013/09/12/d31d2e9e-1a43-11e3-a628-7e6dde8f889d_story.html.

BBC News. 2011. "Libya: French Plane Fires on Military Vehicle." March 19. http://www.bbc.co.uk/news/world-africa-12795971.

Becker, Jo, and Scott Shane. 2012. "Secret 'Kill List' Proves a Test of Obama's Principles and Will." *New York Times*, May 29. http://www.nytimes.com/2012/05/29/world/obamas-leadership-in-war-on-al-qaeda.html.

Cohen, Jon, and Dan Balz. 2011. "Osama bin Laden Killing Gives Obama Quick but Limited Ratings Boost." *Washington Post*, May 3. http://www.washingtonpost.com/politics/osama-bin-laden-killing-gives-obama-quick-but-limited-ratings-boost/2011/05/03/AFhxJegF_story.html.

Corbe, Sylvie. 2013. "France 'Ready to Punish' Syria over Gas Attack." Associated Press, August 27. http://www.huffingtonpost.com/2013/08/27/france-syria_n_3823398.html.

Crowley, Michael. 2011. "Susan Rice, Samantha Power, Rwanda and Libya." *Swampland* blog, TIME.com, March 24. http://swampland.time.com/2011/03/24/susan-rice-samantha-power-rwanda-and-libya/.

Eviatar, Daphne. 2013. "Counterterrorism: Where Are Obama's Policy Changes?" *Reuters*, October 21. http://blogs.reuters.com/great-debate/2013/10/21/counterterrorism-where-are-obamas-policy-changes/.

Feinstein, Lee, and Anne-Marie Slaughter. 2004. "A Duty to Prevent." *Foreign Affairs* 83 (1): 136–150.

Fisher, Max. 2013. "Read the Full Transcript: Kerry's Speech on Syria, Chemical Weapons and the Need to Respond." *Washington Post*, August 26. http://www.washingtonpost.com/blogs/worldviews/wp/2013/08/26/read-the-full-transcript-kerrys-speech-on-syria-chemical-weapons-and-the-need-to-respond/.

Gaddis, John Lewis. 2005. "Grand Strategy in the Second Term." *Foreign Affairs* 84 (1): 2–15.

Greenwald, Glenn. 2012. "Repulsive Progressive Hypocrisy." *Salon*, February 8. http://www.salon.com/2012/02/08/repulsive_progressive_hypocrisy/.

Higgins, Eliot. 2013. "Sy Hersh's Chemical Misfire." ForeignPolicy.com, December 9. http://foreignpolicy.com/2013/12/09/sy-hershs-chemical-misfire/.

Kirkpatrick, David. D., and Kareem Fahim. 2011. "Qaddafi Warns of Assault on Benghazi as U.N. Vote Nears." *New York Times*, March 17. http://www.nytimes.com/2011/03/18/world/africa/18libya.html?pagewanted=all.

Krauthammer, Charles. 1999. "The Clinton Doctrine." CNN.com, March 29. http://edition.cnn.com/ALLPOLITICS/time/1999/03/29/doctrine.html.
Krasner, Stephen. 1999. *Sovereignty: Organized Hypocrisy*. Princeton, NJ: Princeton University Press.
Lake, Anthony. 1994. "Confronting Backlash States." *Foreign Affairs* 73 (2): 45–55.
Lieber, Robert. 2007. *The American Era: Power and Strategy for the 21st Century*. New York: Cambridge University Press.
Mazzetti, Mark, Robert F. Worth, and Michael R. Gordon. 2013. "Obama's Uncertain Path amid Syria Bloodshed." *New York Times*, October 22. http://www.nytimes.com/2013/10/23/world/middleeast/obamas-uncertain-path-amid-syria-bloodshed.html.
New York Times. 2002. "Text of Bush's Speech at West Point." June 1. http://www.nytimes.com/2002/06/01/international/02PTEX-WEB.html.
Nichols, Thomas. 2008. *Eve of Destruction: The Coming Age of Preventive War*. Philadelphia: University of Pennsylvania Press.
Nichols, Thomas. 2014. "Here's What Nuclear Weapons Have to Do with the Ukraine Crisis." *Business Insider*, March 24. http://www.businessinsider.com/tom-nichols-nuclear-weapons-ukraine-crisis-deterrence-2014-3.
Nichols, Thomas, and John Schindler. 2013. "America's Middle East Policy Collapses." *National Interest*, September 16. http://nationalinterest.org/commentary/americas-middle-east-policy-collapses-9073.
Pindell, James W., and Rick Klein. 2007. "Obama Defends Votes in Favor of Iraq Funding: Says He Backs Troops, Not War." *Boston Globe*, March 22. http://www.boston.com/news/nation/articles/2007/03/22/obama_defends_votes_in_favor_of_iraq_funding/.
Obama, Barack. 2011. "Remarks by the President in Address to the Nation on Libya." Washington, DC: White House Office of the Press Secretary, March 28. http://www.whitehouse.gov/the-press-office/2011/03/28/remarks-president-address-nation-libya.
Obama, Barack. 2012. "Remarks by the President to the White House Press Corps." Washington, DC: White House Office of the Press Secretary, August 20. http://www.whitehouse.gov/the-press-office/2012/08/20/remarks-president-white-house-press-corps.
Pifer, Steven. 2014. "Ukraine Crisis' Impact on Nuclear Weapons." CNN.com, March 4. http://www.cnn.com/2014/03/04/opinion/pifer-ukraine-budapest-memorandum/.
Power, Samantha. 2002. *A Problem from Hell: America in the Age of Genocide*. New York: Basic Books.
Rampton, Roberta, and Jeff Mason. 2013. "Obama's Syria Decision: A Walk, a Debate, and a New Approach." *Reuters*, August 31. http://www.reuters.com/article/2013/09/01/us-syria-crisis-obama-decision-idUSBRE98001520130901.
Remnick, David. 2011. "Behind the Curtain." *New Yorker*, September 5.
Rogin, Josh. 2011. "Who Really Said 'Obama Was Leading from Behind'?" The Cable, ForeignPolicy.com, October 27. http://foreignpolicy.com/2011/10/27/who-really-said-obama-was-leading-from-behind/.

Silverstein, Gordon. 1996. *Imbalance of Powers*. New York: Oxford University Press.

Shinkman, Paul. 2013. "Obama: 'Global War on Terror' Is Over." *US News and World Report*, May 23. http://www.usnews.com/news/articles/2013/05/23/obama-global-war-on-terror-is-over.

Somin, Ilya. 2014. "Why the Kosovo 'Precedent' Does Not Justify Russia's Annexation of Crimea." *Volokh Conspiracy* blog. *Washington Post*, March 24. http://www.washingtonpost.com/news/volokh-conspiracy/wp/2014/03/24/crimea-kosovo-and-false-moral-equivalency/.

Strange, Hannah. 2013. "U.S. Raid that Killed bin Laden Was 'an Act of War,' Says Pakistani Report." *Telegraph*, July 9. http://www.telegraph.co.uk/news/worldnews/asia/pakistan/10169655/US-raid-that-killed-bin-Laden-was-an-act-of-war-says-Pakistani-report.html.

Straus, Scott. 2005. "Darfur and the Genocide Debate." *Foreign Affairs* 84 (1): 123–133.

Stojanovic, Dusan. 2014. "How Similar—or Not—Are Crimea and Kosovo?" *Washington Examiner*, March 14. http://www.washingtonexaminer.com/how-similar-or-not-are-crimea-and-kosovo/article/feed/2124453.

Tapper, Jake. "Fmr. Bush National Security Adviser: We Should Have Sanctioned Russia after Georgia Invasion." CNN.com, March 7. http://thelead.blogs.cnn.com/2014/03/07/fmr-bush-national-security-adviser-we-should-have-sanctioned-russia-after-georgia-invasion/.

Telegraph. 2013. "François Hollande: David Cameron Committed 'Schoolboy Error' on Syria." September 11. http://www.telegraph.co.uk/news/worldnews/middleeast/syria/10302082/Francois-Hollande-David-Cameron-committed-schoolboy-error-on-Syria.html.

Todd, Chuck. 2013. "The White House Walk-and-Talk that Changed Obama's Mind on Syria." NBCNews.com, August 31. http://firstread.today.com/_news/2013/08/31/20273128-the-white-house-walk-and-talk-that-changed-obamas-mind-on-syria.

United Kingdom. 2013. "Syria: Transcript of PM's Interview." London: Prime Minister's Office, August 27. GOV.UK.

Washington Post. 2013a. "Full Transcript: Kerry, Hagel and Dempsey Testify at Senate Foreign Relations Committee Hearing on Syria." September 4. http://www.washingtonpost.com/politics/2013/09/03/35ae1048-14ca-11e3-b182-1b3bb2eb474c_story.html.

Washington Post. 2013b. "Transcript: President Obama's Aug. 31 Statement on Syria." August 31. http://www.washingtonpost.com/politics/transcript-president-obamas-aug-31-statement-on-syria/2013/08/31/3019213c-125d-11e3-b4cb-fd7ce041d814_print.html.

Washington Post. 2014. "Full Transcript of President Obama's Commencement Address at West Point." May 28. http://www.washingtonpost.com/politics/full-text-of-president-obamas-commencement-address-at-west-point/2014/05/28/cfbcdcaa-e670-11e3-afc6-a1dd9407abcf_story.html.

The White House. 2011. "Remarks by the President in Address to the Nation on Libya." March 28. http://www.whitehouse.gov/the-press-office/2011/03/28/remarks-president-address-nation-libya.

Wong, Kristina. 2014. "Obama Backs Off Drone Strikes." TheHill.com, May 22. http://thehill.com/policy/defense/206541-obama-backs-off-drone-strikes.

Wroughton, Lesley, and Andrew Osborn. 2014. "Kerry Says He's Seen Raw Data Suggesting Syria Used Chlorine in Attacks." *Reuters*, May 15. http://www.reuters.com/article/2014/05/15/us-syria-crisis-kerry-idUSBREA4E0TN20140515.

Zakaria, Fareed. 2011. "A New Era in U.S. Foreign Policy." CNN.com, August 23. http://globalpublicsquare.blogs.cnn.com/2011/08/23/a-new-era-in-u-s-foreign-policy/.

Zenko, Micah. 2013. "Tracking U.S. Targeted Killings." *Politics, Power, and Preventive Action*, CFR.org, December 31. http://blogs.cfr.org/zenko/2013/12/31/tracking-u-s-targeted-killings/.

PART II

Perspectives on Preventive Drone Strikes

5

Why Drones Are Different

STEPHAN SONNENBERG

Armed drones' demonstrated seductiveness as a weapon of choice for those that have them, coupled with their potential to be deployed in an ever-expanding set of situations, make them a classic "disruptive technology."[1] Accordingly, military planners are increasingly deploying drones not just in traditional warfare contexts, but also—and especially—in scenarios where one party is seeking to prevent a potential attack before it ever materializes. The pioneers of weaponized drone technology continue to use such preventive logic in places like Pakistan, Somalia, Yemen, and elsewhere, even though the deployment of such drone-based warfare in some of these places has been ongoing for over a decade.

In this chapter, I argue that the regulation of drones as a distinct category of weaponry is urgently needed. Drones are *un*conventional weapons systems, and thus do not fall under the current legal framework designed to govern the conventional use of force between armed actors. Without regulation, drone technology is likely to continue to evolve and proliferate, meaning that in the foreseeable future all militaries and many non-state actors will have the ability to deploy some form of weaponized drone technology.

If and when other actors—state and non-state alike—deploy drone technology using the same open-ended preventive war logic that U.S. and other military planners have used to justify the drone strikes in Pakistan, Yemen, and Somalia, one can easily imagine a "new normal" where large parts of the globe will need to learn to cope with the long-term threat of lethal drone strikes by external actors. From the perspective of a military strategist, there is nothing unethical about the introduction of a new disruptive technology, especially from the self-interested and short-term perspectives of nations that command an early technological edge. But human rights activists and humanitarians tend to take a more cautious approach to such revolutionary innovations in war-making technology, largely out of concern for the impact of those new technologies on the civilians who stand to be affected by them.

In what follows, I provide a brief history of legal efforts to humanize warfare, and discuss how modern international humanitarian law applies to the U.S. drone program in Pakistan. Next, I explain why drones should be considered a structurally disruptive technology that tends to seduce military planners into thinking they can deploy armed force without bearing the domestic political costs for doing so, thus making armed conflicts globally more likely. I then discuss how the narrative of drones as low-cost precision weaponry is deceiving and ultimately devastating for civilian populations living in potential conflict zones, and conclude with a call for a regulatory framework that would either ban or seriously curtail the use of armed drones as weapons of war.

A Brief History of International Efforts to "Humanize" Warfare

International Humanitarian Law (IHL) is often referred to as the "law of armed conflict," which has evolved over centuries to regulate "honorable" interstate warfare (see also Glazier, Chapter 6). In its treatise on the law of war, the International Committee of the Red Cross (ICRC) defines IHL as "the branch of international law limiting the use of violence in armed conflicts by: (a) sparing those who do not or no longer directly participate in hostilities; [and] (b) restricting it to the amount necessary to achieve the aim of the conflict, which—independently of the causes fought for—can only be to weaken the military potential of the enemy." From this definition flow the important core principles of IHL: "the distinction between civilians and combatants, the prohibition to attack those *hors de combat*, the prohibition to inflict unnecessary suffering, the principle of necessity, and the principle of proportionality" (Sassòli, Bouvier, and Quintin 2011, 1).

Implicit in IHL's definition are important limitations to what IHL seeks to regulate. IHL does not, for example, "prohibit the use of violence; . . . protect all those affected by an armed conflict; . . . [distinguish] based on the purpose of the conflict; . . . [or] bar a party from overcoming the enemy." It also "presupposes that the parties to an armed conflict have rational aims and that those aims as such do not contradict IHL" (Sassòli, Bouvier, and Quintin 2011, 2). Thus, the cost of taming the worst excesses of war, it seems, is to presumptively deem as legitimate a state's sovereign right to resort to violent force if and when it feels the necessity to do so.

This realization often comes as a surprise to many laypersons unfamiliar with the details and history of IHL. Certainly, contemporary lore about IHL is often framed in much more floral language. The ICRC website, for example, proclaims that Henry Dunant's book *A Memory of Solferino*[2] (which is typically credited as the first articulation of the need to create the Red Cross movement and the various humanitarian treaties that the movement helped foster) "has moved many people and still does today." The website goes on to triumphantly quote the reaction of one pair of French authors after having read the book, namely that "one finishes this book cursing war" (ICRC Resource Center 2012). The text of Dunant's book, however, wastes not a single word calling for the abolition or even the curtailment of war. Instead, Dunant brushes aside the "hopes and aspirations" of pacifists (including one with whom he would later share the first Nobel Peace Prize in 1901), and instead accepts without commentary a future where wars are "hardly" avoidable and indeed where "battles will only become more and more murderous" ([1862] 1986, 116, 128).

In contrast to Dunant's pragmatic embrace of the horrors of war, some of his contemporaries, including military and political leaders, believed it possible to "humanize" warfare. The focus of these "humanitarian" diplomats, military strategists, and activists was not merely limited to taking better care of soldiers wounded in war. Instead, their energies focused on stigmatizing and banning the most inhumane forms of military weaponry.

In 1868, for example, the Declaration of St. Petersburg proclaimed that "the progress of civilization should have the effect of alleviating as much as possible the calamities of war," and sought to ban "the employment of arms which uselessly aggravate the sufferings of disabled men, or render their death inevitable."

This late nineteenth-century tension between those who believe that the best way to promote peace is to *raise* the costs and savagery of war (the deterrence school of thought) and those who believe that the prospects for peace are maximized by efforts to *tame* the conduct of warfare (the humanist school of thought) remains to this day. In 1898, Czar Nicholas II of Russia proposed a conference with the aim of finding "the most effectual means of insuring to all peoples the benefits of a real and durable peace, and, above all, of putting an end to the progressive development of the present armaments" (Rescript of the Russian Emperor [1898] 1909). Referring to Russia's trouble attracting foreign investment for "peaceful" infrastructure development projects such as the construction of the Trans-Siberian railway, the young czar pointed out:

> The economic crisis, due in great part to the system of armaments à l'outrance (to excess), . . . are transforming the armed peace of our days into a crushing burden. . . . It appears evident, then, that if this state of things were prolonged, it would inevitably lead to the very cataclysm which it is desired to avert, and the horrors of which make every thinking man shudder in advance. (Rescript of the Russian Emperor [1898] 1909)

Three months later, the czar's representative seemed aware of the already tepid support for collective disarmament among many of the major powers of the day. In a follow-up circular to the czar's proposal, this representative lamented that despite paying lip-service to the czar's efforts, "several Powers have undertaken fresh armaments, striving to increase further their military forces, and in the presence of this uncertain situation" (Russian Circular [1899] 1909, ¶3). The czar's trepidation was not unfounded: in his instructions to the U.S. delegation to the 1899 Hague Convention, U.S. Secretary of State Jon Hay described the czar's proposals to limit the production, use, and further improvements of firearms, explosives, and other destructive agents as "lacking in practicability":

> The expediency of restraining the inventive genius of our people in the direction of devising means of defense is by no means clear, and, considering the temptations to which men and nations may be exposed in a time of conflict, it is doubtful if an international agreement to this end would prove effective. The dissent of a single powerful nation might render it altogether nugatory. The delegates are, therefore, enjoined not to give the weight of their influence to the promotion of projects the realization of which is so uncertain. (Instructions to the International (Peace) Conference at The Hague [1899] 1909)

Despite this lack of full enthusiasm for the czar's vision, the 1899 conference, as well as a successor conference in 1907, elaborated on the aspirational principles laid forth in St. Petersburg.[3]

With the benefit of historical hindsight, of course, the sad reality is that Mr. Hay's pessimism presaged events on the horizon. Within a matter of years, the spirit of the Hague conferences would be supplanted by one of the most brutal and violent periods of warfare that Europe—and the world—had ever witnessed. The belligerents fighting during World Wars I and II violated literally *every* provision laid forth at the Hague conferences.

After the First World War, international diplomats, struggling to make sense of this carnage and trying desperately to reintroduce some sense of order into the international global order, tried to bring international humanitarian law back in line with their prewar aspirations to end total war. In the brief reprieve to the carnage between World Wars I and II, the League of Nations was given the institutional mandate to promote reductions in national armaments across the league, and to promote pacific settlements to disputes that might otherwise lead to war. In 1928, a treaty was signed in Geneva reiterating the need to ban chemical weapons, and extending that ban to cover biological weapons (Protocol for the Prohibition of the Use in War of Asphyxiating Gas, and of Bacteriological Methods of Warfare 1928). In Paris that same year, a treaty specifically banning war itself was signed, this time with the clear support of the U.S. government (Kellogg-Briand Pact 1928). Alas, as history clearly shows, these efforts proved just as "nugatory" as Mr. Hay had predicted in 1899.

The contemporary lessons of this brief historical survey are important to consider. What makes one set of international legal prohibitions on the use of force effective, whereas others are quickly abandoned during times of war? The ban on chemical and biological weapons, for example, managed to retain its relevance even during the ravages of the Second World War (with some crucial exceptions, especially if one considers the genocidal use of such chemicals by the German government against its own citizens and civilians in conquered territories). So, too, the ban on "dum-dum" (or expanding) bullets during warfare was largely respected. But the Kellogg-Briand prohibition on the use of warfare as a tool of foreign policy was dropped like a hot potato with the onset of World War

II. So, too, was the ban on the use of projectiles and explosives launched from balloons, especially if one considers the German Blitzkrieg and the Allied bombing raids over Germany towards the end of the war a clear violation of the spirit of that earlier prohibition. Most worryingly of all from a humanitarian perspective, the Second World War was waged in a manner that, in many cases, targeted civilian populations directly, either as "collateral damage" to military priorities or as the direct intended targets of an assault.

The end of the Second World War brought about another burst of international legislative efforts to humanize warfare. The first strand of this law, focusing on the legality of armed aggression per se, is typically referred to as *jus ad bellum*. These norms are set forth in the 1945 UN Charter.[4] In addition to laws regulating when it is legal or illegal to use armed force, a second strand of international humanitarian law focuses on reinforcing regulations to guide military activity *during* conflict (*jus in bello*). This body of law applies regardless of the *jus ad bellum* legality of an armed attack, and it applies equally to all belligerent parties. In brief, *jus in bello* provisions require states to ensure that efforts have been made to distinguish active participants in hostilities from noncombatants, that measures have been taken to minimize the "collateral damage" incidental to an attack, that the attack be justified by military necessity, and that the attack does not needlessly impact day-to-day civilian life in the affected region. As is evident throughout this volume, and elsewhere, the debate continues over whether drones represent a net plus or net minus for civilian populations (Eviatar, Chapter 7; Fair, Chapter 8). In the section that follows, I explore these questions by drawing on my first-hand experience researching the impacts of America's drone war in northwest Pakistan.

Drones' Compatibility with International Humanitarian Law

In September of 2012, I co-authored a report on the impact of the U.S. unmanned aerial vehicle (UAV, or colloquially "drone") program in North Waziristan, one of Pakistan's remote and difficult to access Federally Administered Tribal Areas and Frontier Regions (FATA). The report, entitled "Living under Drones,"[5] drew on more than 130 interviews with victims and witnesses to U.S. drone strikes, as well as experts, journalists, and policymakers in Pakistan and elsewhere.[6] It "raise[d] serious concerns about the compliance of particular strikes, and targeted killing trends and practices, with IHL" (Stanford/NYU Clinics 2012,113), demonstrating the negative impacts the U.S. drone program was having, particularly on civilians living in the affected areas. It noted that certain strikes—in particular those documented cases where drone strikes hit mosques, schools, funerals, or a gathering of village elders convened to resolve a community dispute—violated the principle of proportionality, or the idea that a military strike should not cause civilian damage disproportionate to the strategic importance of the intended target. It also documented certain targeting practices that resulted in a high number of deaths of first responders, and a resulting policy shift among some humanitarian responders to no longer provide urgent medical care to those wounded as a result of a drone strike. Furthermore, the report observed that the U.S. "signature strike" policy, whereby an individual would be targeted based on his or her pattern of behavior, was a serious threat to the principle of distinction, or the idea that militaries must make efforts to distinguish between lawful targets and civilians. Little was known in 2012 about what criteria the United States used to launch a "signature strike" (Stanford/NYU Clinics 2012, 12). Estimates in 2012 were that approximately 98 percent of the strikes in Pakistan could be classified as signature strikes (Bergen and Braun 2012; Landay 2013). The report also warned of the risks of drone proliferation to other state and non-state actors (for more recent reports on this issue, see Davis et al. 2014 and Zenko 2014), and speculated that the introduction of drone technology had enabled U.S. policymakers to significantly expand the list of targets for extrajudicial killing[7] beyond what would ever have been conceivable in an era prior to the advent of drones (more recently, Abizaid and Brooks 2014).

The critiques made in the "Living under Drones" report were specific to the U.S. drone program as it existed in Pakistan in 2012. Since 2012, the United States has amended its policy on the use of drones in ways that may have tempered some of the most troubling aspects of the program that were taking place when the report was published. Nonetheless, serious legal questions about the United States' use of its considerable arsenal of drones persist. Many of these concerns could potentially be addressed through additional policy changes or by better oversight of the drone program by policymakers, something that politicians in other countries wishing to acquire drone technology have proposed.

One such proposal, mockingly described by one skeptical German newspaper as the "German drone, good drone" argument, was promoted in 2013 by Thomas de Mazière, former German minister of defense, as part of an overall push to expand the German drones program (Birnbaum 2013). In response to widespread concerns among politicians and civil society activists that the German military would soon be running its own version of the U.S. "suspect-hunting" drones program, de Mazière insisted that Germany's policy oversight and post-1945 military culture would make the German program fundamentally more consistent with IHL norms (Birnbaum 2013).

Perhaps de Mazière was right, and concrete legal traditions and policy choices could allay some or all of rights activists' concerns by ensuring that the use of drones is more consistent with the existing provisions of international humanitarian law. But was he correct in describing drones as merely a "next generation of the airplane" (Birnbaum 2013)? Other commentators seem to concur with that description of drone technology. For instance, Stuart Casey-Maslen, head of research at the Geneva Academy of International Humanitarian Law and Human Rights, describes drones flatly as "only platforms" (2012, 614), and focuses on their payload as the object necessitating analysis. Human Rights Watch also states that "drones themselves . . . are not illegal weapons under the laws of war—they can be used lawfully or unlawfully depending on the circumstances" (2011).

Indeed, many IHL scholars and human rights activists are on record as actually *welcoming* drones as an advance in the field of IHL, since they permit states to better distinguish combatants from noncombatants and to determine a strike opportunity that would minimize the potential for civilian collateral damage (Human Rights Watch 2011). The remainder of this chapter will respond to those arguments and suggest reasons why IHL scholars and human rights activists should scrutinize drones per se (not just their payloads).

Are Drones a Structurally Disruptive Military Technology?

The "Living under Drones" report, and others like it that have been issued since 2012, have received widespread media and civil society attention. Repeatedly during interviews or conference presentations on the subject, my co-authors and I would be asked how drones themselves are different from other weapons. "What," we would be asked, "makes a Hellfire missile fired by a drone any different from a Hellfire missile fired by an Apache helicopter?"

The question has important implications. If indeed an armed drone as it exists today (that is, still involving human operators "in the loop" of each individual decision to kill a suspected target) is nothing other than a sophisticated delivery vehicle for the missiles it carries, then the current coalition of anti-drone activists will no longer logically be able to restrict their criticism only to drones. Instead, those who continue to feel that drones are morally suspect will need to refocus their critique to bring it to bear on preventive warfare (see, for example, Parliamentary Assembly of the Council of Europe 2007, as well as Emery and Brunstetter, Chapter 10), or perhaps on preventing robotic or robot-assisted warfare itself (see, for instance,. Lin 2010; Sharkey 2010; Vallor 2014). Such critiques are not without merit, and others have made them persuasively. But they tend to be much more technical and legalistic than the powerful "gut" sense of unease with present-day drone technology that currently animates many anti-drone activists.

If, on the other hand, existing drone technology is in some way distinguishable from other weapons delivery systems, the current controversy over the use of drones per se seems far more relevant. And if drones are changing the way military strategists think about war (Abizaid and Brooks 2014), including the preventive use of force, it might suggest the need to review or adapt current international legal norms regulating the use of military force. This suggests that the moment is now ripe for the international community to think about ways to prevent the unchecked development and proliferation of drone technology (see Davis et al. 2014; Zenko 2014).

The Siren Song of the Drones: War without the Costs?

The brief history of efforts to constrict countries' right to wage war has shown that legal prohibitions on nation's right to wage war have typically failed. Instead, *domestic* political and economic considerations have tended to be the most effective barriers to war. As the *Economist* wryly warned in 1995, "True, people tend to recoil at the prospect of shedding their own blood, . . . but even kind-hearted citizens may not worry much when someone from another part of the country goes off to fight, especially if that person has volunteered to do so" (1995, 18). How might that same "kind-hearted citizen's" self-interested opposition to war be even further diluted if he or she were promised that wars could be fought exclusively by means of drones? If political opposition to warfare has more to do with a concern for avoiding *domestic* costs and suffering (as opposed to the avoidance of suffering abroad), the introduction of drone technology suggests a much higher likelihood that sovereigns will resort to armed force.

Drone technology is particularly seductive from the perspective of a policymaker wishing to justify war to skeptical constituents. In part, this has to do with modern society's infatuation with technology, as well as administration officials' use of carefully sculpted metaphors to mask the true impacts of the drone program. "Surgical precision" is one such metaphor (see Friedersdorf 2012). To a layperson, the narrative of "surgically precise" drone strikes seems plausible given the breakneck pace of new technological innovations that have revolutionized life in the twentieth and twenty-first centuries. Indeed, modern drones do offer militaries the capacity to strike individual targets with previously unimaginable precision. The narrative of surgical precision evokes the promise of medicine—namely, the ability of a physician to physically excise a tumor or other problem before the illness has the opportunity to corrupt the rest of the patient's body. The implicit suggestion is that one or a few bad individuals can be efficiently killed by means of a drone, and that doing so will "cure" the host nation and restore it back to health. It also draws on the imagery of a surgeon, a professional who is obligated by the principle of the Hippocratic Oath to do no harm and to act with the best interests of the patient in mind.

The reality in countries currently experiencing drone strikes, however, demonstrates the absurdity of this narrative. First, the surgically precise narrative presumes that the precise location of the "tumors" or "insurgents" can be found with ease. But as the U.S. practice in Pakistan demonstrates, the accuracy of even the most technologically advanced drones programs is only as good as the intelligence identifying possible targets. In many instances the intelligence used to identify targets in

Pakistan proved to be deeply flawed. Traditional means of intelligence gathering proved inadequate and prone to local cultural, political biases, and outright greed and corruption. Certain "high-value" individuals were reported killed numerous times before the intended target was confirmed assassinated, leaving a trail of unintended and often unidentified victims dead or injured. According to official reports following drone strikes surveyed in *Rolling Stone* magazine, Baitullah Mehsud, for example, was allegedly "killed" four times prior to his actual death. Likewise, Ilyas Kashmiri was killed at least three times by U.S. drone strikes, each "kill" contested by local sources (Stanford/NYU Clinics 2012, 129, 128). Our research team spoke with a survivor of the second strike that allegedly killed Kashmiri, in which the survivor lost both legs and one eye. He was fifteen at the time the strike occurred in September of 2009.

In 2014, news broke that a specialized unit in the NSA had been using what one anonymous insider described as "unreliable metadata" to feed the CIA and the military with possible targets for such lethal strikes (Scahill and Greenwald 2014). These tactics were allegedly easily subverted when targets, increasingly aware that they were being tracked via their cell phone SIM cards, began to develop strategies to confuse and subvert the NSA surveillance. Indeed, a document found in an abandoned Islamist hideout in Mali purporting to be a "tip-sheet" on how to evade U.S. drone strikes in Yemen included one entry that "the leaders or those sought after . . . should not use communications equipment because the enemy usually keeps a voice tag through which they can identify the speaking person and then locate him" (Mohammed 2011, 2). This, and twenty-one other "tips" detailed in the Mali document, demonstrated how relatively simple measures taken by militants could subvert the United States' most sophisticated surveillance and intelligence gathering technologies and dramatically increase the risk of death and injury to civilians in the area.

Second, the narrative of surgical precision assumes that a "tumor" can be neatly excised from an organism without causing irreparable harm to the rest of that organism. Again, the experience of the U.S. drones program gives pause to those who still believe this to be possible. The extensive drone program in North Waziristan has not made that region a safer place to live for civilians. Nor has it made Pakistan as a whole a more peaceful country. Quite the contrary: North Waziristan remains one of the most inaccessible and dangerous places on the globe today, and terrorist and extremist networks have reportedly spread around the country and infiltrated every major city in Pakistan (Stanford/NYU Clinics 2012). The lives of those who were wrongfully killed or injured by a U.S. drone strike in many hundreds of households have been irreparably disrupted, with devastating economic, social, and psychological effects that were detailed in the "Living under Drones" report. Furthermore, as has been widely described both in the report and elsewhere, the U.S. drone program may today be one of the primary recruitment tools for extremist groups seeking new supporters. The idea that Pakistan is on its way to better "health" by virtue of the U.S. drone program seems utterly misinformed.

Finally, the narrative of surgical precision relies on the image of a skilled and ethical surgeon. Surgeons are bound by the Hippocratic Oath to act with only the best interests of the patient in mind. Military decision-makers, on the other hand, are bound by an equally solemn oath to act only in the interests of their own country. Thus, unless a humanitarian motive is the sole motive for war (a questionable proposition *whenever* it is put forward by military leaders), the idea that anything but the strict national interest of the attacker country would guide drones' deployment seems sadly naïve.

International lawyers would assert, however, that no matter how "surgically precise" a drone strike might be, this precision is of no relevance when it comes to the legality of initiating an armed attack against another sovereign nation (*jus ad bellum*). If drones are indistinguishable from other weapons of conventional warfare, international law would suggest that the systematic use by one country of drones to carry out attacks in another country is sufficient to constitute an illegal act of aggression under international law unless one of three exceptions apply: (1) the country targeted consents to that attack; (2) the attack is justified under the self-defense exception to the use of armed force; or (3) the attack is carried out in accordance with a U.N. Chapter 7 mandate authorizing the use of force. Any country using drones to effectuate attacks in another country would therefore need to justify that attack with reference to one of the three above-cited exceptions. Unless even the most technologically precise strikes can be justified under one of the three exceptions to the general prohibition on the use of armed force contained in the U.N. Charter, even such precise drone strikes would still constitute an illegal act of aggression. For example, the United States has claimed a preventive self-defense justification for its drone strikes in Pakistan. If—as we argued in our 2012 report—those claims fail, the U.S. drone campaign in Pakistan must be considered an illegal armed attack against the sovereign nation of Pakistan. The only exception would be if the United States were acting with the consent of the Pakistani authorities. Indeed, in 2013, former president of Pakistan Pervez Musharraf admitted to having granted to the United States permission to carry out drone strikes in Pakistan where "a target was absolutely isolated," with "no chance of collateral damage," and "only on a few occasions" (Robertson and Botelho 2013). Wikileaks revelations detailed similar acceptance by Pakistani authorities to U.S. drone strikes (U.S. Embassy Cables [2008] 2010). But besides this very isolated evidence of consent by Pakistani authorities, the Pakistani government, notably Pakistan's democratically elected government representatives, have steadfastly protested the U.S. drone strikes and declared them to be acts of illegal aggression against Pakistan.

Even beyond such legal arguments, however, there is also a practical risk in holding drone strikes to a different—and lower—*jus ad bellum* standard. If countries with advanced drone technology are allowed to consider targeted killings of named (and even unnamed) individuals abroad as something short of an armed attack, then other nations—even those without advanced drone technology—are likely to carry out similar attacks in *their* national interest and claim for themselves the same privilege. This would be the start of a system of global targeted assassinations in the name of "national security," all operating very much at cross purposes with international stability, basic norms of asylum, and restraint on the use of force in international relations.

Drones also offer another, equally flawed, promise to policy makers wishing to go to war: that of reduced political cost. It is true that drones make it possible to carry out military strikes without putting soldiers or other military personnel in harm's way. As the noted expert on drones Peter W. Singer famously stated, the most dangerous part of a drone operator's day is his (or her) commute to and from work (Zucchino 2014). But the deaths of members of the attacker country's armed forces are not the only costs of war. Drones guarantee a zero death rate only on the side of the attacking party's armed forces, and only in situations of asymmetric warfare. The technology thus discriminates heavily against countries or armed groups who are not wealthy enough to finance their own drone programs capable of exacting revenge on the attacking country's armed forces. Such a situation generally leads to large-scale global armaments races. Furthermore, drone programs still require enormous resources to operate, meaning that even if the human toll of war might decrease, the financial burden may remain the same.

Finally, widespread use of drones across national borders still carries a significant and dangerous reputational cost. This cost is particularly accentuated vis-à-vis conventional weapons with humans more visibly in charge, perhaps due to the inescapable imagery conjured up by an emotionless, infinitely capable, all-seeing robot poised to kill without remorse at any moment. In other words, the lack of a visible human presence in a drone renders it psychologically "inhumane."

At the time of this writing, only a few countries have used drones to carry out lethal attacks internationally, with the United States as the most prolific user by far. In 2013, the Pew Global Attitudes Project identified only three countries worldwide (Kenya, Israel, and the United States itself) where political majorities supported the U.S. drones program. It is unclear to what extent the widespread unpopularity of the U.S. drone program (in some cases approaching 90 percent of respondents surveyed) influenced overall perceptions of America or Americans abroad. In Pakistan, the only country surveyed by Pew where the United States was carrying out drone strikes, only 5 percent of those surveyed in spring 2013 approved of the U.S. drones program, and only 11 percent rated America as a whole favorably. Noteworthy, too, is that Pakistani respondents to the Pew survey were one of the few national samples to express a dislike not only of U.S. *official* policy, but also of Americans per se. Of course, Pakistani attitudes about Americans are influenced by many more factors than only the U.S. drones program, so causation cannot be inferred from these figures.

That said, it is highly likely, based also on this author's own impressions interacting with Pakistani citizens, that the continued acceptance of the drones program by the American electorate has strongly factored into the personal animosity many Pakistanis feel not just toward American policymakers, but indeed toward the American people. The problem, of course, is that even if citizens in the United States consider drone warfare to be "riskless warfare" (Singer 2009, 322), civilian populations at the receiving end of such campaigns rarely share this perception. My colleagues and I spoke to civilians in northwest Pakistan at the receiving end of America's first such "riskless war" and found that those civilians acted and behaved as would any civilian during a time of *real* war—doing anything they could to protect themselves and their family, and creating psychological constructs to bring sense to the chaos around them. As one interviewee told us, "Before the drone attacks, we didn't know [anything] about America. Now everybody has come to understand and know about America. . . . Almost all people hate America" (Stanford/NYU Clinics 2012, 133).

From a *jus ad bellum* perspective, this is deeply troubling. It is a new and unforeseen challenge for international lawyers to attempt to discourage military strategists' recourse to war if most citizens of the countries launching armed attacks do not even know or consider themselves to be at war. From the perspective of the communities targeted by drone strikes, of course, the knowledge (or ignorance) of the citizens of an attacking country is small comfort, and the fully predictable response of these communities to behave as though they have been the target of an act of armed aggression, thus increasing the likelihood of war and further retributive violence in the future.

"Bug Splats" and Tough Commutes: When Home and Battlefields Merge

As described above, the legality of the decision by a country's leaders to go to war is only one aspect of the international laws governing war. The second aspect is *jus in bello*, or how countries behave once they are *in* an armed conflict. Here as well, the introduction of drones as a means of fighting wars represents a structural shift in the incentives that policymakers face, thus again rendering war more dangerous from a humanitarian perspective.

The military has yet to understand the psychological impact of this hitherto unprecedented situation. Psychologists are only beginning to realize that drone operators are vulnerable to a host of psychological strains typically associated with Post-Traumatic Stress Disorder or other psychological aftereffects typically associated with frontline deployment (Zucchino 2012). Furthermore, it is unclear whether the act of waging war from behind computer screens still allows drone pilots and their supervisors to find the proper balance between military objectives and humanitarian concerns to ensure that each individual strike complies with IHL norms.

Scholars of warfare (and war crimes) have pointed out that "distancing" is one of the most effective strategies soldiers use to justify the commission of war crimes and other atrocities (see, for example, Slim 2008). One common example of such distancing is to mask the humanity of individuals killed with euphemisms. Indeed, drone operators have been quoted as referring to their victims as "bug splats" (a reference to the resemblance of a drone kill to a bug splat when projected via the drone's live video feed).

Furthermore, the drone operator's distancing from his or her target is not merely a term of art—it is a real, physical distancing. Drones allow warfare to be conducted literally from anywhere in the world, without necessitating a physical deployment to the front lines for most of the pilots and support staff needed to keep the drones aloft. Drone pilots are able to commute between home and their "duty stations" every day, thus waging war almost without interruption of their regular civilian lives. The scenario allows for an almost impenetrable unidirectional distancing, where the targets of the drone operator's actions will almost certainly never be able to understand, interact with, or influence those who are targeting them.

Renditions of drone operators of what it is like to kill individuals using a drone paint a worrying picture of the effects of this dehumanization process:

> Operating a drone is "almost like playing the computer game Civilization"—something straight out of "a sci-fi novel." After one mission, in which [a drone pilot] navigated a drone to target a technical college being occupied by insurgents in Iraq, Martin felt "electrified" and "adrenalized," exulting that "we had shot the technical college full of holes, destroying large portions of it and killing only God knew how many people."

Only later did the reality of what he had done sink in: "'I had yet to realize the horror,'" Martin recalls (Hastings 2012).

Defenders of the drone program caution outsiders not to take the video game analogy too far. Drone operators, they assure the public, still have all the necessary ethical and legal oversight that operators in any other air strike would have to ensure that their targeting practices are consistent with international humanitarian law (Hastings 2012). But such assurances cannot negate the fact that a drone pilot's contact with the war zone is entirely virtual, pixelated, and contained in a safe environment structured to validate the importance of each individual "kill."

A second major problem with drones is that they are designed only to monitor or to kill. The technical characteristics of drones as they exist today make it impossible to use them to capture armed combatants. Policymakers who rely on drones thus implicitly abandon any effort to capture or disable armed combatants. The *New York Times* in 2012 speculated that the choice to kill rather than detain suspected enemy combatants may have been the result of the Obama administration's desire to avoid the allegations of torture and other human rights violations that had plagued his predecessor's detention policy (Becker and Shane 2012). The United States fleet of drones in 2012 constituted approximately one third of the U.S. fleet of aircraft, and that percentage is projected to grow as the list of tasks for unmanned vehicles continues to grow (Ackerman and Shachtman 2012). Former CIA director Leon Panetta described drones as "the only game in town" (Shachtman 2009). Indeed, not only their development, but also their deployment has skyrocketed during the Obama presidency. Thus, in the United States at least, the turn to drones as the weapon of choice in the global fight against al-Qaeda and affiliated forces represents a definitive turn away from the traditional IHL preference for capturing or disabling enemy combatants over killing them.

Most fundamentally, the widespread deployment of drones has eroded the boundary between the "theater of war" and the "theater of peace." Efforts to "tame" the practice of war in the late nineteenth and early twentieth centuries already recognized the need to carve out a civilian "safe zone" from the ravages of war. Starting with early attempts at the turn of the century to exempt undefended villages and towns from aerial or artillery bombardment, and continuing with the more formalized provisions contained in the Geneva Conventions regarding civilians and former belligerents who have laid down their arms, IHL places a high value on the principle of distinction. Drones eviscerate any sense of security that civilians might have in zones where drones operate. The "Living under Drones" report, for example, documented the serious livelihood impacts of weaponized drones hovering above. These fears were not irrational: innocent bystanders *were* often killed or injured by drone strikes. Thus, unless the targeting practices of those who operate drones dramatically improve, it is conceivable that the theater of war and the "home front" will merge for both the drone operators and the communities in which drone targets are allegedly found.

With the likelihood of increased proliferation of drone technology in the near future, soldiers and war-essential civilians living thousands of miles removed from the frontlines in otherwise "peaceful" locations might find themselves the target of a drone attack by a hostile nation, just as workers in German factories learned during the Second World War that they were considered legitimate targets by Allied bombers.

Regulating the Drones: Dreams of a Luddite?

If it is true that drones technology constitutes a structurally disruptive technology with the potential to erode some of the most important legal safeguards against the inhumanity of war, it stands to reason that we should regulate and constrain the unfettered military use of drones.

At the time of this writing, such arguments may sound as though they are directed largely against the interests of the United States, Israel, and the United Kingdom (the only three nations known in 2014 to have used drones to kill individuals in an armed conflict). But the number of countries to have acquired weapons-capable drones is already much higher than three and is projected to grow even more in the coming years. Hamas and Hezbollah have reportedly already attempted to use basic drones to attack Israel, and more non-state armed groups, and more sophisticated drone technology, are likely to follow.

Shying away from the argument now risks jettisoning the argument entirely once other nations acquire and implement their own military drone programs. Undoing past proliferation of a "disruptive technology" is far less likely to succeed than attempting to stop such proliferation in the first place. Despite numerous setbacks during the tumultuous twentieth century, the core norms of IHL are now well established. Scholars have postulated that states are motivated to abide by IHL for a mixture of both self-interested and humanitarian reasons, but ultimately it must be in states' self-interest to respect basic IHL principles.

How then, can those incentives to comply with existing laws of war be reinforced? Game theory suggests an effective strategy for "winning" a binary collaborate-or-defect multiple iteration game—often referred to as a prisoner's dilemma). This classic game theory scenario is not unlike the situation of two militaries contemplating whether to use drones to "cheat" traditional *jus ad bellum* and *jus in bello* safeguards against the use of force. If all militaries abide by IHL norms, militaries will be able to fight and win wars "honorably," and civilians will be protected by a system designed to avoid the inhumanity of total warfare. But if one side defects or chooses to "play dirty" (perhaps in search of some short-term gain or military advantage), the tit-for-tat strategy suggests that the other side rationally should also follow suit, leading to a worse outcome for all parties involved. The most effective strategy for any party hoping to "win" in a multiple-round prisoner's dilemma is the so-called "tit-for-tat" strategy (Axelrod 1980). This strategy suggests a simple winning recipe to achieve maximum mutual collaboration: (1) never be the first to defect; (2) be provocable; (3) be forgiving; (4) be clear in your policy.

Traditionally, the stability of IHL has relied on the existence of strong incentives to make military planners' universal adherence to IHL more "sticky." Military, political, and legal motivations helped provide that stickiness. Military men and women deployed in a conflict zone have a strong incentive to respect the provisions of IHL, especially if they believe that their opponents in the conflict would extend the same protections to them if they were to be captured during combat. Furthermore, occupying militaries that abided by IHL safeguards tended to enjoy greater political support from civilians under their control. Thus, if militaries or fighting forces depend on the support of local populations to achieve their military objectives (the hearts and minds doctrine), their perceived adherence to IHL is crucial to that success. Furthermore, the threat of legal repercussions in either domestic or international trials has also served to temper individual soldiers' enthusiasm for engaging in criminal behavior during wartime. These various incentives to comply with IHL have traditionally resulted in a relatively stable, if still sometimes imperfect, equilibrium point wherein most states (and an increasing number of non-state armed actors) chose voluntarily to abide by the provisions of IHL.

Throughout this chapter, I have argued that the introduction of the structurally disruptive drone technology has already disrupted this tenuous equilibrium. In terms of Axelrod's (1980) framework, therefore, the United States has already chosen to defect from the so-called win-win scenario. According to the tit-for-tat strategy, any defections from a "mutual cooperation game" would be rationally met with similar defections by other parties. With the United States and a few other states having begun to incorporate drones into their military arsenals, it is reasonable to expect an arms race among other countries to acquire or develop their own drone technology. And when the first movers to adopt drones employ them in ways that offend other nations' understanding of IHL, the game-theory prediction suggests that those other nations would soon also abandon preexisting normative frameworks in favor of U.S. precedent.

Thus, if the United States continues to claim the "legal right to kill any person it determines is a member of al-Qaeda or its associated forces, in any state on Earth, at any time, based on secret criteria and secret evidence, evaluated in a secret process by unknown and largely anonymous individuals—with no public disclosure of which organizations are considered 'associated forces' (or how combat status is determined or how the United States defines 'participation in hostilities'), no means for anyone outside that secret process to raise questions about the criteria or validity of the evidence, and no means for anyone outside that process to identify or remedy mistakes or abuses" (Abizaid and Brooks 2014, 13), then so too will other nations claim the right to use drones in accordance with those same spurious standards. Civilian life at the end of this tit-for-tat race to the bottom promises to be truly nasty and brutish (and not only for citizens of lesser-developed countries).

This scenario is not inevitable, however. Axelrod's two final rules for winning at a multiple iteration prisoner's dilemma game provide clues to those who would like to avoid such an arms race and reinforce existing IHL standards. Any one individual state acting in isolation will be unlikely to cause any significant reversal of the global race to the bottom described above. But several states acting in unison, proclaiming publicly their determination to adhere to clear and verifiable higher standards for their use of drone technology—perhaps as an explicit counterexample to the United States' more rights-violating model—would have a credible chance of preventing a race to the bottom. There is thus an urgent need for a clear international framework governing the use of drones in armed conflicts (for an alternative view, see Eviatar, Chapter 7). Furthermore, there needs to be a clear indication that states are willing to adhere to such norms, even in light of other states' negative precedents (see Plaw and Reis, Chapter 9) and their presumed initial dismissive attitude towards any such new standards.

There are several possible avenues for the regulation of drones. States can agree to limit the development of new and more technologically advanced drones. Banning further weapons development might be difficult to implement in societies enamored with the idea of unstoppable technological progress. But it is not unprecedented. The nuclear test ban, for example, froze nations' nuclear testing programs in an effort to halt the various ongoing nuclear weapons development programs underway in many parts of the world.

Regulation might also focus on establishing clear guidelines for the use of drones that would bring them firmly in line with the spirit of existing norms governing armed conflict. Many international scholars are convinced that it is possible to use drones in a way that would make them analytically identical to other conventional weaponry. But safeguards are poorly understood even by many international legal scholars, and thus need to be made explicit, lest the existing U.S. policy stand as the sole precedent for other nations to emulate.

Finally, global policy makers could decide to ban the drone entirely in situations where one country does not consider itself to be formally at war with another. In other words, it would preclude the use of drones based on preemptive or preventive use of force justifications. This would avoid the worst of the *jus ad bellum* problems described above, and reserve the new drone technology only for situations where drones do not constitute the sole means of waging war. Others have also called for such a ban (see, for example, Eviatar, Chapter 7).

The development of IHL today represents the nonlinear product of a centuries-long compromise among military strategists, politicians, civil society actors, and civilians. This same set of actors must today again negotiate how drones will be regulated by IHL. In 1997, for example, the

Ottawa Mine Ban Treaty was signed, banning the use of landmines. The treaty regulates what at the time of its negotiation was an existing and widely available military technology, and one that the United States and many other nations at the time vowed they could never give up. But the example today illustrates the efficacy of actors willing to demonstrate their leadership around such issues. In 2014, after fifteen years of vociferous protests and protestations that the United States could never agree to a comprehensive ban on land mines, the Obama administration announced that it would agree to abide by the provisions of the land mine treaty.

This example can serve as a template for committed activists and human rights campaigners wishing to limit the use of drones as a weapon of war. Activists, diplomats, humanitarians, and favorably inclined government actors can initiate an inclusive and transparent convention-drafting process, emulating the process that led to the land mine treaty. This process will presumably begin in a small way and be derided by powerful military planners as naïve and ill-informed. But with time, patience, and a persistent campaign by those involved, momentum can build to persuade even the more recalcitrant states to adhere to more progressive and protective standards consistent with human rights—eventually.

Currently, as they have in the past, the militaries of the most powerful countries in the world are telling the public that the world need not fear lethal drones in the hands of responsible military planners. The question, therefore, is whether we are inclined to believe them.

NOTES

1 A "disruptive technology" is one that begins to carve out a market niche where existing, higher-priced options currently are not being used or are simply too expensive; the technology over time will evolve to threaten and eventually replace the higher-priced technology (Christensen 2011). Military analysts (including some of the other contributors to this edited volume) have compared drones to other major historical military innovations, such as the invention of the crossbow or gunpowder in their respective historical eras, or the introduction of chemical, biological, or nuclear weapons in more recent decades (Abizaid et al. 2014; Eggers et al. 2012; and Nincic, Chapter 3).

2 The book describes a bloody battle between French and Austrian troops and galvanized European civil society into forming the national Red Cross Societies;

it also ultimately led to the codifying of a first treaty relating to the treatment of wounded soldiers in 1864. In 1901, a frail Henry Dunant shared the first Nobel Peace Prize in absentia. He died nine years later in rural Switzerland. (Dunant [1862] 1986).

3 In 1899, for example, signatories declared that "the right of belligerents to adopt means of injuring the enemy are not unlimited" (Convention with Respect to the Laws and Customs of War on Land and Its Annex: Regulations Concerning the Laws and Customs of War on Land, The Hague [July 29, 1899] 1988, §22; hereafter Hague Convention 1899). The 1899 Convention listed a few examples of prohibited behavior, such as "employ[ing] poison or poisoned arms; kill[ing] or wound[ing] treacherously [sic] individuals belonging to the hostile nation or army; . . . declar[ing] that no quarter will be given; . . . or destroy[ing] or seiz[ing] the enemy's property, unless such destruction or seizure be imperatively demanded by the necessities of war" (Hague Convention 1899 §23). The convention also articulated a provision that continues to be the core protection many civilians expect in times of war, namely, that "the attack or bombardment of towns, villages, habitations or buildings which are not defended, is prohibited" (Hague Convention 1899, §25). The same conference also banned for a period of time the launching of explosives or projectiles from balloons, the use of asphyxiating gasses, and the use of so called "dum-dum" bullets (bullets designed to cause maximum damage to human tissue upon entering the body) (Final Act of the International Peace Conference, The Hague [29 July 1899] 1988).

4 Conflicts that threaten international peace and security are to be addressed according to provisions set forth in Chapters 6, 7, or 8 of the Charter. Only Chapter 7 allows the UN Security Council—after having exhausted all nonmilitary means of attempting to resolve the dispute—to authorize the use of military force by member states. The only other permissible uses of armed force by a state are self-defense against an armed attack, and preemptive self-defense against an imminent attack. This is discussed in greater detail in later chapters by Glazier (Chapter 6) and Plaw and Reis (Chapter 9).

5 The report is a collaborative effort by the International Human Rights and Conflict Resolution Clinic at Stanford Law School and Global Justice Clinic at New York University School of Law, 2012; hereafter cited in the text as Stanford/NYU Clinics.

6 Most, if not all, of the report's findings have since been corroborated by subsequent fact-finding efforts, reports by investigative journalists, and unauthorized or court-ordered policy leaks, as well as official acknowledgments or policy reforms by U.S. authorities.

7 UN Special Rapporteur for Extrajudicial, Arbitrary and Summary Executions, Asma Jahangir, described an early U.S. drone strike in Yemen as a clear case of extrajudicial killing (Jahangir 2003).

BIBLIOGRAPHY

Abizaid, General John P., and Rosa Brooks. 2014. "Recommendations and Report of the Task Force on U.S. Drone Policy." Washington, DC: Stimson Center, June. http://www.stimson.org/images/uploads/ task_force_report_final_web_062414.pdf.

Ackerman, Spencer, and Noah Shachtman. 2012. "Almost 1 in 3 U.S. Warplanes Is a Robot." Wired.com, January 9. http://www.wired.com/2012/01/drone-report/.

Atkinson, Rowland, and John Flint. 2001. "Accessing Hidden and Hard-to-Reach Populations: Snowball Research Strategies." *Social Research Update* 33 (Summer). http://sru.soc.surrey.ac.uk/SRU33.pdf.

Axelrod, Robert. 1980. "Effective Choice in the Prisoner's Dilemma." *Journal of Conflict Resolution* 24 (1): 3–25.

Becker, Jo, and Scott Shane. 2012. "Secret 'Kill List' Proves a Test of Obama's Principles and Will." *New York Times*, May 29. http://www.nytimes.com/2012/05/29/world/obamas-leadership-in-war-on-al-qaeda.html.

Bergen, Peter, and Megan Braun. 2012. "Drone is Obama's Weapon of Choice" *CNN*, September 6. http://www.cnn.com/2012/09/05/opinion/bergen-obama-drone/index.html.

Birnbaum, Robert. 2013. "Deutsche Drohne, Gute Drohne." *Der Tagesspiegel*, April 25. http://www.tagesspiegel.de/politik/de-maiziere-im-dialog-deutsche-drohne-gute-drohne/8117712.html.

Casey-Maslen, Stuart. 2012. "Pandora's Box? Drone Strikes under *Jus ad Bellum*, *Jus in Bello*, and International Human Rights Law." *International Review of the Red Cross* 94 (597): 602.

Christensen, Clayton. 1997. *The Innovator's Dilemma*. Boston: Harvard Business School Press.

Cohen, Nissim, and Tamar Arieli. 2011. "Field Research in Conflict Environments: Methodological Challenges and Snowball Sampling." *Journal of Peace Research* 48:423.

Convention with Respect to the Laws and Customs of War on Land and Its Annex: Regulations Concerning the Laws and Customs of War on Land. The Hague. (29 July 1899) 1988. In *The Laws of Armed Conflicts*, edited by D. Schindler and J. Toman, 69–93. Dordrecht, The Netherlands: Martinus Nihjoff. http://www.icrc.org/ihl/INTRO/150?OpenDocument.

Davis, Lynne E., Michael J. McNemey, James Chow, Thomas Hamilton, Sarah Harting, and Daniel Byman. 2014. "Armed and Dangerous? UAVs and U.S. Security." Santa Monica, CA: Rand Corporation. http://www.rand.org/pubs/research_reports/RR449.html.

Declaration of St. Petersburg. (November 29, 1868) 1915. *Conventions and Declarations between the Powers Concerning War, Arbitration and Neutrality*. The Hague: Martinus Nijhoff. http://avalon.law.yale.edu/19th_century/decpeter.asp.

Dunant, Henry. (1862) 1986. *A Memory of Solferino*. Geneva: International Committee of the Red Cross.

Economist. 1995. "Democracies and War: The Politics of Peace." April 17, 17–18.
Eggers, William, Laura Baker, Ruben Gonzalez, and Audrey Vaughn. 2012. "Public Sector, Disrupted: How Disruptive Innovation Can Help Government Achieve More for Less." Deloitte GovLab Study. Washington, DC: Deloitte.
Final Act of the International Peace Conference, The Hague. (29 July 1899) 1988. In *The Laws of Armed Conflicts*, edited by D. Schindler and J. Toman, 50–51. Dordrecht, The Netherlands: Martinus Nihjoff. http://www.icrc.org/applic/ihl/ihl.nsf/Treaty.xsp?documentId=315AEBE3F3DA0DF9C12563CD002D6689&action=openDocument.
Friedersdorf, Conor. 2012. "Calling U.S. Drone Strikes 'Surgical' Is Orwellian Propaganda." *Atlantic*, September 27. http://www.theatlantic.com/politics/archive/2012/09/calling-us-drone-strikes-surgical-is-orwellian-propaganda/262920/.
Goldstein, Daniel M. 2014. "Qualitative Research in Dangerous Places: Becoming an 'Ethnographer' of Violence and Personal Safety." *Social Science Research Council—Drugs, Security and Democracy Program. Working Papers on Research Security*, No.1. New York: Social Science Research Council. http://www.ssrc.org/pages/qualitative-research-in-dangerous-places-becoming-an-ethnographer-of-violence-and-personal-safety/.
Hastings, Michael. 2012. "The Rise of the Killer Drones: How America Goes to War in Secret." *Rolling Stone*, April 26. http://www.rollingstone.com/politics/news/the-rise-of-the-killer-drones-how-america-goes-to-war-in-secret-20120416.
Human Rights Watch. 2011. "Q&A: U.S. Targeted Killings and International Law." December 19. http://www.hrw.org/news/2011/12/19/q-us-targeted-killings-and-international-law.
ICRC Resource Center. 2014. *A Memory of Solferino*, June 16. http://www.icrc.org/eng/resources/documents/publication/p0361.htm.
Instructions to the International (Peace) Conference at The Hague. (April 18, 1899) 1909. In *The Hague Peace Conferences of 1899 and 1907, A Series of Lectures Delivered before the Johns Hopkins University in the Year 1908 by James Brown Scott, Technical Delegate of the United States to the Second Peace Conference at The Hague*, Volume 2—*Documents*. Baltimore, MD: Johns Hopkins University Press. http://avalon.law.yale.edu/19th_century/hag99–03.asp.
International Human Rights and Conflict Resolution Clinic at Stanford Law School and Global Justice Clinic at NYU School of Law. 2012. "Living under Drones: Death, Injury, and Trauma to Civilians from U.S. Drone Practices in Pakistan." September. http://law.stanford.edu/wp-content/uploads/sites/default/files/publication/313671/doc/slspublic/Stanford_NYU_LIVING_UNDER_DRONES.pdf.
Jahangir, Asma. 2003. "Civil and Political Rights, Including Questions of Disappearances and Summary Executions." UN Special Rapporteur for Extrajudicial, Summary and Arbitrary Executions, January 13. Geneva: Commission on Human Rights. UN Doc. E/CN.4/2003/3.

Kellogg-Briand Pact. 1928. *United States Statutes at Large*, vol. 46, part 2, 2343. http://avalon.law.yale.edu/20th_century/kbpact.asp.

Landay, Jonathan S. 2013. "Obama's Drone War Kills 'Others,' Not Just al-Qaeda Leaders." *McClatchy Newspapers*, April 9. http://www.mcclatchydc.com/2013/04/09/188062/obamas-drone-war-kills-others.html.

Letter—Webster to Asburton (1842) 1934. In *Treaties and Other International Acts of the United States of America*, edited by Hunter Miller. Volume 4, *Documents 80–121: 1836–1846*. Washington, DC: Government Printing Office. http://avalon.law.yale.edu/19th_century/br-1842d.asp.

Lin, Patrick. 2010. "Ethical Blowback from Emerging Technologies." *Journal of Military Ethics* 9:4.

Mohammed, Abdullah bin. 2011. "The Al-Qaeda Papers—Drones." Associated Press, June 17. http://hosted.ap.org/specials/interactives/_international/_pdfs/al-Qaeda-papers-drones.pdf.

Parliamentary Assembly, Council of Europe. 2007. *The Concept of Preventive War and Its Consequences for International Relations*. Resolution 1578 (adopted October 4). http://www.assembly.coe.int/nw/xml/XRef/X2H-Xref-ViewHTML.asp?FileID=11677&lang=en.

Protocol for the Prohibition of the Use in War of Asphyxiating Gas, and of Bacteriological Methods of Warfare. 1925. Geneva, June 17. http://avalon.law.yale.edu/20th_century/geneva01.asp.

Robertson, Nic, and Greg Botelho. 2013. "Ex-Pakistani President Musharraf Admits Secret Deal with U.S. on Drone Strikes." *CNN*, April 12. http://edition.cnn.com/2013/04/11/world/asia/pakistan-musharraf-drones/index.html.

Rescript of the Russian Emperor. (August 24, 1898) 1909. *The Hague Peace Conferences of 1899 and 1907, A Series of Lectures Delivered before the Johns Hopkins University in the Year 1908 by James Brown Scott, Technical Delegate of the United States to the Second Peace Conference at the Hague*, Volume 2—*Documents*. Baltimore, MD: Johns Hopkins University Press. http://avalon.law.yale.edu/19th_century/hag99–01.asp#1.

Russian Circular. (January 11, 1899) 1909. In *The Hague Peace Conferences of 1899 and 1907, A Series of Lectures Delivered before the Johns Hopkins University in the Year 1908 by James Brown Scott, Technical Delegate of the United States to the Second Peace Conference at The Hague*, Volume 2—*Documents*. Baltimore, MD: Johns Hopkins University Press. http://avalon.law.yale.edu/19th_century/hag99–02.asp.

Sassòli, Marco, Antoine A. Bouvier, and Anne Quintin. 2011. *How Does Law Protect in War? Vol. 1—Outline of International Humanitarian Law*, 3rd ed. Geneva: International Committee of the Red Cross. http://www.icrc.org/eng/assets/files/publications/icrc-0739-part-i.pdf.

Scahill, Jeremy, and Glenn Greenwald. 2014. "The NSA's Secret Role in the U.S. Assassination Program." *Intercept*, February 10. https://firstlook.org/theintercept/article/2014/02/10/the-nsas-secret-role/.

Shachtman, Noah. 2009. "CIA Chief: Drones 'Only Game in Town' for Stopping Al Qaeda." Wired.com, May 19. http://www.wired.com/2009/05/cia-chief-drones-only-game-in-town-for-stopping-al-qaeda/.

Sharkey, Noel. 2010. "Saying 'No!' to Lethal Autonomous Targeting." *Journal of Military Ethics* 9:4.

Singer, Peter. 2009. *Wired for War: The Robotics Revolution and Conflict in the 21st Century*. New York: Penguin.

Singer, Peter. 2013 "The Proliferation of Drones." *IP Journal*, June 19. https://ip-journal.dgap.org/en/ip-journal/topics/proliferation-drones.

Slim, Hugo. 2008. *Killing Civilians*. New York: Columbia University Press.

Tomz, Michael, and Jessica Weeks. 2013. "Public Opinion and the Democratic Peace." *American Political Science Review* 107 (4): 849–865.

U.S. Embassy Cables. (2008) 2010. "Pakistan Backs U.S. Drone Attacks in Tribal Areas." *Guardian*, November 30. http://www.guardian.co.uk/world/us-embassy-cables-documents/167125.

Zenko, Micah, and Sarah Kreps. 2014. "Limiting Armed Drone Proliferation." Council Special Report No. 69, June. New York: Council on Foreign Relations. www.cfr.org/drones/limiting-armed-drone-proliferation/p33127.

Zucchino, David. 2012. "Stress of Combat Reaches Drone Crews." *Los Angeles Times*, March 18. http://articles.latimes.com/print/2012/mar/18/nation/la-na-drone-stress-20120318.

6

The Drone

It's in the Way That You Use It

DAVID GLAZIER

The use of remotely piloted aircraft, or "drones" in traditional military parlance (and thus the preferred term of this author), to conduct "targeted killings"—deliberate attacks on individuals identified as terrorist adversaries—has supplanted Guantánamo as the most controversial issue in the so-called "war on terror." Proponents see drones as an effective means of striking America's terrorist enemies, stressing their precision targeting capabilities and potential to minimize "collateral" civilian harm. But supporters had little to say about their legality until then-State Department Legal Adviser Harold Koh offered a cursory public justification in March 2010, eight years and 135 strikes since their first use.[1] Other Obama administration officials subsequently followed suit.[2]

Drone critics, meanwhile, continue to question both the legality and wisdom of their use, alleging drone strikes kill more civilians than the government acknowledges (Kilcullen and Exum 2009; O'Connell 2012).[3] Critics have been particularly concerned about the targeting of American citizens, most notably "radical Muslim cleric" Anwar al-Awlaki (see, for example, Benjamin 2012). Some suggest that there are inherent issues with drones themselves; Stephan Sonnenberg, for example, argues that drones are "per se" different from other weapons, requiring "unconventional" legal regulation (Chapter 5).

This chapter focuses on applicable international and domestic law, contending that while both sides in the drone debate get some points right, each also errs about key legal issues. It begins by considering the circumstances in which states can use lethal force to address threats before examining targeted killing specifics. Even while conceding the U.S.

claim to be in a post-9/11 armed conflict, it finds important issues with respect to compliance with both domestic and international laws, including which groups and who within those groups are being targeted; where strikes are taking place; and who is conducting them. The chapter concludes by arguing that U.S. drone proponents are advancing short-sighted legal arguments contravening agreed restrictions on the preventive use of force. This risks starting down a slippery slope towards more killing, by more states, as drone technology proliferates, undermining global respect for human rights and the rule of law, and making the world less safe for all.

The Legality of Resorting to Armed Force

There is longstanding agreement that the use of force in self-defense must be both "necessary" to stop the aggression and "proportionate" in scope; retributive strikes, however deserved, are unlawful (see Lubell 2010). For example, both the U.S. Constitution and international human rights law (IHRL) recognize a fundamental human right to life (on IHRL, see Eviatar, Chapter 7), while Article 2 of the UN Charter requires states to resolve disputes through peaceful means. Furthermore, Chapter VII of the Charter allows the Security Council to authorize force to maintain or restore international peace, while Article 51 provides a narrow exception to the prohibition on unilateral state resort to force, recognizing "the inherent right of individual or collective self-defense *if* an armed attack occurs" (emphasis added).

There are substantive differences between the lawful use of force during an ongoing armed conflict and in self-defense. The law of armed conflict (LOAC), also commonly referred to as international humanitarian law (IHL), or *jus in bello*, requires only that use of force be a "military necessity," meaning that it advances achievement of overall conflict aims. It permits targeting "combatants" based on their "status" without concern for whether they pose any immediate threat; diminishing the enemy's combat capacity is considered a sufficiently legitimate military purpose. When using force in self-defense outside an ongoing conflict, however, law governing the resort to force, or *jus ad bellum*, imposes a substantially more stringent "necessity" standard, limiting targeting to those whose killing is "necessary" to halt the attack.

Despite Article 51's explicit armed attack requirement, many states assert a right to engage in preemptive self-defense, using force to counter *imminent* attacks rather than waiting to absorb a first strike, *if* no other means is adequate to stop it. This is justified as having been part of the "inherent" right of self-defense when the Charter was adopted (see Plaw and Reis, Chapter 9). President George W. Bush's administration articulated a more radical position, "adapt[ing] the concept of imminence" to allow preemptive action against threats involving potential mass civilian casualties even if there is uncertainty "as to the time and place of the enemy's attack" (Bush Administration 2002, 15). But that view was never widely accepted by other states and lacks credible claim to reflect current law (Plaw and Reis argue in Chapter 9 that state practice is trending towards more aggressive approaches, but stops short of concluding that the law has changed). Following the 2003 U.S. invasion of Iraq, UN Secretary General Kofi Annan assembled a blue-ribbon panel to consider emerging threats and the right to use force. It agreed that the UN Charter preserved the preexisting right of self-defense, including limited authority to preempt *imminent* attack, but rejected "preventive" acts against less urgent threats. If a threat lacks true imminence, there is necessarily time for referral to the Security Council. The panel also cautioned that

> in a world full of perceived potential threats, the risk to the global order and the norm of nonintervention on which it continues to be based is simply too great for the legality of unilateral preventive action, as distinct from collectively endorsed action, to be accepted. Allowing one to so act is to allow all. (U.N. Secretary General's High-Level Panel on Threats, Challenges, and Change 2005, 55)

Lawful state killing should thus be an exceptional event, limited to (1) defensive armed conflicts; (2) measures authorized by the Security Council responding to a breach of international peace; (3) proportionate self-defense necessary to counter actual or imminent attacks; and (4) capital punishment imposed by the sentence of legitimately constituted courts. Sporadic state efforts to eliminate perceived enemies through "extra-judicial" killing have routinely provoked international condemnation (see Melzer 2008; Mayer 2009).

States have traditionally regarded terrorism as crime, not war; 9/11 is the only terrorist act widely recognized as an armed attack, and the U.S. military response received at least tacit international acquiescence. Domestically, Congress exercised its constitutional power to sanction hostilities. The September 18, 2001, Authorization for the Use of Military Force (AUMF) allowed the president

> to use all necessary and appropriate force against those nations, organizations, or persons he determines planned, authorized, committed, or aided the terrorist attacks that occurred on September 11, 2001, or harbored such organizations or persons, in order to prevent any future acts of international terrorism against the United States by such nations, organizations or persons. (Authorization for Use of Military Force, 2001)

The combination of international recognition of 9/11 as an armed attack and the subsequent AUMF means that *some* drone strikes falling within the credible scope of the authorized conflict could be lawful. Curiously, however, commentators have largely overlooked the specific limits found in this AUMF language. John Kaag and Sarah Kreps, for example, who argue that "the latitude provided in the AUMF sets a dangerous precedent," make no effort to parse out what the statute actually says (2014, 86), while a leading volume on the legality of targeted killing fails to address AUMF limits at all (see Finkelstein, Ohlin, and Altman 2012). But it is critical to note, for example, that the AUMF's language is entirely in the past tense. To fall within its ambit, a group must have already engaged in conduct linked to 9/11 by September 18, 2001 (see Chapter 1, this volume, for additional discussion regarding AUMF applicability). Purported self-defense justifications are extremely problematic given both the absence of armed attacks on the United States since 9/11 and the imminence requirement for preemptive strikes.

Which Groups May Be Targeted?

Although the AUMF addresses only those responsible for 9/11 (al-Qaeda) and those who aided/sheltered them (the Afghan Taliban), President Bush boldly proclaimed that the United States was fighting "every terrorist group of global reach" (Bush 2001). While the first drone strike in Yemen targeted an al-Qaeda figure, the first strike in Pakistan killed militant leader Nek Mohammed—an enemy of that country, not the United States—as a quid pro quo for access to Pakistani airspace (Mazetti 2013; Fair, Chapter 8). In 2008, the U.S. focus expanded to Pakistani Taliban (Tehrik-i-Taliban Pakistan, or TTP) leadership, a group that is distinct from its Afghan counterparts and coalesced only in 2007. Although Christine Fair asserts in Chapter 8 that strikes against groups like the TTP have been more favorably received in Pakistan than commonly portrayed in the West, targeting the TTP had real adverse consequences. The large number of collateral casualties resulting from failed efforts to kill Baitullah Mehsud, for example, fueled regional opposition to U.S. policies. Even more significantly, it led a group formed to oppose the Pakistani government to sponsor attacks on U.S. facilities and the attempted 2010 Times Square bombing. By creating a new enemy, unilateral presidential extension of hostilities to this group places Americans at risk while the power to "declare war" is lodged with Congress to ensure more participation in, and political accountability for, such decisions.

As was noted in Chapters 1 and 4, although President Barack Obama distanced himself from Bush's "global war on terror" rhetoric, he has expanded targeted killing in terms of both numbers of strikes and their overall scope.[4] Using the ill-defined concept of "associated forces," current targets (more fully identified by Daphne Eviatar in Chapter 7) now include not only the TTP, but also groups like al-Qaeda in the Arabian Peninsula (AQAP) and Somalia's al-Shabaab, both founded well after the AUMF enactment. Although Nasir al-Wuhayshi, AQAP's leader until his June 2015 drone killing, had previously served as Osama bin Laden's secretary and might thus have personally fallen within the scope of the AUMF, AQAP itself emerged in January 2009 as the successor organization to al-Qaeda in Yemen, established in 2006. It is more accurately viewed as "a distinct terrorist group with its own hierarchy and decision making apparatus" than as "an extension of al-Qaeda core" (Johnsen 2015, 259).

Self-defense principles give military forces the logical right to combat co-belligerent forces that join their adversary on the battlefield against them. But mere declarations of affinity, or even military cooperation, are insufficient to expand the legal scope of an armed conflict. In World

War II, it was considered necessary for Congress to declare war separately against Germany, Italy, and Japan despite their formal Axis alliance. Congress subsequently declared war separately against Romania and Bulgaria when their forces later "associated" with Germany. Russia began an existential struggle against the Nazis in June 1941, but remained neutral towards their Japanese co-belligerent until August 1945. Unilateral presidential expansion of hostilities to groups falling outside the literal language of the AUMF would thus be both unprecedented and raise serious concerns over separation of powers.

Strikes against al-Shabaab are particularly problematic. It did not exist until 2006, has not attacked the United States, and currently lacks the capability to reach beyond the Horn of Africa. American intervention would thus require that the group be involved in an internal armed conflict against the Somali government, which then requested U.S. assistance. That would satisfy the *jus ad bellum* requirements of international law, but not the constitutional problem posed by a unilateral presidential commitment to hostilities abroad.

Who May Be Targeted within Those Groups?

The question of lawful targeting extends beyond the groups to be targeted; it also requires identifying which individuals within those entities can be killed. These rules are well established for "traditional" conflicts and can reasonably be extrapolated to non-state adversaries. Unfortunately, the United States has refused to classify current opponents into any recognized legal regime. The Bush administration's "enemy combatant" designation, for example, used to justify Guantánamo detention, lacked legal coherency (Maxwell and Watts 2007). Although overlapping, the sets of who may be detained, tried, and killed in armed conflicts are legally distinct. Habeas jurisprudence addressing detention is thus not dispositive with respect to targeting. Nevertheless, commentators frequently misconstrue Geneva Convention criteria on prisoner of war eligibility as defining who can be targeted (see, for example, Kaag and Kreps 2014).

Correct identification of those liable to being killed is crucial given the lethal finality of these decisions. The targeting of individuals not directly involved in hostilities, such as fifty "Afghan drug lords" reportedly considered for elimination by the Obama administration (Mayer 2009) falls outside international legal authority. So who can be lawfully targeted?

Combatants

Every society proscribes deliberately killing other humans, so the fundamental benefit that the LOAC confers on belligerents is the "combatant's privilege," or immunity from ordinary civil law for acts of violence committed during hostilities. This does not render war lawless; it just requires that combatants be judged against the LOAC/IHL rather than ordinary criminal law. But the cost of this privilege is high:

> Combatants may be attacked at any time until they surrender or are otherwise hors de combat, and not only when actually threatening the enemy.... If a combatant is targeted far behind the front lines, no matter how unlikely such targeting may be, she continues to be a legitimate target.... That illustrates the downside of combatancy: A lawful combatant enjoys the combatant's privilege, but also is a continuing lawful target. (Solis 2010, 188)

Criteria for combatant status in an international armed conflict are formally articulated in the Hague Land Warfare Regulations. Article 1 confers the "rights" and "duties" of war upon armies, as well as militia or volunteer forces, meeting requirements:

1. To be commanded by a person responsible for his subordinates;
2. To have a fixed distinctive emblem recognizable at a distance;
3. To carry arms openly; and
4. To conduct their operations in accordance with the laws and customs of war.

"Combatant" is not synonymous with being either a dedicated "trigger puller" or even a uniformed serviceperson, however. Many members of regular armed forces "contribute to the combat effort in ways that have little to do with actually firing weapons—cooks, administrative personnel . . . and so on. They are nevertheless combatants, for they are entitled to fight" (Solis 2010, 188). There is no legal issue with their deliberate killing. But not all uniformed military personnel can be targeted. Those performing medical and religious functions are protected "noncombatants," including doctors, nurses, chaplains, and "a wide range of specialists, technicians, maintenance staff, drivers, cooks, and administrators provided that they are exclusively assigned to the medical staff" (U.K. Ministry of Defence 2004, § 4.2.2.).

Militaries also employ civilians in support roles. Many, such as acquisitions personnel and financial managers, are far removed from conflict theaters; others accompany fighting forces to positions near the front lines where they perform key administrative and logistical functions. As civilians, they are not liable to deliberate targeting, but they do logically accept increased risks of becoming "collateral damage."

Al-Qaeda and Taliban personnel generally lack fixed distinctive emblems and do not follow LOAC rules.[5] The United States thus has no obligation to accord them combatant status, but could elect to do so in the author's opinion in order to gain clear authority to target and preventively detain them for the duration of hostilities. Their lack of uniforms complicates targeting, but it would be reasonable to consider the al-Qaeda personnel who are provided military training and weapons to be "combatants," liable to direct targeting, while those assigned to medical or religious duties, as well as unarmed members performing support roles typically filled by civilians in state employment, would not be. The dichotomy of roles within traditional armed forces highlights the fact that merely identifying someone as a *member* of al-Qaeda or the Taliban is an insufficient basis for targeting that person; it must still be established that his or her role is equivalent to that of a traditional combatant.

Civilians Directly Participating in Hostilities

The actual U.S. approach has been to try to "have it all." The Bush administration declared al-Qaeda "unlawful enemy combatants," claiming authority to kill them while denying them any LOAC/IHL rights or protections. The Obama administration has altered the terminology but follows the same general approach despite lack of support from mainstream legal commentary.

Denying an adversary combatant status logically makes them civilians. This does not grant total immunity from attack, but limits targeting to "such time as they take a direct part in hostilities" according to Article 51 of the 1977 Additional Geneva Protocol I. How broadly this language should be interpreted is an open question. A literal reading suggests that civilians participating sporadically in hostilities would be immune most of the time, controlling their exposure by choosing when and where to attack. But since many al-Qaeda and Taliban personnel are essentially full-time fighters, engaged in preparation and training between operations just like conventional military personnel, it defies logic that their vulnerability should be temporally limited when U.S. forces are continuously liable to attack.

The Israeli High Court of Justice (HCJ) endorsed a broader interpretation with respect to terrorist groups. Acknowledging that fighting outside Israel's borders must be governed by international armed conflict rules, it agreed that terrorists were not combatants given their failure to meet the Hague criteria (HCJ 2006, ¶¶ 16–21, 24–25). But it rejected arguments that "unlawful combatants" were a distinct third category and instead held them to be "civilians taking a direct part in hostilities" and found Additional Protocol I's Article 51 applicable as customary international law (id. ¶¶ 27–31). This is significant since neither Israel nor the United States are parties to that protocol. The court concluded that "direct participation" went beyond just bearing arms, extending to intelligence collection, transporting fighters and ammunition to/from battles, and servicing weapons (id. ¶ 35).

The court also addressed the meaning of "such time," noting the example of an individual who previously participated in single or sporadic attacks, but no longer does. That individual, it said, has regained protected status and cannot be attacked for past activity (id. ¶ 39). The court was concerned, however, about a "revolving door" approach allowing ongoing participants to periodically regain immunity. It concluded that these individuals remain continuously liable to attack since "the rest between hostilities is nothing other than preparation for the next hostility" (id. ¶¶ 39–40).

This court tempered this targeting authority with four prerequisites:

(1) Attacks must be based on "information which has been most thoroughly verified . . . regarding the identity and activity of the civilian who is allegedly taking part in the hostilities";

(2) Lethal attacks are not permissible "if a less harmful" means can be employed;
(3) A "thorough investigation regarding the precision of the identification of the target and the circumstances of the attack" is required after each strike; and,
(4) Collateral harm to innocent civilians must be able to "withstand the proportionality test." (Id. ¶ 40)

There is little basis to dispute the first and fourth criteria. The second requirement, to use less lethal means, would break new ground as a general rule for armed conflicts, but makes sense given Israel's theoretical police authority over territory where it is legally an occupying power. It would thus be inapplicable to most strikes in an active armed conflict (but *would* defeat the necessity of killing under a self-defense justification). The third mandate for ex post investigation of *every* killing also logically derives from Israeli law enforcement authority, just as U.S. police departments investigate all officer-involved shootings. The LOAC mandates commanders to "repress" war crimes, requiring investigation when there is specific reason to believe a crime may have been committed, but it does not require a searching post-mortem after every attack.

In 2009, the International Committee of the Red Cross (ICRC) issued "interpretive guidance" on direct participation in hostilities following years of expert meetings (Melzer 2009). Although not binding law, countries will find reliance on it advantageous since the ICRC's respected humanitarian credentials will provide an effective buffer against criticism. The guidance acknowledges that there must logically be organized noncivilian armed groups on both sides in a conflict (Melzer 2009). It adopts the term "continuous combat function" to describe the essential characteristics of non-state conflict participants who should constantly be targetable:

> Continuous combat function requires lasting integration into an organized armed group acting as the armed forces of a non-State party to an armed conflict. Thus, individuals whose continuous function involves the preparation, execution, or command of acts or operations amounting to direct participation in hostilities are assuming a continuous combat function. (Melzer 2009, 34)

The guidance calls for distinguishing group members "whose function does not involve direct participation in hostilities . . . [who] remain civilians assuming support functions, similar to private contractors and civilian employees accompanying state armed forces." Examples include "recruiters, trainers, financiers and propagandists" (Melzer 2009, 34).

These criteria provide grounds for questioning the legitimacy of some drone targets. Anwar Awlaki, for example, was originally described as a "radical Muslim cleric" without appreciation that as such, he should have benefited from specific legal protections accorded religious functionaries. Public discussion alleged he recruited and motivated terrorists, functions also seeming to place him in a civilian role. Only around the time of his actual killing did the government openly ascribe an "operational" role to him; there is still insufficient publicly available information to ascertain if this was a good faith determination or a rationalization to bring his killing into notional LOAC compliance; Awlaki's own statements fall short of claiming any direct operational role.

Whether nontraditional conflict adversaries are classified as combatants under the Israeli interpretation of direct participation or the ICRC's "continuous combat function," the ultimate outcome should be the same. Fighters and operational commanders should be liable to attack from the time of their affiliation with the adversary's fighting forces until they return to true civilian life or are incapacitated through capture, wounds, or illness. But terrorist group members who perform functions akin to those of civilians or noncombatants may not be deliberately targeted.

Senior Leadership

Although media reports suggest that by 2010 U.S. drone strikes were mostly targeting low-level fighters, their desired focus was top leaders such as Osama bin Laden and Taliban leader Mullah Omar (see Cloud 2010). Senior military commanders are liable to targeting even when functioning at a strategic level far removed from any fighting. The case of political leadership is more complex; most sources agree that the authority to direct military operations is dispositive. Saddam Hussein, who often wore military uniforms and exercised formal control over Iraqi forces, thus was targetable (Solis 2010). English royalty have military titles and ceremonial uniforms, but they lack actual authority to direct operations and thus are civilians. U.S. presidents are civilians under domestic law, but their authority as commander-in-chief makes them lawful targets, whereas other political leaders without actual command authority are not (see Dinstein 2010).

The HCJ concluded that the "entire chain" overseeing terrorist acts are taking direct part in hostilities, including "those who have sent [the individual]," "the person who decided upon the act, and the person who planned it" (HCJ 2006, ¶ 37). The decision does not address "political" leaders, but they logically fall outside the court's implied requirement for actual linkage to specific hostile acts, while the ICRC excludes "political and humanitarian wings" from the scope of continuous combat functions (Melzer 2009, 32).

Logically, then, the test for targetability of a non-state group's leaders should be essentially the same as that for state officials. Mullah Omar, for example, reportedly directed Taliban military strategy and appointed subordinate commanders (Rashid 2001). He should thus have remained targetable if he continued to exercise that authority over residual Taliban forces after he fled Afghanistan. Similarly, Pakistani Taliban leader Hakimullah Mahsud was reportedly "the architect of a series of suicide bombings and raids on markets, mosques and security installations" (Cloud 2010). This would have made him a target for authorized participants in the conflict between the TTP and the government of Pakistan (but *not* for U.S. forces).

Targeting Unknown Individuals via Signature Strikes

The concept of "targeted killing" assumes reliance on "kill lists" identifying specific individuals for elimination. But in reality, many, if not most, drone killings have been "signature strikes" based on "whether the targeted person's . . . 'pattern of life' fits that of a militant in the eyes of a drone operator thousands of miles away" (Benjamin 2012, 127). Signature strikes are not inherently problematic—soldiers have almost never known the identity of those they killed, and have instead had to make decisions to fire on observed characteristics and conduct.

The current signature strikes, however, give reason for concern as to whether lawful U.S. targets are being struck. While the government is legitimately reluctant to reveal specific targeting criteria lest its adversaries take precautions to avoid attack, reportedly almost any armed "military aged male" might be considered a target (Davidson 2012). A running joke in Washington is that when "the CIA sees 'three guys doing jumping jacks,' the agency thinks it's a terrorist training camp" (Davidson 2012). Yet many men are routinely armed in the countries where strikes take place. Perversely, the government of Yemen compensates relatives of innocent drone victims with AK-47 rifles, increasing their risk of being struck in turn! Moreover, the number of armed groups in Pakistan (some, like al-Qaeda, of legitimate concern to the United States, others, like the TTP, primarily a threat to Pakistan, and still others sponsored by Pakistan to strike against India) calls into question how drone operators would limit attacks to groups falling within the legal scope of U.S. hostilities. Signature strikes thus pose significant reason for concern, even if generally permissible under the LOAC.

Does Nationality Implicate Due Process Considerations?

Many critics have been particularly concerned by deliberate targeting of American citizens such as Anwar al-Awlaki. Some suggest that the Constitution entitles U.S. citizens, even if not foreigners, to some formal "due process" before being killed. Medea Benjamin, for example, contends that like any other American citizen, al-Awlaki was entitled under U.S. law to a presumption of innocence and jury trial, even if it had to be in absentia. Al-Awlaki may have been a traitor who had defected to the enemy, but the Constitution requires that he be convicted on the "testimony of two witnesses" or a "confession in open court," not the say-so of the executive branch (Benjamin 2012, 132).

These arguments implicitly assume the applicability of peacetime rules to targeted killing. Yet even if Daphne Eviatar's arguments in this volume (Chapter 7) about the extraterritorial application of the International Covenant on Civil and Political Rights are correct, LOAC/ IHL principles would still govern these situations under the principle of *lex specialis* (which holds that more specialized rules prevail over the general) when drones are employed in an armed conflict. Neither citizenship nor traditional due process concerns should be relevant. The fact that an enemy fighter has the same nationality as those he or she is fighting has never limited the enemy fighter's killing or preventive detention.[6] *Ex parte Quirin*[7] held in 1942 that a citizen employed as a Nazi saboteur could be denied a jury trial, affirming that Americans joining an enemy get no special constitutional protections—even when captured and held in the United States. Although *Hamdi v. Rumsfeld*[8] required that a citizen held in the United States be able to contest his detention, it also reaffirmed *Quirin*. *Boumediene's*[9] holding that aliens in Guantánamo had a right to habeas review clarified that geography, not citizenship, was the determining factor in establishing enemy rights.

Wartime killing has always been an executive branch prerogative, ranging from strategic choices by the commander in chief down to individual soldiers deciding when to pull their triggers. So long as an individual is identified in good faith as an enemy liable to killing under the LOAC, there should be no concern about his or her individual nationality or formal due process requirements. To assert otherwise is to ignore established law and two centuries of U.S. constitutional experience.

Conversely, however, commentators who would allow targeting outside of an armed conflict with *some* requisite due process concede far too much authority to the government. In that situation, both constitutional and IHRL guarantees of the right to life should prevail, and no executive or judicial process, short of an actual criminal conviction and imposition of a death sentence by a formal court of law, can authorize state killing (Alford 2011). The concept of judges issuing targeting authorization is even more constitutionally infirm than "torture warrant" proposals (Kreimer 2003, 278). *If* there is time to pursue either judicial authorization or the detailed internal procedures by which President Obama reportedly approves adding individuals to a kill list (Becker and Shane 2012), then the degree of imminence required for lawful self-defense is lacking, and killing is impermissible.

Distinction and Proportionality

The Principle of Distinction

Compliance with rules governing which groups, and which members of those groups, can be targeted is insufficient to establish the legality of individual drone strikes. Each attack must be directed at a lawful target; that is, it must distinguish between permissible and impermissible targets. Reports of strikes hitting groups lacking any credible basis as lawful targets, including several discussed in the chapters by Daphne Eviatar (Chapter 7) and Stephan Sonnenberg (Chapter 5), call for objective investigation.

The Principle of Proportionality

The LOAC excuses the inadvertent killing of civilians so long as anticipatable "collateral damage" is not "disproportionate" to the expected military advantage. This rule, known as the principle of proportionality, reflects the legal balancing of competing values of military necessity and humanity.

Unintended civilian drone deaths, particularly in Pakistan, are widely seen as undermining international support for U.S. actions and fueling terrorist recruiting. Judging the strikes' merits poses a daunting challenge, particularly since drone casualty figures are hotly contested. Accurate information is extremely hard to obtain from the remote locations, often "off limits" to Westerners, where strikes take place, and local reporters often have reason to spin facts to their advantage (but see Fair, Chapter 8). The public must generally rely on casualty data amassed by private entities, whose totals vary. On the low end, the U.S.-based *Long War Journal* website tallies 360 drone strikes in Pakistan between January 2006 and July 2014, killing "2,772 leaders and operatives for Taliban, Al Qaeda, and allied extremist groups" at a collateral cost of just 158 civilians (Roggio 2015). This low civilian toll is likely influenced by reported Obama administration practices of declaring any "military-age male" casualty as having been a militant (Becker and Shane 2012). But its totals are called into serious question by reports such as Amnesty International's (2013) investigation of nine 2012–2013 strikes in Pakistan. In the midrange, enjoying more international credibility, the London-based Bureau of Investigative Journalism reports 389 strikes since 2004 resulting in 2,342 to 3,785 deaths, including 416 to 957 civilians, at least 168 being children. At the high end, the Pakistan Body Count (2015) reports similar total casualties, but claims 1,294 to 2,531 of these were civilians. Even if the overall civilian death toll is not excessive compared to the number of strikes conducted, the aggregate numbers are not legally significant. Proportionality requires assessing individual attacks based on predictable casualties, requiring detailed knowledge of what decision-makers knew, or should have known, at the time, and the application of impartial judgment as to the reasonableness of their decisions.

The U.S. "targeted killing" program is often said to be based on Israeli experience, but it does not apply the same legal rigor. The HCJ upheld Israeli targeted killing subject to a requirement for a post-attack legal review, including specific focus on the question of proportionality (HCJ 2006), after considering the legality of attacks over five years that killed roughly 300 targeted individuals at a collateral cost of 150 civilian lives. The CIA, in contrast, reportedly conducted fifteen failed attempts to target a single Pakistani Taliban leader, Baitullah Mehsud, killing more than 321 other individuals before finally succeeding (Mayer 2009). Results like these call for careful inquiry to determine whether they constitute law of war violations, potentially even war crimes. It is extremely likely that some U.S. strikes have violated the principles of distinction and proportionality.

Where Can Drones Be Employed?
Battlefields

Significant discussion has focused on the validity of drone employment away from actual "battlefields," notwithstanding the fact that this is a descriptive, not a legal, term. War is not a sporting event with fixed venues; battlefields are identified after the fact as locations where fighting has taken place. The term is not even listed in the ICRC's *Law of Armed Conflict Dictionary* (Verri 1992) or its IHL index (Solf and Roach 1987). Its sole appearance in IHL rules is the requirement to collect wounded personnel "left on the battlefield."[10] Moreover, aerial warfare has never been limited to "battlefields." One need only recall U.S. strategic bombing during World War II, of North Vietnam, or more recent air operations over the Balkans and Libya—two conflicts lacking any U.S. "battlefield" participation at all. "Battlefields" are actually one place most military aviators do not want to be. Hence the fierce debates over retiring the Air Force's only dedicated close air support platform, the A-10 Warthog, and why the politically well-connected Marine Corps is permitted to have its own jets to ensure air cover for its ground personnel. It defies logic for drones to be limited to "battlefields" absent any similar restriction on manned aircraft.

So where can force be used in armed conflict? Militaries generally use the term "theater of war," but legally hostilities may be conducted in/over the territory of participating states and international waters. Operations in the territory of nonparticipating states, in contrast, violate their sovereignty. The U.S. Navy's international law manual plainly declares that "acts of hostility in neutral territory, including neutral lands, neutral waters, and neutral airspace, are prohibited." (2007, 7-2).

The Territory of Conflict Parties

The fight against al-Qaeda and the Afghan Taliban is complicated by the fact that these two non-state groups now lack recognized territory. As long as the United States continues combat operations in support of the Afghan government, it can credibly claim authority to conduct drone strikes there. American politicians lack the courage to assert this, but the United States would be on even sounder legal ground striking those groups within its own territory. As long as it is engaged in armed conflict against them, an al-Qaeda fighter—even an American citizen—would be a lawful U.S. drone target on the streets of New York or Chicago, however unlikely that is to occur.

Third Party (Neutral) States

Strikes in third countries such as Pakistan, in contrast, are far more problematic. Pakistan is not a formal participant in the U.S. conflicts with al-Qaeda or the Afghan Taliban, so its territory is technically neutral. As such, it has an obligation to deny any belligerent use of its territory harmful to other conflict parties to the best of its abilities. This "duty of prevention is not absolute, but according to [its] power" (Stevenson 1970, 937). Nevertheless, the ability of al-Qaeda and Taliban personnel to take shelter on its territory suggests that Pakistan is likely breaching this legal obligation.

Unwilling or Unable?

American drone proponents assert a broad right to strike enemy forces in neutral territory when that state is "unwilling or unable" to prevent misuse of its territory to United States' detriment (Deeks 2012). But this over-simplistic formulation misstates the actual law. The authority to engage in this self-help derives from the right of self-defense. It is insufficient that the local state has failed to take action to the outside power's detriment; the belligerent intervention must meet the strict self-defense requirements of necessity and proportionality, and be responding to an imminent threat—criteria that the United States played a leading role in establishing during the nineteenth century *Caroline* incident,[11] when it was the "Pakistan" of that day.

Canada remained under British rule following U.S. independence; not surprisingly, Canadian separatists received substantial American sympathy. Participants in an 1837 revolt found refuge and support in New York and employed a U.S.-owned steamboat, the *Caroline*, to ferry men and arms across the Niagara River in preparation for an invasion of what is now Ontario. After observing the *Caroline* reinforce the rebels, the British commander determined that it was essential to destroy the vessel to prevent the invasion. That evening, a British force rowed across the river, seized the *Caroline* on the American side, and set it on fire before sending the boat over Niagara Falls to its destruction.

The U.S. government strongly protested this violation of its sovereignty, and tensions lingered over the incident for several years. They were finally resolved by an 1842 exchange of letters between Secretary of State Daniel Webster and British Special Minister Lord Ashburton. Webster proposed, and Ashburton agreed, that preemptive self-defense and violation of neutral sovereignty could be justified only if the "necessity of self-defence, [was] instant, overwhelming, leaving no choice of means, and no moment for deliberation." They also agreed that international law demanded that when the "necessity of the moment" justified entry into foreign territory, the actions carried out there must "be limited to that necessity, and kept clearly within it" (Rouillard 2004, 110–11).

The U.S. government confirmed this understanding in justifying its 1970 incursion into Cambodia, which was in response to North Vietnam's use of that relatively weak state as both a logistical base for infiltration into South Vietnam and a refuge from counterattack. The Nixon Administration reported its actions as self-defense in a letter to the Security Council as required by Article 51, acknowledging legal constraints on the incursion, which would be "limited and proportionate to the aggressive military operations of the North Vietnamese forces and the threat they pose."[12]

State Department Legal Adviser John R. Stevenson subsequently provided amplifying justification, noting that "the United States has a strong interest in developing rules of international law that limit claimed rights to use armed force and encourage the peaceful resolution of disputes" (Stevenson 1970, 935). While he noted that some scholars have taken a broader view of the right to "self-help," Stevenson reviewed historical examples of interventions before concluding that "whatever the merits of these views prior to 1945, the adoption of the United Nations Charter changed the situation by imposing new and important limitations on the use of armed force," including restricting the unilateral use of force to situations of self-defense (Stevenson 1970, 940).

"Unwilling or unable" is thus a legally insufficient standard; action taken against enemy forces in the territory of a nonparticipating (neutral) third state must also satisfy the necessity, proportionality, and imminence standards of *jus ad bellum* rules governing self-defense.[13] While there may be some room to question the precise meaning of these terms as they relate to contemporary state understandings and practice, it is a legal fallacy to propose that they can be dispensed with today (Bethlehem 2012).

Consent

One of the more controversial aspects concerning U.S. targeted killing has been the role of third state consent. This entails both factual and legal questions. Pakistan and Yemen have clearly given the United States confidential permission to conduct at least some strikes carried out on their territory, even if politically compelled to denounce them for domestic popular consumption. But the precise terms of that consent, including what targets fall within its scope, as well as any quid pro quos, remain speculative (on Pakistan, see Fair, Chapter 8).

Moreover, there have been only limited efforts to analyze the legal significance of "consent." Mary Ellen O'Connell (2012) has consistently called attention to this issue, noting that international law limits state authority to employ violence even within its own borders, as well as arguing that a state cannot grant authority to kill its nationals, which it lacks itself. Sir Daniel Bethlehem, former principal Legal Adviser in the U.K. Foreign and Commonwealth Office, offers further clarity, noting that state consent can overcome sovereignty issues, but cannot provide the legal authority for the killing itself. That must still be drawn from public international law governing either armed conflict or self-defense.[14]

The United States did not seek Cambodia's consent to its 1970 incursion because it reasoned that granting it would have legally made Cambodia a formal co-belligerent of the United States and South Vietnam against the North and thereby would have invited further Communist aggression against that weak state (Stevenson 1970). This understanding should make states reluctant to grant consent for U.S. drone strikes against groups that the local state does not wish to engage itself. But it would be consistent with O'Connell's view. By joining the conflict as a co-belligerent, the local state would gain the authority to target opposing fighters based on their status, as compared to the more limited neutral obligation/authority to merely suppress hostile acts by belligerents within its territory. Consenting to lethal foreign action would then be consistent with its own legal authority.

Absent consent, targeted killing in states not already parties to an armed conflict will require satisfaction of the necessity, proportionality, and imminence standards governing self-defense in order to overcome sovereignty issues. With local state consent, the United States may target lawful adversaries beyond the previous boundaries of an armed conflict, presuming satisfaction of the *jus ad bellum* requirements of domestic and international law establishing the legal existence of a conflict against that adversary. While consent is an insufficient basis for killing in the first place, the expansion of an armed conflict to a third state offers the potential advantage of allowing status-based targeting to fall under the lower *jus in bello* "military necessity" standard, as compared to the substantially more restrictive absolute "necessity" standard of the *jus ad bellum*.

Who May Lawfully Employ Drone Systems?

The 2001 AUMF authorized "the use of United States Armed Forces" against those responsible for the 9/11 attacks. Nevertheless, the CIA, rather than the military, has conducted many of the subsequent drone strikes. While the legality of this approach has been debated fairly extensively, most commentators have missed the obvious answer—it violates international law. The 1923 draft Hague air warfare regulations, long recognized as declaratory of customary international law (Roberts and Guelff 2004), are quite explicit. Article 13 proclaims: "Military aircraft are alone entitled to exercise belligerent rights" (Commission of Jurists 1923, 247). The rules require military aircraft to "bear an external mark indicating its nationality and military character" (id. art. 3, 246) and to be "under the command of a person duly commissioned or enlisted in the military service of the state" (id. art. 14, 247). The "crew must be exclusively military" (id.) and must "wear a fixed distinctive emblem . . . so as to be recognizable at distance" if "separated from their aircraft" (id. art. 15, 248). The regulations further state: "No aircraft other than a belligerent military aircraft shall engage in hostilities in any form; the term 'hostilities' includes the transmission during flight of military intelligence for the immediate use of a belligerent" (id. art. 16), meaning even real-time tactical reconnaissance by CIA drones is legally problematic in an armed conflict. The regulations clearly distinguish military aircraft from other "public non-military aircraft," requiring the latter to bear external marks indicating their "public non-military character" if used for customs or police purposes; otherwise they are "treated on the same footing, as private aircraft" in wartime (id. arts. 4–5, 246).

Despite major technological advances in aerial warfare, subsequent treaties have addressed only a few air warfare issues. This lacuna led the Harvard-based Program on Humanitarian Policy and Conflict Research to sponsor the restatement of current customary rules in the 2009 *Manual on International Law Applicable to Air and Missile Warfare* (Program on Humanitarian Policy and Conflict Research 2009, [HPCR]). One of the manual's significant contributions is updating the 1923 rules to reflect drone development. It defines two relevant terms:

> "Military aircraft" means any aircraft (i) operated by the armed forces of a State; (ii) bearing the military markings of that State; (iii) commanded by a member of the armed forces; and (iv) controlled, manned or preprogrammed by a crew subject to regular armed forces discipline. . . .
>
> "Unmanned Combat Aerial Vehicle (UCAV)" means an unmanned military aircraft of any size which carries and launches a weapon, or which can use on-board technology to direct such a weapon to a target.

It then explicitly states: "Only military aircraft, including UCAVs, are entitled to engage in attacks" (HPCR 2009, xxix).

The consequences for violations are not specified by these rules. But as previously discussed, civilians participating in hostilities both lose their immunity from attack and lack combatant immunity from regular domestic law. Every CIA drone killing is thus potentially a murder under the domestic law of the venue where it takes place and the home state of any third-country victim. This liability extends not only to the actual operators, whose identities are likely sufficiently obscure to provide practical protection from prosecution, but also to government higher-ups overseeing the program. Senior CIA officials, National Security Advisors, and presidential staff who are readily identifiable could thus face life-long risk of prosecution (as would Presidents Bush and Obama if denied head of state immunity).

Perversely, however, the U.S. government is concurrently prosecuting its adversaries in the Guantánamo military commissions on the ill-advised legal theory that unprivileged belligerency constitutes a war crime per se. If this theory were correct—a point the United States may be legally estopped from denying in the event of any subsequent foreign prosecutions—then all those responsible for the CIA program both may be tried by *any* country exercising universal jurisdiction over war crimes and may be prosecuted by military tribunals rather than ordinary civilian courts.

Conclusion: Why Does the Law Matter?

Some U.S. drone strikes—those conducted in a discriminate manner by the military against identified al-Qaeda or Afghan Taliban fighters or operational leaders with the consent of the co-belligerent nations where they take place—may enjoy a claim to legality. Strikes against terrorist groups that are not included within the scope of the 2001 AUMF, in contrast, raise concerns about the executive usurpation of authority constitutionally delegated to Congress in order to determine when, and against whom, the United States may fight, *even if* the United States enjoys the blessing of the foreign state where they take place. And any strikes conducted abroad without consent of the nation whose territory is involved violate international rules about sovereignty and prohibitions against the use of force unless the exacting *jus ad bellum* criteria for self-defense of necessity, proportionality, and imminence are satisfied.

Satisfaction of the *jus ad bellum* is only the first step in the analysis, however. Each strike must be planned and conducted in a discriminate manner, and the anticipated military advantage must outweigh predictable civilian harm. Although the United States has never publicly admitted fault in any drone strike to date, and although the overall number of civilian casualties may be low compared to previous conflicts, this fails to establish the legality of any individual attack. There is clearly sufficient information in the public record to identify problematic attacks calling for follow-up inquiry, and potentially, formal accountability.

Since the United States enjoys self-proclaimed status as the "world's only superpower," one might ask why it should worry about conforming its drone use to the rules applicable to lesser states. Americans widely perceive drone use as making them safer, so why quibble about legality?

First, any U.S. monopoly on lethal drone technology is over. Israel and the United Kingdom are already conducting lethal strikes. Seventy countries are now said to possess surveillance drones, while China, Iran, Russia, South Korea, and Taiwan are actively perfecting armed variants, and China has already promised to make export sales (Zenko 2013).

Moreover, international law considers all states sovereign equals, with the same rights and obligations, regardless of their size or military prowess. What the United States does today, other states can assert the right to do in the future. Ukraine provides a case in point; justifications Vladimir Putin offered for Russia's entry into Crimea were, almost verbatim, excerpts from those proffered by Ronald Reagan and George H. W. Bush in support of their military interventions into Grenada and Panama (for additional examples of states using U.S. self-defense justifications to suit their own foreign policy interests, see Fisk and Ramos 2014).

Fuzzy legal assertions, such as those ambiguously mixing *jus ad bellum* self-defense rules and *jus in bello* targeting authority, are particularly rife for future abuse. Divorcing the "unwilling or unable" standard from the co-requisites of necessity, proportionality, and imminence is even more problematic. U.S. conduct today creates an all too real risk of launching the world down a slippery slope in which states routinely fly drones across borders to eliminate military or political opponents virtually at will.

It is unlikely that even the current level of drone activity really serves American interests. As Brian Glyn Williams persuasively details, there are credible arguments both for and against the effectiveness of U.S. drone policy (2013, 169–228). *Some* strikes likely do contribute to American national security. But this is far from establishing that the full scope of U.S. drone activity is lawful or wise. Micah Zenko's (2013) Council on Foreign Relations report, for example, concludes that many U.S. strikes are counterproductive, and recommends ending signature strikes, limiting targets to senior operational leaders or direct participants in ongoing terrorist operations, and clarifying the scope of AUMF authority. While Zenko's arguments are policy-based, they are consistent with the law discussed in this chapter. Both the framers of the Constitution and the government officials who developed contemporary international law were rational actors who carefully preserved the necessary sphere of action for nations to defend their vital interests. States thus already have the legal authority they need to protect their security. Actions falling outside the scope of existing law are, almost assuredly, ultimately both unnecessary and unwise.

NOTES

1 A Justice Department white paper, leaked to the media in February 2013, provided a confused mash-up of justifications derived from the law of armed conflict and self-defense principles (U.S. Department of Justice undated). An earlier Office of Legal Counsel memo (released in mid-2014) was more coherent, but focused more on U.S. criminal law than either domestic or international rules governing the use of force (Department of Justice 2010).
2 The texts of these key speeches and critical analysis can be found in Anderson and Wittes 2015.
3 Indeed, the very term "drone" is now frequently considered pejorative.
4 In terms of the number of drone strikes, there were 350 in Obama's first six years, as compared to 52 under Bush.
5 Al-Qaeda reportedly did have a uniformed brigade when the United States intervened in Afghanistan. See Williams 2008.
6 *In re Territo*, 156 F.2d 142 (9th Cir., 1946), upheld detaining an American citizen fighting for Italy as a prisoner of war in the United States until the formal conclusion of peace.
7 319 U.S. 1 (1942).
8 542 U.S. 507 (2004).
9 553 U.S. 723 (2008).

10 Convention for the Amelioration of the Condition of the Wounded and Sick in Armed Forces in the Field, art. 15, August 12, 1949, 6 U.S.T. 3114, 75 U.N.T.S. 31 (more commonly known as the First Geneva Convention).
11 For further discussion of this incident, see Jones 1977, 20–29; Rouillard 2004, 106–108.
12 Letter to the Security Council, U.N. Doc. S/9781, May 5, 1970, *American Journal of International Law* 64 (5): 933.
13 See Sonnenberg, Chapter 5, and Plaw and Reis, Chapter 6, for further *ad bellum* discussion.
14 Bethlehem made these comments as part of a panel on "Targeted Killings" at the American Society of International Law's 2013 annual meeting in Washington, D.C., with Marco Sassòli, Hina Shamsi, and the author. Video of the panel can be streamed here: http://www.asil.org/resources/2013-annual-meeting-videos. Sir Daniel's comments are found from 53:53–56:00.

BIBLIOGRAPHY

Alford, Ryan Patrick. 2011. "The Rule of Law at the Crossroads: Consequences of Targeted Killing of Citizens." *Utah Law Review* 2011 (4): 1203–1273.
Amnesty International. 2013. "Will I Be Next? US Drone Strikes in Pakistan." London: Amnesty International, October. https://www.amnestyusa.org/sites/default/files/asa330132013en.pdf.
Anderson, Kenneth, and Benjamin Wittes. 2015. *Speaking the Law: The Obama Administration's Addresses on National Security Law*. Stanford, CA: Hoover Institution.
Authorization for Use of Military Force. 2001. *Statutes at Large*, 115:224.
Barrett, Edward 2010. "Rise of the Drones: Unmanned Systems and the Future of War." Washington, DC: House Committee on Oversight and Government Reform Subcommittee on National Security and Foreign Affairs, March 23.
Becker, Jo, and Scott Shane 2012. "Secret 'Kill List' Proves a Test of Obama's Principles and Will," *New York Times*, May 29. http://www.nytimes.com/2012/05/29/world/obamas-leadership-in-war-on-al-qaeda.html.
Benjamin, Medea. 2012. *Drone Warfare: Killing by Remote Control*. New York: OR Books.
Bethlehem, Daniel. 2012. "Self-Defense against an Imminent or Actual Armed Attack by Non-State Actors." *American Journal of International Law* 106 (4): 770–778.
Blumenfield, Laura. 2006. "In Israel, a Divisive Struggle over Targeted Killing." *Washington Post*, August 25.
Boothy, William. 2009. *Weapons and the Law of Armed Conflict*. New York: Oxford University Press.
Bush Administration. 2002. "The National Security Strategy of the United States of America." Washington, DC: The White House. September. http://nssarchive.us/NSSR/2002.pdf.
Bush, George W. 2001. "Address to a Joint Session of Congress and the American People." White House, September 20. http://georgewbush-whitehouse.archives.gov/news/releases/2001/09/20010920-8.html.

Cloud, David S. 2010. "CIA Drones Have Broader List of Targets." *Los Angeles Times*, May 5. http://articles.latimes.com/2010/may/05/world/la-fg-drone-targets-20100506.
Commission of Jurists. 1923. "[Draft] Rules of Aerial Warfare." Reprinted in *American Journal of International Law* (Supplement) 17 (October): 245–260.
Davidson, Amy. 2012. "The President's Kill List." *New Yorker*, May 30. http://www.newyorker.com/news/daily-comment/the-presidents-kill-list
Deeks, Ashley. 2012. "'Unwilling or Unable': Toward a Normative Framework for Extraterritorial Self Defense." *Virginia Journal of International Law* 52 (3): 483–550.
Dinstein, Yoram. 2010. *The Conduct of Hostilities under the Law of International Armed Conflict*. New York: Cambridge University Press.
Dinstein, Yoram. 2011. *War, Aggression and Self-Defense*, 5th ed. New York: Cambridge University Press.
Fisk, Kerstin, and Jennifer M. Ramos. 2014. "Actions Speak Louder than Words: Preventive Self-Defense as a Cascading Norm." *International Studies Perspectives* 15 (2): 163–185.
Finkelstein, Claire, Jens David Ohlin, and Andrew Altman, eds. 2012. *Targeted Killings: Law and Morality in an Asymmetrical World*. Oxford, UK: Oxford University Press.
High Court of Justice. 2006. *Public Committee against Torture in Israel et al. v. Government of Israel et al.*, HCJ 769/02, December 14. http://elyon1.court.gov.il/files_eng/02/690/007/a34/02007690.a34.pdf.
Human Rights Watch. 2003. "Off Target: The Conduct of the War and Civilian Casualties in Iraq." New York: Human Rights Watch. http://www.hrw.org/reports/2003/usa1203/.
Human Rights Watch. 2013. "'Between a Drone and al-Qaeda': The Civilian Cost of U.S. Targeted Killings in Yemen." New York: Human Rights Watch, October 22. http://www.hrw.org/node/119909/section/7.
International Human Rights and Conflict Resolution Clinic at Stanford Law School and Global Justice Clinic at NYU School of Law. 2012. "Living under Drones: Death, Injury, and Trauma to Civilians from U.S. Drone Practices in Pakistan." September. http://law.stanford.edu/wp-content/uploads/sites/default/files/publication/313671/doc/slspublic/Stanford_NYU_LIVING_UNDER_DRONES.pdf.
Jaffe, Greg. 2010. "Combat Generation: Drone Operators Climb on Winds of Change in the Air Force." *Washington Post*, February 28. http://www.washingtonpost.com/wp-dyn/content/article/2010/02/27/AR2010022703754.html.
Jones, Howard. 1977. *To the Webster-Ashburton Treaty*. Chapel Hill: University of North Carolina Press.
Kaag, John, and Sarah Kreps. 2014. *Drone Warfare*. Cambridge, UK: Polity Press.
Kilcullen, David, and Andrew McDonald Exum. 2009. "Death from Above, Outrage down Below." *New York Times*, May 17. http://www.nytimes.com/2009/05/17/opinion/17exum.html?pagewanted=all&_r=0.

Kreimer, Seth F. 2003. "Too Close to the Rack and the Screw: Constitutional Constraints on Torture in the War on Terror." *University of Pennsylvania Journal of Constitutional Law* 6 (2): 278–325.

Johnsen, Gregory D. 2013. *The Last Refuge: Yemen, Al-Qaeda, and America's War in Arabia*. New York: W. W. Norton.

Lubell, Noam. 2010. *Extraterritorial Use of Force against Non-State Actors*. New York: Oxford University Press.

Maxwell, Mark D., and Sean M. Watts. 2007. "'Unlawful Enemy Combatant': Status, Theory of Culpability, or Neither?" *Journal of International Criminal Justice* 5 (1): 19–25.

Mayer, Jane. 2009. "The Predator War." *New Yorker*, October 26.

Mazetti, Mark. 2013. *The Way of the Knife: The CIA, A Secret Army, and a War at the Ends of the Earth*. New York: Penguin.

Mazetti, Mark, Eric Schmitt, and Robert Worth. 2011. "Two-Year Manhunt Led to Killing of Awlaki in Yemen." *New York Times*, September 30. http://www.mytimes.com/2011/10/01/world/middleeast/anwar-al-awlaki-is-killed-in-yemen.html?pagewanted=all.

McIntyre, Jamie. 2002. "U.S. Accepts Blame for Canadians' Deaths." CNN.com/U.S., June 29. http://edition.cnn.com/2002/US/06/29/friendly.fire/index.html.

Melzer, Nils. 2008. *Targeted Killing in International Law*. New York: Oxford University Press.

Melzer, Nils. 2009. *Interpretive Guidance on the Notion of Direct Participation in Hostilities*. Geneva: International Committee of the Red Cross.

O'Connell, Mary Ellen. 2012. "Unlawful Killing with Combat Drones: A Case Study of Pakistan, 2004–2009." In *Shooting to Kill: Socio-Legal Perspectives on the Use of Lethal Force*, edited by Simon Bronitt, Miriam Gani, and Saskia Hufnagel, 263–292. Oxford, UK: Hart Publishing.

Pakistan Body Count. 2015. "Drone Attacks." August 6. http://pakistanbodycount.org/drone_attack.

Program on Humanitarian Policy and Conflict Research. 2013. *HPCR Manual on International Law Applicable to Air and Missile Warfare*. New York: Cambridge University Press.

Rashid, Ahmed. 2000. *Taliban*. New Haven, CT: Yale University Press.

Roberts, Adam, and Richard Guelff. 2004. *Documents on the Laws of War*. New York: Oxford University Press.

Roggio, Bill. 2015. "Charting the Data for U.S. Airstrikes in Pakistan, 2004–2015." *Long War Journal*. http://www.longwarjournal.org/pakistan-strikes.

Rouillard, Louis-Philippe. 2004. "The Caroline Case: Anticipatory Self-Defense in Contemporary International Law." *Miskolc Journal of International Law* 1 (2): 104–120.

Singer, Peter. W. 2009. *Wired for War: The Robotics Revolution and Conflict in the 21st Century*. New York: Penguin.

Solf, Waldemar A., and J. Ashley Roach, eds. 1987. *Index of International Humanitarian Law*. Geneva: International Committee of the Red Cross.

Solis, Gary D. 2010. *The Law of Armed Conflict*. New York: Cambridge University Press.

Stevenson, John R. 1970. "Speech to Association of the Bar of the City of New York." *American Journal of International Law* 64 (4): 933–41.

United Kingdom Ministry of Defence. 2004. *The Manual of the Law of Armed Conflict*. New York: Oxford University Press.

United Nations Secretary General's High-Level Panel on Threats, Challenges, and Change. 2005. *A More Secure World: Our Shared Responsibility*. U.N. Doc A/59/565. http://www.unrol.org/files/gaA.59.565_En.pdf.

U.S. Department of Defense. 1992. *Final Report to Congress: Conduct of the Persian Gulf War*, April. http://www.dod.mil/pubs/foi/operation_and_plans/PersianGulf-War/404.pdf.

U.S. Department of Justice. 2010. "Applicability of Federal Criminal Laws and the Constitution to Contemplated Lethal Operation as against Shaykh Anwar al-Aulaqi." Office of Legal Counsel, July 16. http://fas.org/irp/agency/doj/olc/aulaqi.pdf.

U.S. Department of Justice. Undated. *White Paper: Lawfulness of a Lethal Operation Directed against a U.S. Citizen Who Is a Senior Operational Leader of Al-Qa'ida or an Associated Force*. Washington, DC: Department of Justice. http://msnbcmedia.msn.com/i/msnbc/sections/news/020413_DOJ_White_Paper.pdf

U.S. Navy. 2007. *The Commander's Handbook on the Law of Naval Operations*. Naval Warfare Publication (NWP) 1–14M. Newport, RI: Naval War College.

Verri, Pietro.1992. *Dictionary of the International Law of Armed Conflict*. Geneva: International Committee of the Red Cross.

Whittle, Richard. 2014. *Predator: The Secret Origins of the Drone Revolution*. New York: Henry Holt.

Williams, Brian Glyn. 2008. "The Al-Qaeda We Don't Know: The 055 Brigade," *World Politics Review*, October 26. http://www.worldpoliticsreview.com/article.aspx?id=2821.

Williams, Brian Glyn 2013. *Predators: The CIA's Drone War on al Qaeda*. Dulles, VA: Potomac Books.

Zenko, Micah. 2013. "Reforming U.S. Drone Strike Policies." Council Special Report No. 65. New York: Council on Foreign Relations, January. http://www.cfr.org/wars-and-warfare/reforming-us-drone-strike-policies/p29736?co=C009601.

7

Drones and the Law

Why We Do Not Need a New Legal Framework for Targeted Killing

DAPHNE EVIATAR

The United States' use of unmanned aerial vehicles (UAVs, or "drones") to conduct so-called "targeted killings" of suspected terrorists can be viewed as a use of preventive force designed to prevent sporadic acts of terrorism from escalating into an all-out war.[1] Indeed, drones are politically popular in the United States precisely because the use of lethal force against faraway terror suspects—whose activities could, conceivably, one day threaten us—seems far preferable to sending troops overseas or waiting for an actual threat against the United States to materialize.

But the liberal use of lethal force to kill thousands of suspected terrorists abroad is extremely dangerous, and it does not square with any acceptable interpretation of international law. Indeed, nowhere in either international treaties or international common law is such "preventive" lethal force deemed acceptable outside a zone of armed conflict, unless it is in response to a specific, imminent threat to human life. As discussed in earlier chapters, laws of war and international human rights law have developed precisely to prevent politically popular attitudes at home from swaying state policy in favor of the liberal killing of foreigners suspected of wrongdoing abroad. After all, as David Glazier concludes in Chapter 6, we can easily imagine the damaging precedent such a use of preventive force sets. Even CIA Director John Brennan admits that the United States is "establishing precedents that other nations may follow, and not all of them will be nations that share our interests or the premium we put on protecting human life, including innocent civilians" (2012).

Such concerns have not stopped supporters of "targeted killings" from arguing that drones are a perfectly lawful and effective counterterrorism tool, although in doing so, they often skew or ignore those aspects of international law that do not suit their purposes. As earlier chapters have noted, the Obama administration claims the United States is fighting a broad and ever-changing war that requires the ability to use lethal force, delivered by drones or otherwise, whenever the president determines it is needed wherever the United States' enemies may be found. This follows from the U.S. government's claim that the "war on terror"—now called the "war against the Taliban, al-Qaeda, and 'associated forces'"—has no geographic limits. The United States acknowledges the role of international law in only the vaguest sense, and conveniently ignores the instances when law conflicts with its policy. Even when some individuals acknowledge legal concerns, they contend that the U.S. drone program is necessary to our fight against terrorism.

In my view, although the United States' ongoing war may not have easily definable geographic boundaries, it must have limits. It is clear as a matter of law that a war between a state and an armed group (a "non-international armed conflict," or NIAC, in the parlance of International Humanitarian Law) has some boundaries, even if they are not exactly geographical. International law, which has been carefully developed and is grounded in political and ethical considerations, provides those boundaries. The United States needs to explicitly acknowledge how international law limits its powers so that it can exercise the "soft power" critical to achieving its counterterrorism objectives. In particular, it must be seen as respecting and abiding by international law in order to win the international cooperation it needs and to respond effectively to anti-American sentiment fueled by perceptions of the United States as a rogue power using military might with impunity.[2] It is therefore critical for the United States to publicly assert and demonstrate that it respects and follows existing international law, as interpreted by the vast majority of nations, international bodies, and other widely respected legal authorities.

In this chapter, I review the relevant international law (International Humanitarian Law [IHL] and International Human Rights Law [IHRL]) governing states' use of lethal force outside of their borders and discuss the key questions regarding whether or not the United States is involved in an armed conflict with the "associated forces" it targets in drone strikes in Pakistan and Yemen. I then provide an overview of the two main terrorist groups with which the United States claims to be engaged in armed conflict outside of a clear battlefield and relying heavily on the use of drones, and argue that the United States does not appear to be engaged in a lawful, publicly declared armed conflict with them. This situation does not require us to redefine armed conflict or to create a new legal framework to govern U.S. counterterrorism operations. Nor does it require us to create a new international framework under IHL to specifically govern the development and use of drones, as Stephan Sonnenberg argues in Chapter 5. Rather, I demonstrate that the extraterritorial application of IHRL is sufficient to govern U.S. covert lethal action in Pakistan and Yemen. I conclude with a discussion of the importance of providing enough information about the drone program to demonstrate compliance with international human rights law.

Relevant International Law

International law prohibits lethal action by states unless it meets certain requirements (see Sonnenberg, Chapter 5, and Glazier, Chapter 6). Those requirements are different depending on whether the action takes place within or outside of an armed conflict. In armed conflict, International Humanitarian Law (IHL), also known as the Law of Armed Conflict, applies. Under IHL, each state in an International Armed Conflict (IAC), or war between states, can lawfully target the armed forces of the other state or states party to the conflict. Civilians may not be targeted unless, and only while, they are "directly participating in hostilities" (DPH). Some interpret the law to also allow targeting of civilians who perform a "continuous combat function," such as acting as a farmer by day and a soldier by night (see International Committee of the Red Cross 2009; see also Glazier, Chapter 6, for additional discussion).

In a non-international armed conflict (NIAC), or war between a state and a non-state actor, such as a terrorist group, the state may target only the armed forces of the non-state actors who are party to the armed conflict. The law outside of an armed conflict is completely different, however: everyone is a civilian, and therefore not targetable except in extremely limited circumstances. Although the U.S. government and media frequently refer to "combatants" or "militants" when describing military operations outside a war zone, this is misleading. There are no "combatants" or "militants" outside of a legally recognized armed conflict. Where there is no ongoing armed conflict, International Human Rights Law (IHRL) applies and sets forth strict limits for the use of lethal force.

Modern International Human Rights Law arose out of the creation of the United Nations after World War II. Following the horrors witnessed in that war, the new international body established a Commission on Human Rights and charged it with drafting a document that would set forth the meaning of the "fundamental human rights" and freedoms proclaimed in the UN Charter.[3] Led by Eleanor Roosevelt, the commission drafted the Modern International Human Rights Law, which was enshrined in the Universal Declaration of Human Rights (UDHR) and adopted by fifty-six member states on December 10, 1948. The declaration outlined the basic civil, political, economic, social, and cultural rights that all human beings should enjoy, including "the right to life, liberty, and security of person."

Although many consider the Universal Declaration itself customary international law, two major human rights treaties were drafted afterwards to enforce the rights covered by the declaration. The International Covenant on Civil and Political Rights (ICCPR) focuses on the right to life and on freedom of speech, religion, and voting. The International Covenant on Economic, Social, and Cultural Rights (ICESCR) focuses on rights to food, education, health, and shelter (Piccard 2010).[4] Together with the Universal Declaration, these documents are commonly referred to as the International Bill of Human Rights. There are 163 state parties to the ICESCR and 168 parties to the ICCPR, meaning that those states have agreed to be bound by the treaties' terms.

The basic right to life enshrined in the ICCPR has been interpreted to limit a state's use of lethal force outside of a recognized armed conflict to only those situations where the target of lethal force is presenting an imminent threat to human life that cannot be otherwise ameliorated. The United Nations' "Basic Principles on the Use of Force and Firearms by Law Enforcement Officials" states: the "intentional lethal use of firearms may only be made when strictly unavoidable in order to protect life" (1990, 2). Thus firearms may be used only in self-defense or in defense of others "against the imminent threat of death or serious injury" or "to prevent the perpetration of a particularly serious crime involving grave threat to life" and "only when less extreme means are insufficient to achieve these objectives" (United Nations 1990, 2; see also Gaggioli 2013).

Context Matters: In a Zone of Armed Conflict or Outside of One?

Given the different legal frameworks governing states' use of force, one of the fundamental questions swirling around the United States' secretive use of lethal drone strikes is whether the U.S. government believes it is acting within or outside of an armed conflict. Despite repeated questioning from lawmakers and human rights groups, the United States has never made this clear. Its vague and conflicting explanations of its authority so far suggest that the U.S. government is trying to fashion its own rules to justify its behavior, rather than following established law.

While the Obama administration says it is in an armed conflict against the Taliban, al-Qaeda, and its "associated forces" (see, for example, Brennan 2012) wherever they may be found, the administration has not defined "associated forces." Adding to the broad scope is the fact that there are many armed groups around the world that claim, or at one point had claimed, an affiliation with al-Qaeda. Meanwhile, U.S. officials have repeatedly referred to an "al-Qaeda core"—the group the United States went to war against in 2001—as "decimated" (Nicholas and Nelson 2013; CNN 2012). Indeed, most of the various terrorist groups that have adopted the al-Qaeda name or proclaimed their affinity with al-Qaeda since 9/11 are not targeting the United States, but fighting domestic civil wars or mounting regional insurgencies.

But from a legal standpoint, the United States cannot merely declare, without further explanation, that the armed conflict extends to any insurgent group around the world that claims ideological affiliation with a "core" group that essentially no longer exists. This position ignores settled international law. Under IHL, an armed conflict exists only if hostilities are taking place between a state and another state (IAC) or between a state and a clearly defined non-state group (NIAC) with a central command and control structure. To constitute NIAC, hostilities must reach a level of intensity that involve more than sporadic acts of terrorism. The International Committee of the Red Cross (ICRC) has written extensively on this question and said: "Non-international armed conflicts are protracted armed confrontations occurring between governmental armed forces and the forces of one or more armed groups, or between such groups arising on the territory of a State [party to the Geneva Conventions]. The armed confrontation must reach a minimum level of intensity and the parties involved in the conflict must show a minimum of organisation" (2008, 5).

According to an ICRC opinion paper, "In order to distinguish an armed conflict, in the meaning of common Article 3, from less serious forms of violence, such as internal disturbances and tensions, riots or acts of banditry, the situation must reach a certain threshold of confrontation" (2008, 3). The two relevant criteria are the intensity of the hostilities and the organization of the parties (*Prosecutor v. Tadic* 1995). Hostilities reach a minimum level of intensity when, for example, "hostilities are of a collective character or when the government is obliged to use military force against the insurgents, instead of mere police forces" (International Committee of the Red Cross 2008, 3). Non-governmental groups are considered "parties to the conflict" only if "they possess organized armed forces," such as forces that operate "under a certain command structure and have the capacity to sustain military operations" (International Committee of the Red Cross 2008, 3).

To determine whether, when, and where the United States is in an armed conflict, it is therefore critical to understand (1) who the individual groups are that might constitute "associated forces," (2) their connection to al-Qaeda and/or the Taliban, (3) whether there is a central command and control structure, and (4) what level of hostilities exist between them and the United States. As U.S. officials have acknowledged, al-Qaeda has splintered into many different branches and affiliates. This became particularly obvious in June 2013, when Ayman al-Zawahri, the leader of al-Qaeda, sent a letter directing an al-Qaeda affiliate in Syria, Islamic State in Iraq and Syria, also known as ISIS, to withdraw to Iraq and leave operations in Syria to someone else. ISIS leader Abu Bakr al-Baghdadi responded that his fighters would remain in Syria "as long as we have a vein that pumps and an eye that blinks" (Hubbard 2014; see also Lister 2014). The public splits among al-Qaeda affiliates illustrate the growing decentralization of al-Qaeda and its ideology since 9/11, and particularly since the death of Osama bin Laden following the raid by U.S. Special Forces in May 2011 (Wilson, Whitlock, and Branigin 2011).

Furthermore, although a variety of groups across Africa and the Middle East have taken up violent jihad, their targets are mostly local governments or other governments in the region. Most are not targeting the United States or other Western countries and do not have close ties to a central al-Qaeda organization, or "al-Qaeda core." (Some, like the Islamic State, or ISIS, are openly at odds with al-Qaeda.) Rather, many share a general ideology and, in some cases, have adopted the widely recognized brand name of "al-Qaeda" for purposes of terrorizing a local population and government. "There is really not one al-Qaeda anymore" says Gregory D. Johnsen, the author of *The Last Refuge: Yemen, al Qaeda and America's War in Arabia* (2012, quoted in Hubbard 2014). "It has taken on the local flavor of wherever it is, although none of the groups have really disavowed transnational jihad" (quoted in Hubbard 2014). Charles E. Berger, National Intelligence Fellow at the Council on Foreign Relations and assistant special agent in charge of the Federal Bureau of Investigation, calls it "a loose coalition of separate terrorist groups with their own individual causes" (2014). Berger argues:

> Al Qaeda's propaganda machine went into high gear as military action in Afghanistan and Pakistan pushed it into survival mode. With little more than public pronouncements, other terrorist groups rebranded themselves as Al Qaeda. Counterterrorism officials and media outlets began to refer to these groups as Al Qaeda "branches," "franchises" or simply "Al Qaeda." However, these terms overstate the relationships between these groups and the Al Qaeda organization. Counterterrorism officials will often use the term "Al Qaeda linked" but will rarely define the term. While some affiliates have a few former Al Qaeda core members among its [sic] leadership, others are organizationally distinct. There is some communication between the affiliates; however, these connections are tenuous.

For purposes of applying the laws of war, because these disparate groups do not operate under a centralized command and control structure, they cannot be considered one coordinated party to armed conflict with the United States. The remaining question, then, is whether the United States is engaged in an armed conflict with any one or more of them as separate entities. To answer that, it is necessary to determine who these separate entities are that the United States may or may not consider "associated forces." Do they, either alone or together, meet the criteria for a party to an armed conflict with the United States? Consider AQAP in Yemen or the Pakistani Taliban, for instance.

AQAP in Yemen

Al-Qaeda in the Arabian Peninsula (AQAP) is the al-Qaeda–affiliated group U.S. officials most frequently refer to as representing the greatest threat to the United States. Though formed in January 2009 through a union of the Saudi and Yemeni branches of al-Qaeda, its origins date back to the 1990s, when thousands of mujahedeen returned to Yemen after fighting the Soviet occupation in Afghanistan. Of all of the al-Qaeda affiliates, AQAP is the only one whose members or followers are known to have directly attempted to attack the United States, on approximately four occasions. The last known attempt was in 2012. All of those attempts have failed.

Part of the concern is that AQAP has grown from just a few hundred fighters in 2009 to over a thousand in recent years. Some attribute its growth to an increasingly aggressive U.S. drone campaign in Yemen, which has targeted and killed AQAP members and supporters. Critics of the U.S. drone program claim that these killings create strong anti-U.S. sentiment in Yemen, both because some attacks kill innocent bystanders, along with AQAP fighters, and because even AQAP fighters are also fathers, husbands, and heads of households and communities, upon whom many people depend (see, for example, Johnsen 2013). Their being killed by foreign soldiers without explanation therefore can be exploited by local AQAP leaders to recruit new fighters to their cause.

Pakistani Taliban

The Pakistani Taliban are believed to "cooperate closely with al Qaeda and provide shelter and support for the global terror group's top leaders and operatives," according to the *Long War Journal* (Roggio 2013). Formally known as Tehrik-e-Taliban Pakistan, this group has never been a united fighting force, but rather is an umbrella organization for up to thirty different Islamist militant organizations operating across the tribal area along the Afghan border (Walsh 2014). Formerly directed by leaders from the Mehsud tribe, the group has been torn apart by rivalries since the American drone killings of Baitullah Mehsud (Masood 2009), the Pakistani Taliban founder, in 2009; his deputy, Wali ur-Rehman, in May of 2014; and the second leader, Hakimullah Mehsud (Walsh 2013), in November. Rivalries among various remaining commanders have led to violent infighting within the Pakistani Taliban and reportedly thwarted the Pakistani government's effort to negotiate a peace deal.

The scattered nature of these groups, the fact that almost none of them is targeting the United States, and the apparent lack of any central command structure linking them weigh strongly against the United States being able to treat them, for purposes of international law, as an enemy actor, either separately or united, against which the United States is engaged in an armed conflict. Legal experts disagree about whether "associated forces" even makes sense as a concept in a non-international armed conflict. The idea stems from the concept of "co-belligerency" in international armed conflicts, but many international law experts claim the analogy is false (see, for example, Heyns 2013). As UN Special Rapporteur for Summary, Arbitrary, and Extrajudicial Killings, Christof Heyns, explains,

> The idea that the concept of co-belligerency can be transposed into non-international armed conflicts has been met with resistance because it ignores the significant differences between the various forms of armed conflict and opens the door for an expansion of targeting without clear limits. (2013, 13)

The established legal position, explains Heyns, is that "where the individuals targeted are not part of the same command and control structures as the organized armed group or are not part of a single military hierarchical structure, they ought not to be regarded as part of the same group, even if there are close ties between the groups" (2013, 13). This specification raises the question of whether the United States has joined a previously existing war against any one or more of these groups.

Is the United States at War in Yemen or Pakistan?

Yemen

It is possible that the United States is engaged in an armed conflict in Yemen with AQAP, based not on the group's failed attempts to attack the United States, but on the possibility that the United States has intervened in an armed conflict between AQAP and the Yemeni government.[5] However, the U.S. government has never indicated that it has done so.

Many observers, including Robert Chesney, professor at the University of Texas School of Law and Senior Fellow at the Brookings Institution, have come to assume that the United States is waging war in Yemen, and just has not admitted it. "Simply put, there is and for some time has been a non-international armed conflict underway in Yemen pitting AQAP against the government of Yemen, and the United States is a party to this conflict as well," Chesney has written (2012).

While international law permits the United States to join Yemen's war, such a war may not be lawful under U.S. domestic law because it is not legitimated by the Authorization for the Use of Military Force of 2001 (David Glazier notes the constitutional problems this creates in Chapter 6). It is also not clear whether U.S. targeted killings and other military and CIA activities in Yemen are aimed exclusively at AQAP or whether they target a broader range of insurgents fighting the Yemeni government (Goodman 2013a).

Pakistan

Similarly, it is possible that the United States is engaged in an armed conflict in parts of Pakistan, but increasingly unlikely. On the one hand, the United States has considered the Federally Administered Tribal Areas (FATA) region in Pakistan an extension of the geographic zone of armed conflict in Afghanistan, because fighters frequently cross the border and are harbored by al-Qaeda operatives. That could make them a legitimate target as part of the war in Afghanistan.

On the other hand, the United States may have intervened in Pakistan's war with the Pakistani Taliban and other armed insurgent groups. It remains unclear whether the Pakistani government has consented to such action, and whether that consent continues. Although Pakistan is widely believed to have tacitly supported U.S. drone strikes on its territory in the past, more recent public statements from Pakistani officials declaring that U.S. strikes are illegal, counterproductive, and violate the country's sovereignty cast doubt on whether the country's elected officials consent to ongoing operations (Dawn.com 2012; see also Reuters 2012a and Reuters 2012b). For several months in 2014, the U.S. government suspended drone strikes in Pakistan, reportedly at the request of the Pakistani government, which was attempting to engage in peace negotiations with Taliban leaders. However, drone strikes have since resumed.

Finally, it is not clear whom the United States has targeted in Pakistan. Is it only members of al-Qaeda or the Taliban fueling the war in Afghanistan? We do not know because these are all covert operations. Given the fractured nature of both al-Qaeda and the Pakistani Taliban, it seems likely that U.S. drone strikes have exceeded that focus.

Extraterritorial Application of International Human Rights Law

While the U.S. government has refused to make public any information about whom it is targeting and on what legal basis, it has also made clear that it does not believe its actions are governed by International Human Rights Law, the prevailing international law that virtually all international legal authorities believe applies to the use of lethal force outside of armed conflict. Instead, as I demonstrate below, the United States' position on IHRL is at odds with both the original purpose of the treaty and with the views of nearly every other state in the international system.

The ICCPR, as described above, enshrines and particularizes international human rights norms created by the Universal Declaration of Human Rights. With regard to the geographical scope of states' obligation to protect these rights, the ICCPR states:

> Each state Party to the present Covenant undertakes to respect and to ensure to all individuals within its territory and subject to its jurisdiction the rights recognized in the present Covenant, without distinction of any kind. (United Nations 1976, Article 2 [1])

Yet this language is ambiguous: Does the state party commit to respect and ensure rights only to those individuals who are both within its territory AND subject to its jurisdiction? Or does it commit to respect the rights of all individuals everywhere and "to ensure" those rights only "to all individuals within its territory and subject to its jurisdiction"? Given the overall purpose of the treaty and the fact that states can only, as a practical matter, "ensure" rights to those entitled to the protections of its laws, the latter interpretation makes more sense. Even as a matter of basic grammar it makes more sense, since a State doesn't "respect... all individuals" but does "respect... the rights." Still, despite an enormous amount of careful analysis of the treaty, its purpose, and its negotiation history, which has led the overwhelming majority of other signatory nations and legal authorities to interpret the ICCPR in this manner, the United States has insisted that the treaty applies only within its borders and only to those individuals subject to its jurisdiction.

The question of geographical purview first arose in 1995. When responding to a question regarding the scope of the covenant during the United States' Initial Report to the Human Rights Committee, then-U.S. State Department legal adviser Conrad Harper stated: "The Covenant was not regarded as having extraterritorial application" (U.S. Department of State 2010, 1). Since then, the U.S. government has maintained that Article 2(1) of the ICCPR requires state parties to recognize the rights enumerated in the covenant only for "individuals who are both within the territory of a State Party and subject to that State Party's sovereign authority," so that "the terms of the Covenant apply exclusively within the territory of the United States" (U.S. Department of State 2010, 1–2, 15).

Over the years, however, scholars and even U.S. legal advisors have found this interpretation contrary to the purpose of the treaty itself, as well as unsupported by the historical record of negotiations leading to its ratification. As Santa Clara law professor and Stanford visiting scholar Beth Van Schaack has put it:

> Through a process of cross-fertilization and parallel reasoning, a doctrinal convergence is now discernable with the opinions and other views of authoritative decision-makers representing the range of human rights treaty bodies and tribunals that have confronted the issue. According to this consensus, States owe human rights obligations to all individuals within the authority, power and control of their agents or instrumen-

talities, and can be found responsible whenever they cause harm to such individuals. (2014, 22)

The most convincing argument for why this is the better reading of the ICCPR comes from Yale international law professor Harold Koh, who, as State Department legal adviser in 2010, directed an exhaustive analysis of the treaty conducted by his team of State Department lawyers. The memo they produced concluded that the United States' interpretation of the treaty's geographical scope is wrong, and argued forcefully for the United States to change its position on this matter.[6]

Leaked to the *New York Times* in March 2014 (Savage 2014), the memo reveals that in October 2010, Koh informed the government that he and his team believed the 1995 interpretation of the ICCPR's geographic scope was a misinterpretation of the treaty that not only misread the language of the document on its face, but also ignored its purpose and the history of negotiations that led to it. Despite the U.S. precedent to the contrary, Koh's interpretation should come as no surprise: it would make little sense to interpret the ICCPR as not reaching beyond a state's borders given that it was an outgrowth of the Universal Declaration of Human Rights, which itself was a direct response to the Nazi atrocities both within and outside of Germany's borders. Koh concluded that the United States' 1995 interpretation is "in significant tension with the treaty's language, context, and object and purpose, as well as with interpretations of important U.S. allies, the Human Rights Committee and the ICJ, and developments in related bodies of law" (U.S. Department of State 2010, 4). A more accurate interpretation would read the ICCPR as imposing upon states the "obligations to respect Covenant rights"—that is, "not to violate those rights through its own actions or the actions of its agents"—in situations where "a state exercises authority or effective control over the person or context at issue." Koh concluded: "In my view, the 1995 Interpretation is no longer tenable and the USG legal position should be reviewed and revised accordingly" (U.S. Department of State 2010, 4).

The U.S. government has never explained publicly why it has not followed its own legal adviser's advice and has refused to change its position. Meanwhile, the United Nations' Human Rights Committee, the European Court of Human Rights, the International Criminal Tribunal for the former Yugoslavia (ICTY), the International Criminal Tribunal for Rwanda (ICTR), the Inter-American Court of Human Rights, and the International Court of Justice have interpreted the treaty otherwise: they have all found that such obligations "follow the flag" wherever a state exercises effective control in its overseas operations.[7]

Of all signatories who have expressed views on the issue, only the United States and Israel interpret the ICCPR's applicability otherwise. Significantly, these are the only two states that rely heavily on targeted killings outside their borders. Beth Van Schaack has called the United States' position

> increasingly out-of-step with the established jurisprudence and with arguments being advanced—and conceded—by our coalition partners and other allies. . . . This firm stance confirmed the United States as a persistent objector to any emerging customary norm. . . . [It] ultimately undermines the legitimacy of U.S. arguments in these fora as well as its commitment to the human rights project more broadly. (2014, 24)

Under the circumstances, it is difficult to avoid the conclusion Van Schaack reaches, that "the United States' so-called 'legal position' actually reflects a strategic policy choice to endeavor to evade scrutiny of its extraterritorial exploits on the merits" (2014, 24). Many of the Bush administration's most deplored actions undertaken in the name of its "war on terror"—particularly the torture of suspected terrorists and their supporters in overseas CIA prisons and the creation of the Guantánamo Bay detention camp—reinforce that view. U.S. policymakers may see that choice as strategically supporting their short-term goal of continuing to detain and bomb suspected terrorists overseas. But the long-term consequences for U.S. counterterrorism goals need to be more seriously considered.

We Do Not Need a Middle Way

On January 25, 2002, then-White House counsel Alberto Gonzales famously wrote to President George W. Bush that "the war against terrorism is a new kind of war" that is different from "the traditional clash between nations adhering to the laws of war." He reasoned that "this new paradigm renders obsolete Geneva's strict limitations on questioning of enemy prisoners and renders quaint some of its provisions" (2002, 2). The determination that existing international law did not apply to the U.S. "war on terror" was ultimately used to justify the United States' torture and other abuses of suspected terrorists in U.S. detention sites. It was a fateful decision that discredited the United States' reputation as a nation that respects human rights and the rule of law and undermined its ability to have any meaningful influence in those realms on the world stage going forward.[8]

Similarly, the "preemptive" (Marks 2006, 60)—actually, preventive—war against Iraq based on a suggested link to terrorism and the potential danger of weapons of mass destruction (that did not actually exist) likewise was based on a belief in the need for a new international legal framework, set forth in the 2002 National Security Strategy, which acknowledged the "capabilities and objectives of today's adversaries" (Bush Administration 2002, 15) as opposed to yesterday's. In the wake of these actions, as well as the U.S. reinterpretation of international law that allowed them, the United States continues to struggle to restore its reputation and global influence. It also struggles to re-earn the respect of both allies and potential enemies—who have since used U.S. violations of human rights and international law to promote hatred of Americans and recruit a new generation of terrorists. In the meantime, a debate has arisen among scholars and policymakers about whether the international legal framework developed after World War II and incrementally enhanced ever since has become "outmoded" with the rise of terrorism (Eisenberg 2014).

This is not the first time policymakers and intellectuals have claimed the existence of a new kind of warfare that demands a new legal paradigm. Robert D. Sloane, associate professor of Law at Boston University, writes in the *Michigan Law Review* that "during World War II . . . General Wilhelm Keitel, among other Nazi elites, described the conflict between Germany and the Soviet Union, which Hitler labeled 'Bolshevist terrorism,' in eerily similar terms: as a new kind of ideological warfare that rendered the 1929 Geneva Convention on the Treatment of Prisoners of War obsolete" (2007, 449).

Such an example does not prove that all arguments for changing the legal paradigm are sinister (see, for example, Brooks 2004). But it does demonstrate the need to approach such claims about a new kind of warfare with caution. The Obama administration itself has not explicitly advocated for a new legal paradigm governing war and peace the way the Bush administration did. But it does ignore established distinctions between war and peace—legal definitions of armed conflict—to claim that the United States is at war anywhere that al-Qaeda–linked terrorists may be found, and therefore that the more permissive targeting rules operative in an armed conflict should apply to all U.S. targeted killings. This is, in effect, to argue for a new legal paradigm, one that envisions an endless global war with no geographical or temporal boundaries. This is a dangerous precedent to set. It is not actually creating a new paradigm; instead, it is ignoring a well-established one that developed in order to make war and its destructiveness an exceptional circumstance and peace the norm.

Although drones are a new technology that makes killing in far-away places easier for the country that has them, this does not alter the facts of when an armed conflict exists. The impact of killing is still the same, as is the impact of granting permission to kill on both sides. To suggest that the ability of drone technology to fight terrorists abroad should lead us to a permanent state of war will in the long run endanger Americans as well as our enemies (and the innocents deemed "collateral damage"). It violates not only the letter, but also the spirit of the postwar international legal regime.

As with all laws, treaties should be interpreted according to their plain meaning and in light of their overall purpose. The Vienna Convention on the Law of Treaties (VCLT) (United Nations 1969) specifically warns against parsing the plain language of a treaty in such a way as to avoid its intended meaning. Article 31(1) of the VCLT states: "A treaty shall be interpreted in good faith in accordance with the ordinary meaning given to the terms of the treaty in their context and in the light of its object and purpose" (United Nations 1969, 340). Although the United States is not a party to the VCLT, U.S. courts have routinely applied its rules as a form of customary international law (see, for example, *Fujitsu Ltd. v. Federal Exp. Corp.* 2001).

The purpose of limiting self-defense actions in the UN Charter is, of course, to narrowly limit the instances when it is legitimate for a state to use lethal force against another country or armed group. Thus, the

Charter's preamble sets forth its purposes, including "to save succeeding generations from the scourge of war" and "to reaffirm faith in fundamental human rights, in the dignity and worth of the human person" (United Nations 1945).

Against this backdrop, it is impossible to define the "imminent" threat needed to justify lethal force in self-defense as anything other than a threat of an aggressive action that is "about to happen." U.S. officials have claimed to follow "international law principles" (Preston 2012)—but this claim depends simply on redefining the terms of international law itself. This is not to comply with the law; it is to selectively ignore or redefine international laws the United States does not like. Once again, this sets a dangerous precedent for other nations that may be eager to define an "imminent" threat in equally muddled and perhaps even broader terms (see Fisk and Ramos 2014 for a discussion of recent attempts to redefine the term's meaning).

Acknowledging the extraterritorial applicability of international human rights law will serve U.S. interests in other ways as well. In particular, it is critical to the United States' ability to win stronger support for its counterterrorism operations, as well as to its credibility as a world leader seeking to encourage rights-respecting behavior in other nations. As Harold Koh wrote in his U.S. State Department opinion in October 2010:

> Our prior position has been a source of ongoing international tension, with significant deleterious effects on our international human rights reputation and our ability to promote international human rights internationally. The prior administration was severely criticized in UN fora, by important U.S. allies, by members of Congress, by domestic and international human rights groups, and in the domestic and international media. The 1995 Interpretation is seen as allowing alleged incidents of abusive extraterritorial practices such as torture and "extraordinary rendition," and as immunizing such practices from legal review by preserving the policy option for U.S. personnel to act in a "legal black hole" once they step outside the territorial United States. By contrast, revising our legal position to recognize some application of the ICCPR to U.S. conduct abroad would have a salutary effect on our international reputation. It would significantly advance our international standing and reputation for

respect for the international rule of law, which are primary commitments of this Administration. (4–5)

Beth Van Schaack agrees that shifting its view "will demonstrate the United States' respect for the views of human rights bodies and of its allies, bolster the universality of certain core human rights protections, and do much to bring to a close a historical chapter marred by allegations that the United States was endeavoring to create—and exploit—rights-free zones" (2014, 24).

U.S. targeted killing operations have unfortunately opened a new chapter in which the United States is perceived as ignoring human rights and attempting to create and exploit lawless zones. For the United States to state publicly that international human rights law applies to its attempts to kill people with armed drones outside U.S. territory would be a significant step toward assuring U.S. allies and critics alike that the United States seeks only to kill people outside an armed conflict when lethal action is actually necessary.

Whether or not the strict letter of the law requires it, international support is sufficiently important to U.S. counterterrorism operations that, as a matter of policy, Heyns's recommendations for increased transparency around drone killings—including identification of the legal basis for targeting, the individuals targeted, and the unintended casualties—should be adopted. Even if the targeted killing operations themselves must not be revealed in advance so as not to undermine their efficacy, and specific sources and methods used must remain secret, there is no justification after a strike for not revealing sufficient information for the public to determine whether the strike was lawful. Without that level of transparency, the legal requirement for accountability is meaningless.

Significantly, there is nothing in the law governing either the Department of Defense or the CIA that prohibits disclosure of such information (Lederman 2014). The secrecy is merely a policy choice made by the executive. It is a choice that both the President and his senior advisors have themselves questioned repeatedly. As President Obama acknowledged in May 2014, the United States "must be more transparent about both the basis of our counterterrorism actions and the manner in which they are carried out" (Obama 2014). CIA Director and former counterterrorism advisor John Brennan has similarly said, "I think the rule should be that if we're going to take actions overseas that result in the deaths of people, the United States should take responsibility for that" (DeYoung 2012). Yet the executive branch has consistently failed to take public responsibility by refusing to provide the most basic information about whom it is killing and why.

The United States' demonstration of compliance with international law is extremely important for the global precedent it sets. The United States' current use of drones appears to be an act of "preventive" force beyond what is allowed by the international legal regime. So long as the United States continues to operate in this manner without sufficient explanation, it encourages other states to flout international law as well, risking an escalation of the use of preventive force, whether by drones or otherwise. Sonnenberg's (Chapter 5) application of game theory to this scenario yields the same conclusion.

As the Rand Corporation recently reported in its analysis of the increased reliance on drones and its implications,

> The United States has an interest in how others use armed UAVs, and will need to address how its own use of these systems can be fit into a broader set of international norms so as to discourage their misuse by others. There is evidence that U.S. leadership—and failure to lead—can matter in shaping international behavior. (Davis et al. 2014, 1)

David Glazier (Chapter 6) makes a similar assertion: "What the United States does today, other states can assert the right to do in the future"—noting that Vladimir Putin's justifications for invading Crimea were "almost verbatim excerpts" from those proffered by Ronald Reagan and George H. W. Bush to support military interventions in Grenada and Panama. Confidence in the legitimacy of United States' use of lethal force is critical to the United States regaining the public trust it has lost over the years, particularly since the mistreatment of detainees and other human rights abuses the United States committed after the September 11 attacks.

Acknowledging the applicability of international human rights law to the use of drones would not preclude development of new regulations governing these particular weapons-delivery systems, which Sonnenberg argues for in his chapter. But any new regulations must take account of and explicitly seek to bring nation-states into compliance with existing international human rights law if they are to represent a step forward in the legal protection of civilians in armed conflict. And as Sonnenberg notes, states must make clear that they intend to comply with these rules.

What we need now is *not* a new set of laws or a new, more expansive reading of the current framework for analyzing the legality of U.S. counterterrorism operations—a phenomenon that Avery Plaw and João Franco Reis (Chapter 9) note is already occurring. What we need now is for the United States to demonstrate publicly that it is complying with existing, long-standing, still-critical international human rights law. Such a demonstration would provide a much-needed boost to the United States' reputation and credibility around the world, as well as enhance its power and influence. So long as the United States refuses publicly to comply with widely accepted international human rights law, it will remain an easy target for propagandists who can point to the United States' past use of torture and now its secret killing campaigns to whip up anti-U.S. sentiment for their own political purposes. This is what we have seen occur, for example, in Pakistan, where ongoing unrest and anti-American sentiment continue to threaten U.S. troops and interests in the region. Belief in and support for U.S. policy are critical to global cooperation, to stemming the tide of anti-American sentiment that fuels terrorism, and ultimately, to the efficacy of the drone program itself. A firm and public declaration that the United States honors and abides by international human rights law wherever the U.S. government takes action would go a long way toward dispelling the sort of anger toward the United States that places both U.S. troops and American citizens everywhere at risk.

To continue to undertake secret lethal operations without that public trust, meanwhile, undermines the core principles of American democracy that the U.S. government proclaims to stand for. On a practical level, it undermines U.S. global counterterrorism efforts as well. In delivering the commencement speech at West Point Military Academy in 2014, President Obama said: "Our intelligence community has done outstanding work . . . but when we cannot explain our efforts clearly and publicly, we face terrorist propaganda and international suspicion, we erode legitimacy with our partners and our people, and we reduce accountability in our own government." It was an eloquent statement of the problem. Now the United States needs to act on the solution.

NOTES

1. I use "drones" in this chapter because it is the term most widely used and understood by the public, by policymakers, and by the U.S. military.
2. Surveys conducted by the Pew Research Center have found that most non-Americans oppose U.S. drone strikes. For example, a March 2013 survey in Pakistan found that 68% of Pakistanis disapproved of U.S. drone strikes targeting suspected terrorists in their country, and only 5% approved. Slightly more Pakistanis (21%) approve if the United States carries out its strikes in conjunction with the Pakistani government. Almost three-quarters (74%) of Pakistanis said the drone strikes killed too many innocent people. A 2012 Pew Research Center survey of 19 countries plus the United States found that in 17 of them, more than half disapproved of the U.S. conducting drone strikes to target extremists. The policy was particularly unpopular in majority Muslim nations, but it also faced disapproval in Europe and other regions as well. The strongest disapproval was registered in Greece (90%), Egypt (89%), Jordan (85%), Turkey (81%), Spain (76%), Brazil (76%), and Japan (75%).
3. As discussed elsewhere in the volume, the UN Charter also authorizes a state to use lethal force outside of a recognized armed conflict in self-defense (see United Nations 1945, chap. 7), in response to an attack. Some, including the United States government, interpret Article 51 of the UN Charter also to permit the use of force in order to prevent an imminent attack on the state. As with the situation in armed conflict, the use of lethal force in self-defense outside of armed conflict must also satisfy the requirements of necessity and proportionality (Military and Paramilitary Activities in and against Nicaragua (*Nicaragua v. United States of America*) 1984; Schachter 1984; see also Schrijver and van den Herik 2010). The use of force, therefore, must be limited to what is required to repel an attack. The use of force may not be designed to preempt an anticipated future threat. The UN Charter also requires that lethal force used in self-defense be reported to the Security Council.
4. The United States, however, has ratified only the ICCPR, and has done so with various "reservations," "understandings," and "declarations" designed to limit the applicability of the convention to the United States.
5. As of this writing, the state of the Yemeni government was uncertain because President Abed Rabbo Mansour Hadi had been ousted by a group of Shiite Houthi rebels.
6. To note, the United States has not yet done so.
7. The European Court of Human Rights establishes binding precedent for the forty-seven member States of the Council of Europe. The ICTY and ICTR are inter-

national tribunals established by the UN Security Council to adjudicate crimes in former Yugoslavia and Rwanda, respectively. The Inter-American Court of Human Rights issues judgments on the American Convention on Human Rights and is part of the human rights enforcement mechanism of the Organization of American States, of which the United States is part, although the United States has signed, but not ratified the convention and does not accept the jurisdiction of the Inter-American Court. The International Court of Justice is the judicial arm of the United Nations. The Human Rights Committee issues nonbinding interpretations of the ICCPR for that treaty's 167 states party, including the United States.

8 We saw one recent example of how U.S. conduct has undermined its credibility in enforcing human rights norms abroad when Russian President Vladimir Putin took the unusual step in July 2014 of sanctioning the U.S. soldiers who abused prisoners at the Abu Ghraib prison in Iraq as part of his claim that the United States' accusations that Russia violates human rights are hypocritical. See Rogin 2014.

BIBLIOGRAPHY

Abbot, Sebastian. 2012. "AP IMPACT: New Light on Drone War's Death Toll." Associated Press, February 26. http://news.yahoo.com/ap-impact-light-drone-wars-death-toll-150321926.html.

Afzal, Madiha. 2013. "Drone Strikes and Anti-Americanism in Pakistan." Washington, DC: Brookings Institution, February 7. http://www.brookings.edu/research/opinions/2013/02/07-drones-anti-americanism-pakistan-afzal.

Authorization for the Use of Military Force. 2001. *Statutes at Large*, vol. 224, sec. 115.

BBC News. 2010. "Al-Qaeda Offshoot Claims Cargo Bombs." November 5. http://www.bbc.com/news/world-middle-east-11703355.

Berger, Charles E. 2014. "The Balkanization of al-Qaeda." *National Interest*, February 21. http://nationalinterest.org/commentary/the-balkanization-al-qaeda-9912.

Brennan, John. 2012. "The Ethics and Efficacy of U.S. Counterterrorism Strategy." Washington, DC: Wilson Center, April 30. http://www.wilsoncenter.org/event/the-efficacy-and-ethics-us-counterterrorism-strategy.

Brooks, Rosa Ehrenreich. 2004. "War Everywhere: Rights, National Security Law, and the Law of Armed Conflict in the Age of Terror." *University of Pennsylvania Law Review* 153 (2): 675–761.

Bush Administration. 2002. "The National Security Strategy of the United States of America." Washington, DC: The White House. September. http://nssarchive.us/NSSR/2002.pdf.

Center for Civilians in Conflict and Human Rights Clinic at Columbia Law School. 2012. *The Civilian Impact of Drones: Unexamined Costs, Unanswered Questions*. New York: Columbia Law School; Washington, DC: Center for Civilians in Conflict. http://civiliansinconflict.org/uploads/files/publications/The_Civilian_Impact_of_Drones_w_cover.pdf.

Chesney, Robert. 2012a. "Reactions to the ACLU Suit: There Is Armed Conflict in Yemen, and the U.S. Is Party to It." *Lawfare*, July 18. http://www.lawfareblog.com/2012/07/reactions-to-the-aclu-suit-there-is-armed-conflict-in-yemen-and-the-us-is-party-to-it/.

Chesney, Robert. 2012b. "The United States as a Party to an AQAP-Specific Armed Conflict in Yemen." *Lawfare*, January 31. http://www.lawfareblog.com/2012/01/yemen-armed-conflic/.

CNN. 2012. "Obama: 'Al Qaeda Has Been Decimated.'" November 1. https://www.youtube.com/watch?v=GQjztrnJzCM.

CNN. 2013. "It's Not Over, Somali Terrorists Say after Mall Attack that Killed 67." *CNN World*, October 2. http://www.cnn.com/2013/10/02/world/africa/kenya-mall-attack-shabaab-warning/index.html.

Davis, Lynn E., Michael J. McNerney, James S. Chow, Thomas Hamilton, Sarah Harting, and Daniel Byman. 2014. "Armed and Dangerous? UAVs and U.S. Security." Santa Monica, CA: RAND Corporation. http://www.rand.org/pubs/research_reports/RR449.html.

Dawn.com. 2012. "President Zardari Asks U.S. to End Drone Strikes, Remove Mistrust." September 15. http://dawn.com/2012/09/16/president-zardari-urges-us-to-immediately-cease-drone-strikes/.

Dawn.com. 2013. "Nawaz Says He Urged Obama to End Drone Strikes in Pakistan." October 24. http://www.dawn.com/news/1051311.

DeYoung, Karen. 2012. "CIA Veteran John Brennan Has Transformed U.S. Counterterrorism Policy." *Washington Post*, October 24. http://www.washingtonpost.com/world/national-security/cia-veteran-john-brennan-has-transformed-us-counterterrorism-policy/2012/10/24/318b8eec-1c7c-11e2-ad90-ba5920e56eb3_story.html.

DeYoung, Karen, and Greg Miller. 2014. "U.S. Curtails Drone Strikes in Pakistan as Officials There Seek Peace Talks with Taliban." *Washington Post*, February 4. http://www.washingtonpost.com/world/national-security/us-curtails-drone-strikes-in-pakistan-as-officials-there-seek-peace-talks-with-taliban/2014/02/04/1d63f52a-8dd8-11e3-833c-33098f9e5267_story.html.

Dilanian, Ken. 2014. "Debate Grows over Proposal for CIA to Turn Over Drones to Pentagon." *Los Angeles Times*, May 11. http://touch.latimes.com/#section/-1/article/p2p-80167118/.

Eisenberg, Richard B. 2014. "On the 'Future of War.'" *Air Force General Counsel Blog*, January 17. http://afgeneralcounsel.dodlive.mil/2014/01/17/on-the-future-of-war-2/.

Fair, Christine C., Karl Kaltenthaler, and William J. Miller. 2012. "The Drone War: Pakistani Public Opposition to American Drone Strikes in Pakistan." Social Science Research Network, December 23. http://papers.ssrn.com/sol3/papers.cfm?abstract_id=2193354.

Feinstein, Dianne. 2013. "Feinstein Statement on Intelligence Committee Oversight of Targeted Killings." Washington, DC, February 13. http://www.feinstein.senate.gov/public/index.cfm/press-releases?ID=5b8dbe0c-07b6-4714-b663-b01c7c9b99b8.

Fisk, Kerstin, and Jennifer M. Ramos. 2014. "Actions Speak Louder than Words: Preventive Self-Defense as a Cascading Norm." *International Studies Perspectives* 15 (2): 163–185.

Freedom of Information Act. 1996. *U.S. Code*, vol. 5, sec. 552.

Fujitsu Ltd. v. Federal Exp. Corp. 2001. 247 F.3d 423, 433 (2d Cir.).

Gaggioli, Gloria. 2013. "The Use of Force in Armed Conflicts: Interplay Between the Conduct of Hostilities and Law Enforcement Paradigms." *International Council of the Red Cross*, November. http://www.icrc.org/eng/assets/files/publications/icrc-002-4171.pdf.

Goldsmith, Jack. 2014. "Questions about CIA v. DOD Drone Strikes." *Lawfare*, May 13. http://www.lawfareblog.com/2014/05/questions-about-cia-v-dod-drone-strikes/.

Gonzales, Alberto. 2002. "Decision Re Application of the Geneva Convention on Prisoners of War to the Conflict with Al Qaeda and the Taliban." January 25. http://www.cfr.org/terrorism-and-the-law/application-geneva-convention-prisoners-war-conflict-al-qaeda-taliban/p11893.

Goodman, Ryan. 2013a. "A New War?: The United States Involvement in Yemen's Internal Armed Conflict [Updated]." *Just Security*, October 28. http://justsecurity.org/2013/10/28/war-united-states-yemen-internal-armed-conflict-insurgency/.

Goodman, Ryan. 2013b. "No more foreign wars? Yet America is fighting in Yemen's civil war." *The Guardian*, 22 October. http://www.theguardian.com/commentisfree/2013/oct/22/foreign-wars-america-fighting-yemen.

Goodman, Ryan. 2014. "What's Missing in *New York Times*' 'Latest' Version of U.S. Military Role in Yemen." *Just Security*, May 12. http://justsecurity.org/10349/whats-missing-york-times-latest-version-military-role-yemen/.

Gorman, Siobhan, and Adam Entous. 2011. "CIA Plans Yemen Drone Strikes." *Wall Street Journal*, June 14. ttp://online.wsj.com/article/SB10001424052702303848104576384051572679110.html.

Heyns, Christof. 2013. "Report of the Special Rapporteur on Extrajudicial, Summary or Arbitrary executions." New York: United Nations General Assembly, September 13. http://justsecurity.org/wp-content/uploads/2013/10/UN-Special-Rapporteur-Extrajudicial-Christof-Heyns-Report-Drones.pdf.

Holder, Eric. 2012. "Attorney General Eric Holder Speaks at Northwestern University School of Law." Washington, DC: U.S. Department of Justice, March 5. http://www.justice.gov/iso/opa/ag/speeches/2012/ag-speech-1203051.html.

Hubbard, Ben. 2014. "The Franchising of Al Qaeda." *New York Times*, January 25. http://www.nytimes.com/2014/01/26/sunday-review/the-franchising-of-al-qaeda.html?_r=0.

Ignatius, David. 2011. "Rewriting Rumsfeld's Rules." *Washington Post*, June 3. https://www.washingtonpost.com/opinions/rewriting-rumsfelds-rules/2011/06/02/AGHIXPIH_story.html.

International Committee of the Red Cross. 2008. "How Is the Term 'Armed Conflict' Defined in International Humanitarian Law?" Geneva, March. http://www.icrc.org/eng/assets/files/other/opinion-paper-armed-conflict.pdf.

International Committee of the Red Cross. 2009. "Direct Participation in Hostilities: Questions & Answers." Geneva: ICRC Resource Centre, February 6. http://www.icrc.org/eng/resources/documents/faq/direct-participation-ihl-faq-020609.htm#a2.

International Human Rights and Conflict Resolution Clinic at Stanford Law School and Global Justice Clinic at NYU School of Law. 2012. "Living under Drones: Death, Injury, and Trauma to Civilians from U.S. Drone Practices in Pakistan." September. http://law.stanford.edu/wp-content/uploads/sites/default/files/publication/313671/doc/slspublic/Stanford_NYU_LIVING_UNDER_DRONES.pdf.

International Judicial Monitor. 2006. "General Principles of International Law: Treaty Interpretation." *International Judicial Monitor* 1(4). http://www.judicialmonitor.org/archive_0906/generalprinciples.html.

International Justice Project. "United States of America's Reservations to the ICCPR." http://www.internationaljusticeproject.org/juvICCPR.cfm.

Jamjoom, Mohammed. 2014. "Official: Extensive U.S. Involvement in Anti-Terror Operation in Yemen." *CNN World*, April 23. http://www.cnn.com/2014/04/22/world/meast/yemen-terror-operation-dna/.

Johnsen, Gregory D. 2012. *The Last Refuge: Yemen, al-Qaeda, and America's War in Arabia*. New York: W. W. Norton.

Johnsen, Gregory D. 2013. "How We Lost Yemen." *Foreign Policy*, August 6. http://www.foreignpolicy.com/articles/2013/08/06/how_we_lost_yemen_al_qaeda?wp_login_redirect=0.

Joscelyn, Thomas. 2010. "AQAP Claims Responsibility for Cargo Planes Plot." *Long War Journal*, November 6. http://www.longwarjournal.org/archives/2010/11/aqap_claims_responsi.php#.

Laub, Zachary, and Jonathan Masters. 2014. "Al-Qaeda in the Islamic Maghreb (AQIM)." New York: Council on Foreign Relations, January 8. http://www.cfr.org/terrorist-organizations-and-networks/al-qaeda-islamic-maghreb-aqim/p12717.

Lederman, Marty. 2014. "Major Development Concerning Transparency of the Use of Force in Yemen." *Just Security*, May 27. http://justsecurity.org/10821/major-development-transparency-force/.

Lister, Tim. 2014. "Al Qaeda 'Disowns' Affiliate, Blaming It for Disaster in Syria." *CNN World*, February 4. http://www.cnn.com/2014/02/03/world/meast/syria-al-qaeda/.

Mackey, Robert. 2014. "Artists Try to Prick Conscience of Drone Operators with Giant Portrait of Orphan in Pakistani Field." *New York Times*, April 9. http://thelede.blogs.nytimes.com/2014/04/09/artists-try-to-prick-the-conscience-of-drone-operators-with-giant-portrait-of-orphan-in-pakistani-field/?_php=true&_type=blogs&_php=true&_type=blogs&_r=1.

Marks, Stephen P. 2006. "International Law and the 'War on Terrorism': Post 9/11 Responses by the United States and Asia Pacific Countries." *Asia Pacific Law Review* 14 (1): 43–74.

Masood, Salman. 2009. "Taliban in Pakistan Confirm that Their Leader Is Dead." *New York Times*, August 25. http://www.nytimes.com/2009/08/26/world/asia/26pstan.html?adxnnl=1&adxnadx=1398027694-sfdNGtYom59Mwq5FGX1XSw&_r=1&.

Military and Paramilitary Activities in and against Nicaragua (*Nicaragua v. United States of America*), Jurisdiction and Admissibility. 1984. 392 ICJ REP. June 27, 1986.

Moore, John B. 1906. *A Digest of International Law as Embodied in Diplomatic Discussions, Treaties and Other International Agreements*, vol. 2. Washington, DC: Government Printing Office.

National Defense Authorization Act for Fiscal Year 2005. 2004. *U.S. Code*, vol. 1811, sec. 118.

Nicholas, Peter, and Colleen McCain Nelson. 2013. "Obama Has New Take on Terror Threat." *Wall Street Journal*, August 7. http://online.wsj.com/news/articles/SB10001424127887323838204578654391039980004.

Nossiter, Adam. 2014. "Keeping Al Qaeda's West African Unit on the Run." *New York Times*, April 29. http://www.nytimes.com/2014/04/30/world/africa/keeping-al-qaedas-west-african-unit-on-the-run.html?_r=0.

#NotABugSplat. 2014. "A Giant Art Installation Targets Predator Drone Operators." http://notabugsplat.com/.

Obama, Barack. 2009. Classified National Security Information, Executive Order 13526. *Federal Register*, 75:1013.

Obama, Barack. 2011. "Letter from the President on the War Powers Resolution." Washington, DC: Office of the Press Secretary. http://www.whitehouse.gov/the-press-office/2011/06/15/letter-president-war-powers-resolution.

Obama, Barack. 2014. "Remarks by the President at the United States Military Academy Commencement Ceremony." West Point, New York. Washington, DC: White House Office of the Press Secretary, May 28. http://www.whitehouse.gov/the-press-office/2014/05/28/remarks-president-west-point-academy-commencement-ceremony.

O'Connell, Mary Ellen. 2012. "Unlawful Killing with Combat Drones: A Case Study of Pakistan, 2004–2009." In *Shooting to Kill: Socio-Legal Perspectives on the Use of Lethal Force*, edited by Simon Bronitt, Miriam Gani, and Saskia Hufnagel, 263–292. Oxford, UK: Hart Publishing.

Parks, Colonel W. Hays. 1989. "Memorandum on Executive Order 12333 and Assassination." November 2. http://www.hks.harvard.edu/cchrp/Use%20of%20Force/October%202002/Parks_final.pdf.

Pew Research Center. 2012. "Global Opinion of Obama Slips, International Policies Faulted." Washington, DC, June. http://www.pewglobal.org/2012/06/13/global-opinion-of-obama-slips-international-policies-faulted/.

Pew Research Center. 2013. "Obama and Drone Strikes: Support but Questions at Home, Opposition Abroad." Washington, DC: March. http://www.pewresearch.org/fact-tank/2013/05/24/obama-and-drone-strikes-support-but-questions-at-home-opposition-abroad/.

Phillippi v. CIA. 1976. 6 F.2d. 1009 (D.C. Cir.).

Piccard, Ann. 2010. "The United States' Failure to Ratify the International Covenant on Economic, Social and Cultural Rights: Must the Poor Be Always with Us?" *Scholar: St. Mary's Law Review on Minority Issues* 13 (Winter): 231.

Preston, Stephen W. 2012. "Remarks of CIA General Counsel Stephen W. Preston at Harvard Law School." Washington, DC: Central Intelligence Agency, Office of Public Affairs, April 10. https://www.cia.gov/news-information/speeches-testimony/2012-speeches-testimony/cia-general-counsel-harvard.html.

Priest, Dana, and William M. Arkin. 2011. "'Top Secret America': A Look at the Military's Joint Special Operations Command." *Washington Post*, September 2. http://www.washingtonpost.com/world/national-security/top-secret-america-a-look-at-the-militarys-joint-special-operations-command/2011/08/30/gIQAvYuAxJ_story.html.

Prosecutor v. Tadic. 1995. Case no. IT-94-1-I. Decision on the Defence Motion for Interlocutory Appeal on Jurisdiction, ¶ 70. Bosnia and Herzegovina, Serbia: International Criminal Tribunal for the Former Yugoslavia, October 2.

Reagan, Ronald. 1981. United States Intelligence Activities, Executive Order 12333. *Federal Register*, 46:59941.

Reuters. 2012a. "Pakistan Says U.S. Not Listening: Drone Strikes Must Stop." April 26. http://www.reuters.com/article/2012/04/26/us-pakistan-minister-drones-idUSBRE83P0AM20120426.

Reuters. 2012b. "Pakistan Condemns 'Illegal' U.S. Drone Strikes." *Express Tribune*, June 4. http://tribune.com.pk/story/388730/pakistan-condemns-illegal-us-drone-strikes/.

Roggio, Bill. 2013. "Pakistani Taliban Threaten U.S. for Killing Leader in Drone Strike." *Long War Journal*, November 9. http://www.longwarjournal.org/archives/2013/11/pakistani_taliban_th_1.php#ixzz31XLzXLze.

Roggio, Bill. 2014. "U.S. Drone Strike Kills 6 AQAP Fighters in Central Yemen." *Long War Journal*, May 12. http://www.longwarjournal.org/archives/2014/05/us_drone_strike_kill_27.php.

Rogin, Josh. 2014. "Putin Sanctions Abu Ghraib Soldiers." *Daily Beast*, July 19. http://www.thedailybeast.com/articles/2014/07/19/putin-sanctions-abu-ghraib-soldiers-and-guantanamo-officials.html.

Rumsfeld, Donald H. 2004. "Testimony Prepared for Delivery to the National Commission on Terrorist Attacks Upon the United States." Washington, DC: U.S. Department of Defense, March 23. http://www.defense.gov/speeches/speech.aspx?speechid=105.

Savage, Charlie. 2014. "U.S. Seems Unlikely to Accept that Rights Treaty Applies to Its Actions Abroad." *New York Times*, March 6. http://www.nytimes.com/2014/03/07/world/us-seems-unlikely-to-accept-that-rights-treaty-applies-to-its-actions-abroad.html.

Schachter, Oscar. 1984. "The Right of States to Use Armed Force." *Michigan Law Review* 82 (5–6): 1620–1646.

Schachter, Oscar. 1991. *Developments in International Law: International Law in Theory and Practice*, 2nd ed. Leiden, The Netherlands: Brill Academic Publishers.

Schrijver, Nico, and Larissa van den Herik. 2010. "Leiden Policy Recommendations on Counter-terrorism and International Law." Leiden: The Netherlands: Leiden University, Grotius Centre for International Legal Studies, April 1.

Shane, Scott, and Eric Schmitt. 2012. "Qaeda Plot to Attack Plane Foiled, U.S. Officials Say." *New York Times*, May 7. http://www.nytimes.com/2012/05/08/world/middleeast/us-says-terrorist-plot-to-attack-plane-foiled.html.

Sloane, Robert D. 2007. "Prologue to a Voluntarist War Convention." *Michigan Law Review* 106 (December): 443–485.

Stanford University. 2012. "Al-Nusra Front." *Mapping Militant Organizations*. Stanford, CA. http://www.stanford.edu/group/mappingmilitants/cgi-bin/groups/view/493.

Szoldra, Paul. 2013. "Al-Shabab: A Guide to the Terror Group behind the Deadly Kenyan Mall Attack." *Business Insider*, September 22. http://www.businessinsider.com/who-are-al-shabab-2013-9.

Taj, Farhat. 2009. "Drone Attacks—A Survey." *News*, March 5. http://www.thenews.com.pk/TodaysPrintDetail.aspx?ID=165781&Cat=9&dt=3/4/2009.

Tucker, Patrick. 2015. "Every Country Will Have Armed Drones within 10 Years." *Defense One*, May 6. http://www.defenseone.com/technology/2014/05/every-country-will-have-armed-drones-within-ten-years/83878/.

United Nations. 1945. "Chapter VII: Action with Respect to Threats to the Peace, Breaches of the Peace, and Acts of Aggression." *Charter of the United Nations*. New York: United Nations. http://www.un.org/en/documents/charter/chapter7.shtml.

United Nations. 1969. Vienna Convention on the Law of Treaties (with Annex). May 23. TIAS no. 18232. http://www.un.org/en/sections/un-charter/chapter-vii/index.html.

United Nations. 1976. International Covenant on Civil and Political Rights. New York,, March 23. UNTS no. 171. http://www.ohchr.org/en/professionalinterest/pages/ccpr.aspx.

United Nations. 1990. "Basic Principles on the Use of Force and Firearms by Law Enforcement Officials." Havana, Cuba. http://www.ohchr.org/EN/ProfessionalInterest/Pages/UseOfForceAndFirearms.aspx.

U.S. Code. 2014. Vol. 10, sec. 119.

U.S. Code. 2014. Vol. 50, sec. 3093.

U.S. Congress. Senate. Select Committee on Intelligence. 2013. Open Hearing: Nomination of John O. Brennan to be Director of the Central Intelligence Agency. 113th Cong., 18th sess., February 17. http://www.intelligence.senate.gov/130207/posthearing.pdf.

U.S. Department of Justice. Undated. *White Paper: Lawfulness of a Lethal Operation Directed against a U.S. Citizen Who Is a Senior Operational Leader of Al-Qa'ida or an Associated Force*. Washington, DC: Department of Justice. http://msnbcmedia.msn.com/i/msnbc/sections/news/020413_DOJ_White_Paper.pdf

U.S. Department of State. Office of the Legal Adviser. 2010. *Memorandum Opinion on the Geographic Scope of the International Covenant on Civil and Political Rights*. Washington, DC: Department of State. http://justsecurity.org/wp-content/uploads/2014/03/state-department-iccpr-memo.pdf.

U.S. Senate. 2013. "Feinstein Statement on Intelligence Committee Oversight of Targeted Killings." Washington, DC, February 13. http://www.feinstein.senate.gov/public/index.cfm/press-releases?ID=5b8dbe0c-07b6-4714-b663-b01c7c9b99b8.

Van Schaack, Beth. 2014. "The United States' Position on the Extraterritorial Application of Human Rights Obligations: Now Is the Time for Change." *International Law Studies, U.S. Naval War College* 90:20–65.

Walker, Andrew. 2012. "What is Boko Haram?" Washington, DC: United States Institute of Peace, May 30. http://www.usip.org/publications/what-boko-haram.

Walsh, Declan. 2013. "In Pakistan, Drone Strike Turns a Villain into a Victim." *New York Times*, November 3. http://www.nytimes.com/2013/11/04/world/asia/in-pakistan-death-by-drone-turns-a-villain-into-a-martyr.html.

Walsh, Declan. 2014. "Fractured State of Pakistani Taliban Calls Peace Deal into Question." *New York Times*, April 20. http://www.nytimes.com/2014/04/21/world/asia/pakistani-taliban.html.

War Powers Resolution. 1973. *U.S. Code*, vol. 50, sec. 1541–1548.

Williams, Brian Glyn. 2010 "The CIA's Covert Predator Drone War in Pakistan, 2004–2010: The History of an Assassination Campaign." *Studies in Conflict and Terrorism* 33 (10): 871–892.

Wilson, Scott, Craig Whitlock, and William Branigin. 2011. "Osama bin Laden Killed in U.S. Raid, Buried at Sea." *Washington Post*, May 2. http://www.washingtonpost.com/national/osama-bin-laden-killed-in-us-raid-buried-at-sea/2011/05/02/AFx0yAZF_story.html.

8

Studying Drones

The Low Quality Information Environment of Pakistan's Tribal Areas

C. CHRISTINE FAIR

Pakistan captures the attention of those interested in U.S. drone policy because it has experienced far more drone strikes than any other country from 2004 onward. Even though the U.S. Central Intelligence Agency (CIA) and the U.S. Air Force developed the capacity to weaponize remotely piloted aerial vehicles (RPVs)[1] well before the events of September 2001,[2] it took the events of 9/11 to galvanize the Bush administration to finally approve the use of armed drones in what became the U.S.-led "Global War on Terror" (Frisbee 2004). Analysts believe that the CIA first employed a weaponized drone on February 4, 2002, in an effort to kill Bin Laden near the city of Khost, in Afghanistan's Paktia province. In what now seems unusual, Secretary of Defense Donald Rumsfeld, using the passive voice of government obfuscation, acknowledged the strike: "A decision was made to fire the Hellfire missile. It was fired" (Sifton 2012). The United States then expanded the use of armed RPVs to kill alleged terrorists and insurgents in Pakistan's Federally Administered Tribal Areas (FATA).[3] Subsequently, and consistent with the expanding scope of the global war on terror, the U.S. intelligence and military agencies employed armed RPVs in Yemen, Somalia, Libya, and elsewhere. However, the most notorious of these theaters is the FATA in Pakistan, which has been the site of the vast majority of U.S. armed RPV strikes since they began after 9/11.

As the covert use of RPVs in Pakistan proliferated, so has the body of writing on the program and its consequences. Unfortunately, many of these analyses fall short because they pay inadequate attention to the specificities of the program in Pakistan. In this chapter, I first provide important contextual information that should foreground any study of RPV usage in Pakistan. Next, I discuss some of the problems with popular notions of "Pakistani sovereignty" that undergird commentary about the RPV program there. I then evaluate Pakistan's willingness and ability to do more to protect the international community from the terrorist groups ensconced in its territory. Following this, I critique recent human rights advocacy reports, highlighting the methodological, and even ethical, problems that undermine their examination of drones. I conclude with a number of thoughts on how scholars can improve the quality and thus reliability of their work to understand the impact of the RPV program in Pakistan and perhaps other low information environments.

The Covert Armed RPV Program in Pakistan

The first RPV strike in Pakistan's FATA was what some U.S. officials call a "good will kill" to eliminate a notorious Pakistani militant leader Nek Mohammad. Mark Mazzetti (2013) recounts how on a hot June day in 2004, Nek Mohammad was killed by a RPV while lounging in his mud compound in South Waziristan. He had been speaking on a satellite phone with one of the numerous reporters who frequently interviewed him. The Pakistani military claimed responsibility for the strike and the militant's demise. This was the first lie in what would become a concatenation of unsustainable fictions.

In fact, the CIA had executed the man, even though he was not an al-Qaeda operative and did not target the United States or its forces in Afghanistan. Nek Mohammad was an enemy of the Pakistani state, responsible for killing Pakistani troops and humiliating the army after making and then breaking a peace accord with the Pakistani army. The targeted killing was the CIA's first "good will kill." It inked, with Mohammad's blood, a secret bargain between the CIA and Pakistan's military and intelligence agency (the Interservices Intelligence Directorate [ISI]) that would grant the CIA access to Pakistan's air space and thus use RPV strikes to kill America's enemies. Bound by conditions of the covenant with the ISI, the CIA's RPVs would be constrained to narrow "flight boxes" in the FATA. This was to ensure that U.S. spies would not have access to "places where Islamabad didn't want the Americans to go: Pakistan's nuclear facilities, and mountain camps where Kashmiri militants were trained for attacks against India" (Mazzetti 2013, 109). The ISI also insisted that the United States operate all RPV flights in Pakistan under the CIA's covert-action authority, often referred to as "Title 50" operations (Wall 2011). This meant that the United States could never acknowledge that such strikes were taking place and "Pakistan would either take credit for individual kills or remain silent" (Mazzetti 2013, 109). Also, FATA's unique and archaic governance structure would facilitate this plausible deniability and obfuscate any details of the program. President Musharraf, who brokered the deal with President Bush, believed maintaining the ruse would be easy, telling a CIA operative during the negotiations that "in Pakistan, things fall out of the sky all the time" (Mazzetti 2013, 109).[4]

In the early years after the Nek Mohammad killing, the United States used armed RPV attacks sporadically. Between 2004 and 2007, there were only 9 attacks. Then, however, the Bush administration became increasingly convinced that RPV attacks were an effective way to defeat the militants in Pakistan's tribal areas, and in 2008 alone, the Bush White House launched 33 strikes. When Barack Obama became the U.S. president, he became ever more reliant upon armed RPV strikes to achieve his strategic objective of defeating al-Qaeda. In 2009, there were 53 RPV strikes, in 2010, the "year of the drone," there were 118 RPV attacks, and in 2011, there were 70 RPV attacks (Bergen and Tiedemann 2010). According to data from the New America Foundation (2013), there were 48 RPV strikes in 2012, and 13 in 2013 as of May.. Despite the attention to the RPV program in international media, the program in Pakistan is still technically covert. Accurate information about the program is thus very difficult to obtain, and even accounts in peer-reviewed journals contain many errors. U.S. government officials are generally prohibited from even acknowledging any particular RPV strike in Pakistan, despite the fact that RPVs are heavily reported in Pakistani and international media, albeit without reliable and confirmable details (Savage 2013).[5]

The Question of Pakistani Sovereignty

Officials in both the Bush and Obama administrations have justified the CIA RPV program by referencing both domestic and international law. As discussed in earlier chapters (see Chapter 1 and 6, for example), the 2001 Authorization for the Use of Military Force (AUMF) authorizes U.S. counterterrorism operations to target and kill members of the Afghan Taliban and al-Qaeda and its affiliates wherever they may be. Congress passed the AUMF just days after the 9/11 attacks. This statute permits the U.S. president "to use all necessary and appropriate force" to pursue those parties responsible for the 9/11 terrorist attacks. With respect to international law, the Obama administration justifies the program with reference to the right to self-defense, as laid out in Article 51 of the United Nations Charter. The Obama administration asserts that because the United States is in a state of armed conflict with al-Qaeda and associated forces, it is entitled to target them under the doctrine of self-defense (Masters 2013).

Critics reject this legal rationale (see Sonnenberg, Chapter 5, Glazier Chapter 6, and Eviatar, Chapter 7, for varying degrees of this argument). The Stanford-New York University Law School Clinics' report, "Living under Drones" (International Human Rights and Conflict Resolution Clinic, 2012), which examines the impact of U.S. RPV policy on civilians living in northwest Pakistan, voices skepticism that killings carried out today can be justified by the AUMF of 2001. The authors of the report also take considerable issue with the mobilization of the UN principle of self-defense to justify the attacks. The authors question whether the American RPV program violates Pakistan's sovereignty, an issue that hinges upon whether or not Pakistan has consented to the program and whether the United States is lawfully acting in self-defense. On the issue of Pakistani consent, different authors and organizations take different positions. Some are willing to concede that elements of the Pakistani state assented to the RPV attacks at least in the past, even if the state of current cooperation is unknown (Mazzetti 2013; Sanger 2012; IHR 2012). The International Crisis Group suspects that elements of the Pakistani state remain complicit and rebuff those who take Pakistani public denouncements at face value.

In contrast, Ben Emmerson, the UN special rapporteur on the Promotion and Protection of Human Rights and Fundamental Freedoms while Countering Terrorism, made categorical statements that the U.S. RPV program violated Pakistan's sovereignty. After his three-day visit to Pakistan in March 2013, Emmerson announced that there was no evidence of a U.S.-Pakistan agreement on RPV use and that Emmerson's official position was buttressed by "a thorough search of government records," an exceedingly unlikely scenario (International Crisis Group 2013, 20). The International Crisis Group expressed dismay that Emmerson "ignored evidence not only of tacit Pakistani consent during the Musharraf regime, as disclosed by then-Prime Minister Gilani in 2008 and again in 2010 and subsequently confirmed by Musharraf himself, but also of continued cooperation after Musharraf's removal in mid-2008, including the presumed role of Shamsi and Shahbaz airbases" (2013, 20).

While Pakistani officials deny any such agreement, American officials who can speak on this matter often reaffirm its existence, as Daniel Markey, a member of the Secretary of State's Policy Planning Staff from 2003 to 2007, does. In an interview with Ritika Singh, he explains:

> Musharraf's consent represented both that of the Pakistani military and its civilian government. Not only did he grant his consent, but initially, the Pakistani military tried to take credit for these kinds of attacks—claiming that they weren't the work of drones, but Pakistani air strikes. This wasn't a very credible claim on Pakistan's part, but it worked for a while because the strikes were initially much less frequent than they are now. And the misdirection helped the Pakistani government weather the domestic backlash. (Singh 2012)

Musharraf did not follow through on any of his public complaints, confirming the mutual understanding that such protests were political drama for domestic consumption. As Markey explains, "One can only assume. . . . that the private messages from the Pakistani government were different from their public messages" (Singh 2012). David Sanger, the chief Washington correspondent for the *New York Times*, suggests that this permission continued at least until 2011. Investigating the rules of RPV deployment in Obama policy, one of Sanger's interlocutors explained that with respect to host-nation permission, "a country must expressly invite the United States to use RPVs to strike targets inside its territory—which was the case with Pakistan until the traumas of 2011. . . . or they must be employed in a country that is 'unwilling or unable to suppress the threat.'"[6] Sanger further cites "a senior intelligence officer who is responsible for overseeing the program [who] insists that the United States sticks to those rules" (2012, 258).

Although Nawaz Sharif and his Pakistan Muslim League-Nawaz (PML-N) campaigned on an anti-drone platform, he tempered his opposition upon becoming prime minister in May 2013 (Shah 2012). Despite Sharif's insistence that the United States halt the use of RPVs, Pakistani former officials confirm that Pakistan's military and intelligence agencies support the strikes. The preferences of the military and intelligence agencies seem again and again to trump those of Pakistan's elected officials. Speaking of this civil-military discord, Husain Haqqani, Pakistan's former ambassador to the United States, explained:

> The Pakistani ISI actually resisted U.S. efforts to keep its own government in Islamabad informed. . . . The ISI did not like Pakistani civilian officials finding out anything about their dealings with the United States about armed Predator drones, but the U.S. government wanted the civilian leadership to remain in the picture. . . . [The ISI was in the habit of] protesting against the drones publicly while privately negotiating over whom the drones would target. (Quoted in Hirsh 2013)

Such recent reporting vindicates the suspicions of those analysts and organizations, such as the International Crisis Group, who have long suspected that some parts of the Pakistani state are complicit. Writing from an authoritative position on Pakistan's domestic politics and civil-military relations, the International Crisis Group observed that "even after the National Assembly . . . passed resolutions like the one in April 2012 that declared cessation of U.S. drone strikes an official policy objective, Pakistan has not yet taken any concrete steps to challenge the program. It has not, for instance, lodged a formal complaint with the UN Security Council" (2013, 29–30). Even if Pakistan's official position cannot be clarified, the Pakistani government continues to "deconflict" the airspace; in other words, they continue to ensure that RPVs operating in Pakistan do not collide with other aircraft—civilian or military—in the area. It should be recalled that RPVs do not simply "sneak in, bomb, and sneak out." Rather, a mix of RPVs hovers at different levels of altitude in Pakistan for hours and even days.

For those unfamiliar with Pakistan, the public statements by politicians condemning the RPVs may be adequate evidence that the Pakistani state does not facilitate, much less approve of, these attacks.

However, analysts who are more familiar with Pakistani politics understand that elected officials do not exercise control over national security policy. In fact, when democracy returned in 1990 following the death of dictator General Zia ul Haq and the electoral victory of Benazir Bhutto's Pakistan's People Party, the army "allowed Ms. Bhutto" to become the prime minister, provided that she agreed not to interfere in the affairs of the armed forces (Khan 2012, 227-228).

This remains the case. In July 2013, the official commission established by the Pakistan government to investigate the U.S. raid on Osama bin Laden's safe haven in Pakistan concluded that while setting defense policy constitutionally falls under the purview of the civilian government, "in reality. . . . defense policy in Pakistan is considered the responsibility of the military and not the civilian government even if the civilian government goes through the motions of providing inputs into a policymaking process from which it is essentially excluded" (Abbottabad Commission 2013, 159).[7]

Is Pakistan Willing to and Capable of Acting against Targets on Its Soil?

Closely related to the issue of Pakistan's sovereignty is the question of Pakistan's ability and willingness to exercise the rule of law and take action against those militants operating in and from Pakistan. On this matter the Stanford-NYU Law School Clinics' report concedes that "in the absence of Pakistani consent, U.S. use of force in Pakistan may not constitute an unlawful violation of Pakistan's sovereignty if the force is necessary in self-defense in response to an armed attack—either as a response to the attacks of September 11, 2001, or as anticipatory self-defense to mitigate threats posed by non-state groups" in the FATA (IHR 2012, 106-107). The report further points out that for this use of force to be lawful in Pakistan, Pakistan must also be shown to be "unwilling or unable to take [the appropriate steps, itself, against the non-state group]"(IHR 2012, 107). The report thus casts doubt upon whether contemporary RPV attacks can be justified by reference to the events of 9/11. The authors are also doubtful about the resort to "anticipatory" self-defense because it is unlikely that the majority of the drone strikes have averted attacks that are "instant, overwhelming, and leaving no choice of means, and no moment of deliberation"(IHR 2012, 107–108; see also Sonnenberg, Chapter 5; Glazier, Chapter 6; Eviatar, Chapter 7).

Indeed, recent reporting casts doubt upon the U.S. claims that RPV strikes target al-Qaeda and Taliban operatives or their associates to prevent imminent attack on the United States and its interests. Recent reporting by Jonathan Landay, based upon a privileged review of primary source materials, indicated that as many as "265 of up to 482 people who the U.S. intelligence reports estimated the CIA killed during a 12-month period ending in September 2011 were not senior al-Qaeda leaders but instead were 'assessed' as Afghan, Pakistani and unknown extremists. Drones killed only six top al-Qaeda leaders in those months, according to news media accounts" (2013). This is consistent with interviews with American and Pakistani officials, who concede that the U.S. RPVs are killing "Pakistani terrorists," such as Pakistani Taliban leaders (e.g. Nek Mohammad in 2004, Baitullah Mehsood in 2009, Waliur Rehman in 2013, among numerous others). What motivation does the United States have to eliminate Pakistan's enemies, who pose no significant imminent threat to the United States? Simple. The United States has a supreme interest in Pakistani domestic security and stability and seeks to help Pakistan achieve this, even while the two countries remain mired in other differences.

It is from this perspective that the issue of Pakistan's sovereignty becomes very difficult to assess. It is well established that Pakistan has cultivated Islamist militancy from the earliest days of the state. The state has employed Islamist militants to prosecute Pakistan's proxy war with India over the disputed disposition of Kashmir and India's rising position in the international system since 1947. Pakistan has instrumentalized political Islam in Afghanistan since the mid-1950s and Islamist militancy there since the early 1970s. Yet it is also the case that in recent years, some of Pakistan's erstwhile allies have mobilized to target the state. The most prominent of these is a network of commanders who operate under the banner of the Tehreek-e-Taliban-e-Pakistan (Pakistani Taliban, or TTP) and who have set their sights upon dismantling Pakistan's democracy. They have killed tens of thousands of Pakistanis, including women and children, military and paramilitary personnel, police and other law enforcement entities, bureaucrats and political figures alike (Fair 2011a).

While well-cited reports such as those by the afore-noted Stanford-NYU collaboration and that of Columbia University Law School (2012) do not focus upon this question of Pakistani sovereignty and intent, the International Crisis Group (ICG) report engages with it directly. The group considers that Pakistan's military has a record of forging deals to appease some of Pakistan's Taliban groups and that these initiatives "have jeopardised the safety of the communities those groups terrorise, including Shia and Barelvi communities and women" (2013, 30). The International Crisis Group also observes that the military denies access to independent observers in FATA, precluding them from collecting proof of human rights violations by militants. The ICG believes that "the military's support to Afghanistan-oriented proxies, such as the al-Qaeda linked Haqqani network, as well as local Taliban groups, such as those headed by Maulvi Nazir and Hafiz Gul Bahadur, invites U.S. drone strikes in the first place" and argues that "any successful and comprehensive counter-terrorism policy in FATA would have to address all these challenges candidly" (2013, 30). The ICG also explains that whereas the national elected leadership had tried to be more assertive on the issues of oversight of counterterrorism and counterterrorism policies, their role remains limited, with the dominant role played by the army.

The restriction of the RPV program to Pakistan's FATA (which is comprised of seven tribal agencies and six frontier regions) is important to understanding why the RPV policy is problematic in Pakistan and the difficulty in assessing Pakistan's will and capacity to do more to contend with terrorists within its territory. The FATA is governed by a colonial instrument called the Frontier Crimes Regulation, or FCR, which effectively renders residents of FATA to be "second class citizens" (Fair 2014a). As a consequence, foreign journalists are prohibited from travelling to FATA without the approval of the Ministry of Interior and/or an escort from the military and intelligence services. Even ordinary Pakistanis cannot legally visit the area unless they themselves have family ties there. Thus, it is extremely difficult to obtain accurate information from what has long been something of an informational black hole. These restrictions serve the Pakistani state's interests because it has long used FATA to host a dizzying array of Islamist militant groups operating in Afghanistan, India, and even Pakistan itself (Haqqani 2005; Rubin

2002; Hussain 2005; Swami 2007). Some of Pakistan's most hardened Islamist militants have found sanctuary in FATA as a consequence.

Several aspects of the FCR have enormous and nearly universally unacknowledged implications for the U.S. use of armed RPVs in FATA. Under the FCR, an entire family or clan can be punished just because one member has granted terrorists sanctuary in their home. This clause has been used to justify the Pakistani air strikes and draconian army operations that have caused enormous civilian casualties and forced displacement. As of March 2013, the United Nations reported that there are still some 758,000 persons who are internally displaced due to ongoing security operations in FATA as well as parts of Khyber Pakhtunkhwa (UNOCHA 2013). Part of the unrecognized legitimizing discourse surrounding the use of armed RPVs in FATA is the unfortunate fact that residents of FATA are second-class citizens, and the legal regime under which they are governed permits the state to ignore individual innocence and guilt. The United States exploits this predicament, and Pakistan perpetuates it by sustaining a legal regime that discriminates between the citizens of the so-called "settled areas," where the constitution applies, and "lesser" citizens under the rule of the FCR.

There is another, equally unappreciated aspect of the tribal areas: because FATA is governed under the FCR, it has no police forces; instead, paramilitary, military, and tribal militia forces keep order. As a result, the arrest of militants, collection of evidence, and subsequent prosecution in Pakistan's courts are not viable options in FATA. (In contrast, high-value targets captured in the rest of Pakistan are tried under Pakistani law or, in some cases, remanded to the United States.) While law and order approaches may be infinitely preferable to the use of armed RPVs, successive Pakistani governments have closed this route by choosing to defer from bringing the area and its people fully under Pakistan's constitution (White 2008). Thus, the only alternatives to doing nothing to combat the militants in FATA are devastating and indiscriminate Pakistani military operations or special forces raids into Pakistani territory by Afghanistan-based troops (Pak Institute for Peace Studies 2008, 2009, 2011).[8]

Questions about Pakistan's will and capacity to protect its own citizens from the ravages of the various terrorist organizations located in FATA, coupled with the state's insistence upon maintaining the FCR, have important humanitarian implications for the residents of FATA. Is it even the case that the U.S. armed RPV program is the biggest source of human insecurity in FATA? Given that more Pakistanis in FATA die from Pakistani military operations and terrorist attacks alike, why does the use of armed RPVs in FATA, clearly conducted with coordination with Pakistani military and intelligence agencies, attract the attention of international humanitarian organizations? The implication is that a person's death is less noteworthy unless it can be attributed to an armed RPV.

Methodological Issues in the RPV Civilian Impacts Debate

There is no question that U.S. RPV strikes have killed innocent persons. What is at stake is how many of the persons killed are in fact innocent civilians (Singh 2012). Numerous organizations, such as the New America Foundation, the *Long War Journal* of the Foundation for the Defense of Democracies, the Bureau of Investigative Journalism, Columbia Law School, among others have all sought to track RPV strikes and their outcomes. As well-intended as these efforts may be, the data are most certainly flawed. When one compares accounts of the same strike in the various databases, there is important disagreement about who was targeted and with what outcomes. Sometimes there is even disagreement about *where* the strike took place. For example, Amnesty International (2013) reported that an elderly woman (Mamana Bibi) was killed while tending her crops in her village in Ghundi Kala. However, according to the New America Foundation (2012), this attack took place in the village of Tapi. Such divergence occurs "because news accounts, upon which these databases rely, sometimes disagree about these details, and for reasons described above, it is impossible to independently verify which—if any—account is accurate" (Fair 2014b).[9]

While it is *difficult* to independently confirm media reports of RPV strikes in FATA, it is not impossible as some claim (for example, Stanford-NYU Law School Clinics; Amnesty International). Sebastian Abbot of the Associated Press did just that when he dispatched Waziristan-based stringers to independently investigate ten of the reportedly deadliest strikes from the previous year and a half. The team spoke with about eighty villagers at the sites, and, contrary to the widespread perception that civilians—rather than militants—are the principle victims, the team was told that a "significant majority [70 percent] of the dead were combatants" (Abbott 2012). Furthermore, those figures were driven by one very deadly attack on March 17, 2011. When the Associated Press team excluded that extraordinary attack, they found that nearly 90 percent of the people killed were militants, according to the villagers interviewed (Abbot 2012).

Although there is considerable uncertainty about how many innocent persons U.S. RPV strikes have killed, even those who have long opposed the use of armed RPVs now concede that civilian casualties may not be the single most salient objection to the program. The New America Foundation, based upon trend analysis of its own data on civilian casualties, accepts that "it seems clear the civilian casualties have now dropped dramatically, thanks to more precision weaponry and greater care—and the casualties are far lower than if conventional bombs were dropped"(Sanger 2012, 250). Nonetheless, the specter of civilian casualties animates much of the opposition to RPVs. Unfortunately, some of the critical methodological, analytical, and empirical shortcomings of specific reports on drones and human rights are less than desirable.

In "Living under Drones," published jointly by the law school clinics of Stanford University and New York University, the authors attempt to document the civilian cost of the U.S. RPV program in Pakistan's tribal agency of Waziristan. When authors veer away from their core legal expertise and into social sciences, they make several fundamental and avoidable empirical blunders beginning with their problematic convenience sample. A fundamental problem with the study is that it is funded and facilitated by an organization opposed to the use of armed RPVs in Pakistan and elsewhere. The authors explain that "in December 2011, Reprieve, a charity based in the United Kingdom, contacted the Stanford Clinic to ask whether it would be interested in conducting independent investigations into whether, and to what extent drone strikes in Pakistan conformed to international law and caused harm and/or injury to civilians"(International Human Rights 2012, i). It is important to note that Reprieve, and its Pakistani partner organization, the Foundation for Fundamental Rights (FFR), have been vigorous opponents of the RPV program and have argued forcefully for its termination. Thus, at the inception of this project, the law schools were asked to conduct research on behalf of an organization that is fundamentally opposed to RPVs. This represents a fundamental conflict of interest, which the authors do not seem to recognize.

The researchers compounded this ethical problem by allowing Reprieve and FFR to provide the research team with logistical support in Pakistan. In fact, the FFR "assisted in contacting many of the potential interviewees, particularly those who reside in North Waziristan, and in the difficult work of arranging interviews" (International Human Rights 2012, i). The group made no attempt to describe the outcome of a "typical" RPV encounter; rather, the authors sought out persons who self-identified as some form of RPV victim. The authors note that their analysis is based upon a meager 130 "interviews with victims and witnesses of RPV activity, their family members, current and former Pakistani government officials, representatives from five major Pakistani political parties, subject matter experts, lawyers, medical professionals, development and humanitarian workers, members of civil society, academic, and journalists" (International Human Rights 2012, 2).[10]

The authors concede that they did no interviews in North Waziristan or any of the other agencies comprising the FATA. Rather, they conducted their interviews during two separate trips to Pakistan in March and May 2012. All of the interviews took place in the twin cities of Islamabad and Rawalpindi, Peshawar, and Lahore. The authors claim that they conducted interviews with sixty-nine "experiential victims," who claimed to be "witnesses to drone strikes or surveillance, victims of strikes, or family members of victims from North Waziristan" (International Human Rights 2012, 2).

The authors of the report readily concede that the "majority of the experiential victims interviewed were arranged with the assistance of the Foundation for Fundamental Rights, a legal nonprofit based in Islamabad that has become the most prominent legal advocate for drone victims in Pakistan.... Some interviews also included a researcher from either Reprieve or the Foundation for Fundamental Rights" (International Human Rights 2012, 3). The role of this organization in selecting and interviewing respondents raises numerous ethical and empirical concerns, not the least of which is social desirability bias (Spector 2004).[11] Even though the interviewees were not compensated, they were provided with travel arrangements by FFR. This situation also creates opportunities for respondent coercion. The respondents may fear that should they offer accounts that differ from FFR/Reprieve's preferred anti-drone position, they may be unable to return home or not be selected for future interviews, which may provide the opportunity for travel. To summarize, the sample on which the researchers based their conclusions is at best a non-random convenience sample, fraught with dependent-variable selection bias as well as respondent social desirability bias (and in the worst case, possible coercion).

The authors could have sought to provide a countervailing perspective by including the views of pro-drone Pakistanis, which are captured in surveys and in Pakistani editorials (Fair, Kaltenthaler, and Miller, 2014; Zalmay 2013; Yousefzai 2012; "Silent, Fearful Support for U.S. Drones in Tribal Pakistan" 2013; "Drop the Pilot" 2013; Taj 2012; Hussain 2010). Researchers who have actually interacted with residents of Wazirstan, some of whom are from the tribal areas themselves, have found that many residents in FATA vigorously support the U.S. armed RPV program and even compare them to *ababil*, the holy swallows mentioned in the Koran (Surat-al-Fil, or Verse of the Elephant). In that incident, Allah dispatches the *ababil* to repel a Yemeni warlord (Abraha) and his army of elephants, which invaded Mecca, by dropping black stones upon the invaders (Taj 2010). For many persons in FATA, there are few other means to target those militants who are terrorizing parts of the tribal areas and the rest of Pakistan. There are no police or other law enforcement entities in the tribal areas. The Pakistani security forces conduct ground offensives, artillery bombardment, and air strikes that kill many innocents[12] (Watson Institute 2014) and displace millions (International Crisis Group 2010; UNOCHA 2014b; Internal Displacement Monitoring Centre 2014).[13] The Stanford-NYU collaborative effort—and virtually every other advocacy-driven report for that matter—fails to ponder these other concurrent uses of force, even peremptorily. It is not enough simply to say that most Pakistanis oppose drones. This is especially the case when this majority does not experience any of the externalities of the drone program, the varied militants in the tribal areas and their predations, or the haphazard efforts of the Pakistani military to deal with the militants selectively while maintaining a governance architecture (FCR) that is conducive to sustaining the militancy. Arguably, the views of those persons most exposed to the externalities of these realities deserve a privileged place in the policy, ethics, and political narratives surrounding these realities.

The authors of the Stanford-NYU report acknowledge that there is fear of retribution "from all sides—Pakistani military, intelligence services, non-state armed groups—for speaking with outsiders about the issues raised in this report" (International Human Rights 2012, 4). Despite this admission, the authors were surprisingly willing to take every utterance by their interviewees at face value. The authors explain:

> The research team has made extensive efforts to check information provided by interviewees against that provided in other interviews, known general background information, other reports and investigations, media reports, and physical evidence wherever possible. Many of the interviewees provided victims' identification cards and some shared photographs of victims and strike sites, or medical records documenting their injuries. We also reviewed pieces of missile shrapnel. (International Human Rights 2012, 5)

Alas, the team did not include forensic or munitions experts and thus cannot verify that purported damage to human life or property was due to RPVs. There is no chain of custody surrounding these artifacts that can demonstrate that they came from a drone or that the fragment—or its parent missile—is responsible for the event in question. As lawyers, the team well understands these evidentiary issues, but ignores them in this report.

These concerns are significant. Pakistani media have reported individuals and groups who have circulated fraudulent photos of persons who they alleged were injured by drones but were not ("Right-Wing Bigots Circulating Fake Pictures of Drone Victims to Deflect Attention From Taliban's Attack on Malala" 2012; Khan 2012). Given pervasive corruption in Pakistan, fake birth and death certificates can be easily acquired for a small fee. There are also numerous other possible explanations for ordinance debris and injuries. After all, drone strikes occur where terrorists conduct operations and secure safe haven (Pakistan Commission for Human Rights, 2012). Pakistan military and paramilitary organizations have also operated in the tribal areas, devastating the agencies in which they operate. Not only can terrorists and Pakistan's security forces account for some of the alleged injuries and debris; they may also account for the post-traumatic stress disorder and other disruptions to ordinary life that the authors attribute solely to the omnipresent RPVs.[14]

Other studies make similar mistakes. In 2013, Amnesty International conducted its own study of drones in Pakistan and published its findings in a report provocatively titled "Will I Be Next? U.S. Drone Strikes in Pakistan," with a photograph of young girl looking wistfully at the camera. The title and cover of the report imply that drones are so indiscriminate that any child has a reasonable fear of dying in an RPV attack. Amnesty International explained that it interviewed some sixty "survivors of drone strikes, relatives of victims, eyewitnesses, residents of affected areas, members of armed groups and Pakistani officials" between late 2012 and September 2013 (2013). The Amnesty International team did conduct interviews in some parts of North Waziristan, in addition to Khyber Pakhtunkhwa, Islamabad, and Rawalpindi, and they did made efforts to account for the violence perpetrated by state and non-state actors in the region. Despite these methodological improvements over the Stanford-NYU effort, Amnesty International focused mainly on two of the most controversial RPV strikes, which were clearly outliers. The organization made no effort to contextualize the RPV strikes in relation to all other drone strikes in Pakistan, even though the organization notes that it reviewed "all 45 reported U.S. drone strikes in Pakistan from January 2012 to August 2013" (2013, 18).

Amnesty International, like other advocacy organizations, put an enormous amount of weight upon the testimony of the alleged eye witnesses, including children as young as fifteen, eight, seven, and even five years of age.[15] In the report's study of the killing of a sixty-eight-year-old woman referred to as "Mamana Bibi," they rely heavily upon the testimony of Zubair Rehman, one of her teenaged grandsons. While the report details the ages of her other grandchildren cited in the report, nowhere does it state Zubair's age (he is depicted in a photo with his father, and it appears as if he is a teenager). The reliance upon Zubair's testimony is problematic because he reports seeing things that suggest that the aircraft that killed his grandmother could not have been an RPV. He claims, "The drone planes were flying over our village all day and night, flying in pairs sometimes three together. We had grown used to them flying over our village all the time" (2013, 19). There are at least two problems associated with this testimony, if it is accurate. RPVs cannot fly in formation as he suggests.[16] Equally disconcerting is Amnesty International's claim, without reference to any particular witness, that "Mamana Bibi was blown into pieces by at least two Hellfire missiles fired concurrently from a U.S. drone aircraft" (2013, 19). As presumed evidence, Amnesty International published photos that allegedly depict "debris from the missiles fired from a U.S. drone aircraft that killed Mamana Bibi" (2013, 22). However, David Axe, an American military correspondent, notes of this report that the "mangled metal pieces could just as easily have come from a TOW missile or another munition launched by a Pakistani military plane or helicopter" (2013).[17]

If Zubair Rehman's testimony is accurate, his account suggests that his grandmother was killed not by an RPV, but by Pakistani fighter aircraft (such as F-16s), which do fly together in formation and can launch munitions simultaneously as Zubair suggests. Indeed, Pakistan's military is quite active in the FATA. In November 2011, Pakistan's then air chief marshal, Rao Qamar Suleman, stated that "in the first two years of counter-insurgency operations, the air force conducted more than 5,500 strike sorties, dropped 10,600 bombs and hit 4,600 targets" (Trimble 2011).

A staple of nearly every advocacy-driven report on drones, including the reports by Amnesty International and the Stanford-NYU Law School clinics, is to opine that residents of the tribal areas are traumatized by the incessant sound of RPVs buzzing overhead. Amnesty International quotes an interviewee who says, "When the drone plane comes and we hear the sound of 'ghommm' people feel very scared. The drone plane can launch missiles at any time" (2013, 29). Another interlocutor explains that "everyone is scared and they can't get out of their house without any tension and from the fear of drone attacks. People are mentally disturbed as a result of the drone flights.... We can't sleep because of the planes' loud sound. Even if they don't attack we still have the fear of attack in our mind" (2013, 31). This theme also figures prominently in the report of the Stanford-NYU Law School Clinics. While these claims are popular, they are untenable. Most of the RPVs that the United States uses in the FATA are for surveillance. Not only are these RPVs flying in altitudes where they tend to be inaudible,[18] but such a noisy platform would be a useless for surveillance (U.S. Air Force 2010a, 2010b; Kelly 2002).

Conclusions: Can We Do Better?

There is virtually no likelihood that the information environment in Pakistan will improve. It is also unlikely that interest in the RPV program will disappear. The questions that the varied reports addressed herein will not go away. It is imperative that all persons studying the use of RPVs in Pakistan—or elsewhere—attempt to be as rigorous as possible. There are a number of important improvements that researchers and organizations can and should implement.

First, as the work of Sebastion Abbot attests, it is possible to interview persons in the vicinity of drone blasts. As his work has demonstrated, judicious use of trained and professional stringers with local knowledge and language skills provide insights that are invaluable. This does add costs, but it also adds accuracy.

Second, authors should be aware of the problematic ways in which they draw their samples. Some advocacy organizations want to focus upon the most salacious and outrageous of outcomes. This is irresponsible and disingenuous. Consumers of their reports are entitled to know how typical or atypical a particular event is.

Third, all researchers need to treat all interlocutors' accounts with skepticism for the various reasons noted above. The implied assumption of these advocacy-driven accounts is that so-called eyewitnesses are always truthful, but American government officials are always untruthful. Organizations should demur from relying upon the testimony of children. Countless research into the problems with children as witnesses shows that they have difficulty with accurate recall and are vulnerable to suggestion. These problems attenuate with age. Scholars who do research on children must obtain special dispensation from ethical review boards, although this is typically hard to obtain because the risk to the child that can result from participating in social science research may exceed the potential value of his or her participation.

Fourth, simple photographs of injuries and scrap metal are not evidence, and they should be not treated as such. If the same lawyers who included them in their report were litigating the defense, they would be first to cry foul about their inclusion as evidence.

Fifth, these organizations need to be more conversant with the legal, social, and political dynamics of the countries they wish to understand. As noted above, these details matter not only for the contexts of these studies, but also for the interpretation of events that these authors observe.

Sixth, they must become more knowledgeable about the weapons system they are studying. The Stanford-NYU effort does reproduce many technical specifications. Despite this, the authors proliferate the canard of the incessant buzzing of drones hovering above hapless civilians living in distress below. As this chapter shows, there is an enormous amount of open source commercial information about drones, and numerous avionics experts who can be consulted. Amnesty International could have saved itself some embarrassment had it run Zubair's claims by persons knowledgeable about different aircraft platforms and their capabilities.

Seventh, advocacy organizations should consult with munitions, forensics, and even satellite imagery experts. These professionals can provide some assessment of the probability that a given crater is due to an event described by an eyewitness. Different kinds of munitions have different explosive characteristics and may inflict different kinds of injuries, depending upon the distance from the point of explosion, exposure to the munition, and other details about the alleged victim and his or her location with respect to a given blast.

Finally, advocacy organizations should work with social scientists to improve their methods. For example, why not choose case studies of drone strikes based upon a random sample of drone strikes, as social science norms dictate? The answer is clear: a random sample of RPV strikes may not be as outrageous as the particular strikes that attract the advocacy organizations. While it is impossible to do a "random sample" of drone witnesses for any number of reasons, the exclusion of pro-drone Pakistanis is simply unethical.

If advocacy organizations do not want to adhere to the best standards of social science possible, they should simply concede that they are trafficking in public outrage and stop referring to their efforts as "research" and "analysis." The public should treat these efforts accordingly.

NOTES

1. While RPVs are popularly known as "drones," in this chapter I mostly use the term "RPV" to remind the reader that these vehicles are indeed piloted, albeit remotely. Many human rights organizations, in contrast, prefer to use the expression "drones" in their publications, which gives the impression that these are killing machines with little human oversight. Most Pakistanis use the expression "drone," and, for this reason, I also use this expression as context dictates.
2. This chapter is based upon C. Christine Fair, "Drones, Spies, Terrorists and Second Class Citizenship in Pakistan" (2014); and Christine Fair, "The Problems with Studying Civilian Casualties from Drone Usage in Pakistan: What We Can't Know" (2011).
3. The peculiarity of this region and the laws that govern it are discussed herein.
4. There are two kinds of covert armed RPV attacks that are used in Pakistan and elsewhere: personality strikes and signature strikes. While the covert armed RPV program has garnered domestic and international criticism generally, the increasing use of "signature strikes" has been particularly controversial because such strikes are targeted at "men believed to be militants associated with terrorist groups, but whose identities aren't always known" (Entous, Gorman, and Barnes 2011). Whereas personality strikes require the operator to develop a high level of certainty about the target's identity and location, based on multiple sources such as "imagery, cell phone intercepts and informants on the ground" (Miller 2012), operators may "initiate a signature strike after observing certain patterns of behavior"(CLS 2012, 32–33). Simply put, "The CIA had approval from the White House to carry out missile strikes in Pakistan even when CIA targeters weren't certain about exactly who it was they were killing" (Mazzetti 2013, 290). In principle, when conducting signature strikes, the United States assesses whether the individuals in question exhibit behaviors that match a pre-identified "signature" (that is, a pattern of observable activities and/or personal networks) that suggests that they are associated with al-Qaeda and/or the Pakistani or Afghan Taliban organizations (Zenko 2013). Because the identity of the target is unknown, even during and after the strike, innocent civilians may die in the strikes—a likelihood that both current and former U.S. government officials concede (IHR 2012). While the Bush administration employed personality strikes from 2004 and signature strikes from 2008 in Pakistan, the Obama administration redoubled the use of both types until the end of 2013, when RPV strikes sharply declined in Pakistan (Entous, Gorman, and Barnes 2013). Whereas the United States routinely offers apologies and compensation for civilian casualties in Iraq and Afghanistan, the United States has no mechanism to recognize civilian harm much less make amends for the same in Pakistan.
5. Acting U.S. ambassador to Pakistan met with anti-drone Code Pink activists in November 2011. Even discussing the existence of the program and the possible outcomes of the strikes caused Hoagland to remark, "I probably just, you know,

got into big trouble with what I just said" ("Acting U.S. Ambassador to Pakistan Met with Code Pink, Discussed 'Classified' Drone Casualty Counts"). In April 2013 the Obama administration offered its first detailed justification of a program it had previously refused to discuss.
6 The "traumas" included several incidents that raised tensions between the United States and Pakistan. For example, on January 27, 2011, a CIA agent, Raymond Davis, killed three Pakistani civilians in Lahore (Walsh 2011). Later in the year, in November, a NATO "attack" caused the deaths of at least twenty-four Pakistani soldiers in the Mohmand tribal region of Pakistan and led to a two-month suspension of RPV strikes in Pakistan (Masood and Schmitt 2011).
7 In essence, the United States is colluding with the military over the expressed wishes of Pakistan's elected officials and exploiting the peculiar situation that prevails with respect to civil-military relations in Pakistan. In Pakistan, it is the military and its intelligence agencies that make all pertinent national security decisions, not the elected officials. The U.S. government officially professes to support democratization in Pakistan and civilian control over the military. Unfortunately, this is just one of several cases in which the United States has pursued priorities that undermine longer term prospects for civilian control over the military and democratization. However, ultimately, the United States does not control who makes the most germane decisions, even if it is in a position to benefit from the prevailing situation in which the army is in control. (As such, the case of Pakistan provides counterevidence to Thomas Nichols's argument about the erosion of sovereignty, Chapter 4).
8 Militants in the FATA operate against international forces in Afghanistan and are reportedly responsible for killing some 43,000 Pakistanis since 9/11 (Pak Institute for Peace Studies). Like many databases, that of the Pak Institute for Peace Studies is not always clear about what sorts of attacks it tallies and what criteria it uses to code different kinds of violence. The following numbers are from the institute's annual reports from 2008 and 2011: 7,107 Pakistanis were killed in 2011; 10,003 in 2010; 12,632 in 2009; 7,997 in 2008; 3,448 in 2007; 907 in 2006; and 216 in 2005, for a total of 42,310.
9 While the United States typically takes the blame for the near total information blackout about who is targeted and with what outcomes, Pakistan is perhaps equally if not more culpable. From the inception of the program, Pakistan insisted that it be covert and restricted to FATA, which is difficult to access. As explained above, Pakistanis cannot go to FATA unless they have family ties. It is almost impossible for foreigners to go to FATA legally without the approval and even escort of Pakistan's military and intelligence agencies. (The author went on one such trip to South and North Waziristan in August 2010.) Pakistani newspapers report casualty figures based upon Taliban self-reports or even Pakistani government officials' statements, and these figures are in turn picked by international papers. There is rarely any attempt to independently confirm details, and often the details of these accounts are contradictory about numbers of casualties and other aspects of the strikes, such as the exact location and the persons who were targeted.

10 This sample is still relatively large compared to similar work by other advocacy organizations such as Amnesty International. Its October 2013 report is discussed below.
11 Moreover, given that Reprieve and FFR are staunch drone foes, readers should be dubious that the organizations would provide an unbiased selection of interview subjects for the study. (After all, would anyone be persuaded by the findings of a study of the health effects of cigarette smoking funded and facilitated by a company that produces and/or sells cigarettes?) However, even if the organizations in question had intended to provide an unbiased sample, the methodology of selecting interviewees obviates any such intention.
12 The Watson Institute at Brown University estimates that Pakistan's armed forces have killed at least 52,000 Pakistanis (combatant and noncombatant) and injured more than 50,000 since 2004.
13 By July 2009, there were some 2.8 million persons who were internally displaced due to military operations in the Malakand region of Khyber Pakhtunkhwa. In September 2009, the military undertook operations in Khyber agency that displaced between 56,000 and 100,000 persons in less than one month. Later, in October 2009, the Pakistan military commenced military operations in South Waziristan, causing 428,000 residents, or more than half of the agency's population, to flee. In March 2010, the military began a second major military offensive in Orakzai and Kurram, which displaced another 328,000. After the summer 2014 Pakistan military operation in North Waziristan, at least 457,000 persons were displaced, according to the United Nations Office for Coordination of Humanitarian Affairs (UNOCHA). Since 2004, the Internal Displacement Monitoring Centre estimates that some five million persons have been displaced from Pakistan's northwest. Pakistan's IDP population peaked in 2009, when some three million persons were displaced.
14 Needless to say, the continuous buzzing of drones is also more of a myth than a reality. A surveillance platform that created such noise would not be very efficacious. Moreover, drones typically fly at altitudes that render them invisible and inaudible.
15 Many of the authors of these reports are lawyers. As lawyers understand, the testimony of children is extremely unreliable (Ceci and Bruck 1995; Stolzenberg and Pezdek 2013; Quas et al. 1999). For a view that is more optimistic about the utility and accuracy of children witnesses, see (Pozzulo and Warren 2003).
16 Predators and Reapers, the two armed unmanned systems that the United States use in Pakistan, cannot fly in pairs much less triplets because their pilots, who are seated in cubicle-like pods thousands of miles away from the theater, do not have the visibility that permits them to fly in close formation with other drones, or any other kind of aircraft for that matter. While the Air Force is seeking to develop "sense and avoid" technology that would permit RPVs to fly as described, that technology is still in the experimental phase. According to a February 4, 2012 article in the *Air Force Times*, "The military's

unmanned aircraft do not have the ability to 'sense and avoid' nearby aircraft.... The Air Force can't install sense-and-avoid equipment on the Predators because the necessary gear—radars, infrared cameras, transponders—is either too big or consumes too much power. A Predator equipped with sense-and-avoid equipment 'can't carry anything else,' said Dave Bither, Mav6's vice president for strategic development. 'Right now, the technology is a generation away.'"

17 Pakistan's air force uses Lockheed Martin F-16s as well as AH-1 Cobra attack helicopters, which also fly in groups of two and even four.

18 The Reaper has a ceiling altitude of 50,000 feet. The Predator has a ceiling of 25,000 feet. While these aircraft can fly at higher altitudes, they tend to fly at much lower altitudes in theater. For example, according to one report, the Predator is "most effective at about 10,000 feet, within range of most anti-aircraft fire" (Kelley 2002). This comports with the claim of Army General Raymond Odierno who said that the "Predator flies at about 10,000 feet," adding that "it's so high up [the insurgents] have trouble hearing it" (Axe 2013).

BIBLIOGRAPHY

Abbot, Sebastian. 2012. "AP Impact: New Light on Drone War's Death Toll." Associated Press, February 26. http://news.yahoo.com/ap-impact-light-drone-wars-death-toll-150321926.html.

Abbottabad Commission. 2013. *Report of the Abbottabad Commission.* http://s3.documentcloud.org/documents/724833/aljazeera-bin-laden-dossier.pdf.

"Acting U.S. Ambassador to Pakistan Met with Code Pink, Discussed 'Classified' Drone Casualty Counts." 2011. *Daily Caller,* November 5. http://dailycaller.com/2012/11/05/acting-us-ambassador-to-pakistan-met-with-code-pink-discussed-classified-drone-casualty-counts/#ixzz2EJ4GxBjr.

Amnesty International. 2013. "Will I Be Next? US Drone Strikes in Pakistan." London: Amnesty International, October. https://www.amnestyusa.org/sites/default/files/asa330132013en.pdf

Axe, David. 2013. "Dear Amnesty International, Do You Even Know How Drones Work?" *War Is Boring* blog, October 22. https://medium.com/war-is-boring/dear-amnesty-international-do-you-even-know-how-drones-work-56b89ee705b2.

Bergen, Peter, and Katherine Tiedemann. 2010. *The Year of the Drone: An Analysis of US Drone Strikes in Pakistan, 2004–2010.* Washington, DC: New America Foundation.

Ceci, Stephen, and Maggie Bruck. 1995. *Jeopardy in the Courtroom: A Scientific Analysis of Children's Testimony.* Washington, DC: American Psychological Association.

Columbia Law School Human Rights Clinic (CLS). 2012. *The Civilian Impact of Drones: Unexamined Costs, Unanswered Questions.* New York: Center for Civilians in Conflict and Human Rights Clinic at Columbia Law School. http://civiliansinconflict.org/uploads/files/publications/The_Civilian_Impact_of_Drones_w_cover.pdf.

Currier, Cora. 2013. "How Does the U.S. Mark Unidentified Men in Pakistan and Yemen as Drone Targets?" *ProPublica*, March 1. http://www.propublica.org/article/how-does-the-u.s.-mark-unidentified-men-in-pakistan-and-yemen-as-drone-targ.

"Drop the Pilot." 2013. *Economist*, October 19. http://www.economist.com/news/asia/21588142-surprising-number-pakistanis-are-favour-drone-strikes-drop-pilot.

Entous, Adam, Siobhan Gorman, and Julian E. Barnes. 2011. "U.S. Tightens Drone Rules." *Wall Street Journal*, November 4. online.wsj.com/article/SB10001424052970204621904577013982672973836.html?mod=WSJ_hp_LEFTTopStories.

Entous, Adam, Siobhan Gorman, and Julian E. Barnes. 2013. "U.S. Tightens Drone Rules." Cora Currier, "Everything We Know So Far About Drone Strikes." *Propublica*, February 5. http://www.propublica.org/article/everything-we-know-so-far-about-drone-strikes

Fair, C. Christine. 2011a. "The Militant Challenge in Pakistan." *Asia Policy* 11 (January): 105–137.

Fair, C. Christine. 2011b. "The Problems with Studying Civilian Casualties from Drone Usage in Pakistan: What We Can't Know." *Monkey Cage*, August 17. http://themonkeycage.org/2011/08/17/the-problems-with-studying-civilian-casualties-from-drone-usage-in-pakistan-what-we-can%E2%80%99t-know/.

Fair, C. Christine. 2014a. "Drones, Spies, Terrorists, and Second-Class Citizenship in Pakistan." *Small Wars and Insurgencies* 23 (1): 205–235.

Fair, C. Christine. 2014b. "Ethical and Methodological Issues in Assessing Drones Impacts in Pakistan." *Monkey Cage*, October 6. https://www.washingtonpost.com/blogs/monkey-cage/wp/2014/10/06/ethical-and-methodological-issues-in-assessing-drones-civilian-impacts-in-pakistan/

Fair, C. Christine, Karl Kaltenthaler, and William J. Miller. 2014. "Pakistani Public Opposition to American Drone Strikes." *Political Science Quarterly* 129 (1): 1-33.

Frisbee, Lieutenant Colonel Sean M. 2004. "Weaponizing the Predator UAV: Toward a New Theory of Weapon System Innovation." Montgomery, AL: School of Advanced Air and Space Studies, Air University, Maxwell Air Force Base, June. http://dtlweb.au.af.mil/webclient/DeliveryManager?pid=31241.

Haqqani, Husain. 2005. *Pakistan: Between Mosque and Military*. Washington, DC: Carnegie Endowment for International Peace.

Hirsh, Michael. 2013. "Pakistan Signed Secret 'Protocol' Allowing Drones." *National Journal*, October 23. http://www.nationaljournal.com/white-house/pakistan-signed-secret-protocol-allowing-drones-20131023.

Hussain, Irfan. 2010. "Howling at the Moon." *Dawn*, January 9. http://www.dawn.com/news/833723/howling-at-the-moon.

Hussain, Rizwan. 2005. *Pakistan and the Emergence of Islamic Militancy in Afghanistan*. Burlington, VT: Ashgate.

International Crisis Group. 2010. "Pakistan: The Worsening IDP Crisis." *Crisis Group Asia Briefing* no. 111, September 16. http://www.crisisgroup.org/~/media/Files/asia/south-asia/pakistan/B111%20Pakistan%20-%20The%20Worsening%20IDP%20Crisis.pdf.

International Crisis Group. 2013. "Drones: Myths and Reality in Pakistan." *Asia Report* no. 247 (May): Brussels/Islamabad.

Internal Displacement Monitoring Centre. 2014. "Pakistan." Geneva, June. http://www.internal-displacement.org/south-and-south-east-asia/pakistan/.

International Human Rights and Conflict Resolution Clinic at Stanford Law School and Global Justice Clinic at NYU School of Law. 2012. "Living under Drones: Death, Injury, and Trauma to Civilians from U.S. Drone Practices in Pakistan." September. http://law.stanford.edu/wp-content/uploads/sites/default/files/publication/313671/doc/slspublic/Stanford_NYU_LIVING_UNDER_DRONES.pdf.

Landay, Jonathan S. 2013. "Obama's Drone War Kills 'Others,' Not Just al-Qaeda Leaders." *McClatchy Newspapers*, April 9. http://www.mcclatchydc.com/2013/04/09/188062/obamas-drone-war-kills-others.html.

Kelly, Matt. 2002. "Pilotless Spy Plane Plagued by Shortcomings, Despite Successes in Afghanistan Conflict." Associated Press, February 5. http://napavalleyregister.com/news/pilotless-spy-plane-plagued-by-shortcomings-despite-successes-in-afghanistan/article_753257b5-d3c7-5fc7-b8ec-40bfc5140678.html.

Khan, Feroz H. 2012. *Eating Grass*. Stanford, CA: Stanford University Press.

Khan, Sarah. 2012. "Islamist Bigots Circulating Fake Pictures of Drone Victims to Deflect Attention from Taliban's Attack on Malala." *Let Us Build Pakistan*, October 12. http://criticalppp.com/archives/229851.

Masood, Salman, and Eric Schmitt. 2011. "Tensions Flare between U.S. and Pakistan after Strike." *New York Times*, November 26. http://www.nytimes.com/2011/11/27/world/asia/pakistan-says-nato-helicopters-kill-dozens-of-soldiers.html?_r=0.

Masters, Jonathan. 2013. "Backgrounder: Targeted Killing." New York: Council on Foreign Relations, May. http://www.cfr.org/counterterrorism/targeted-killings/p9627.

Mazzetti, Mark. 2013. *The Way of the Knife: The CIA, a Secret Army, and a War at the Ends of the Earth*. New York: Penguin.

Miller, Greg. 2012. "CIA Seeks New Authority to Expand Yemen Drone Campaign." *Washington Post*, April 18. http://articles.washingtonpost.com/2012-04-18/world/35453346_1_signature-strikes-drone-strike-drone-program.

Mujumdar, Dave. 2012. "AF, Firms Seek UAV Flights in Civil Air Space." *Air Force Times*, February 4.

New America Foundation. 2012. "Drone Wars Pakistan." Washington, DC, October 24. http://securitydata.newamerica.net/drones/pakistan/043-2012.

New America Foundation. 2013. "The Drone War in Pakistan." Washington, DC, June. http://natsec.newamerica.net/drones/pakistan/analysis.

Pak Institute for Peace Studies. 2008. *Pakistan Security Report 2008*. Islamabad: PIPS.

Pak Institute for Peace Studies. 2009. *PIPS Security Report 2009*. http://san-pips.com/index.php?action=books&id=main.

Pak Institute for Peace Studies. 2011. *Pakistan Security Report 2011*. Islamabad: PIPS.

Pak Institute for Peace Studies. 2012. "Civilian Casualties in Armed Conflicts in Pakistan: Timeline 2012." http://san-pips.com/index.php?action=reports&id=tml2.

Pakistan Commission for Human Rights. 2013. *State of Human Rights in 2012*. Lahore: HRCP.

Pozzulo, Joanna D., and Kelly L. Warren. 2003. "Descriptions and Identifications of Strangers by Youth and Adult Eyewitnesses." *Journal of Applied Psychology* 88 (2): 315–323.

Quas, J. A., G. S. Goodman, S. Bidrose, M. Pipe, S. Craw, and D. S. Ablin. 1999. "Emotion and Memory: Children's Long-Term Remembering, Forgetting, and Suggestibility." *Journal of Experimental Child Psychology* 72 (4): 235–270.

Rubin, Barnett R. 2002. *The Fragmentation of Afghanistan*. New Haven, CT: Yale University Press.

Sanger, David E. 2012. *Confront and Conceal: Obama's Secret Wars and Surprising Use of American Power*. New York: Crown.

Savage, Charlie. 2013. "Top U.S. Security Official Says 'Rigorous Standards' Are Used for Drone Strikes." *New York Times*, April 30. http://www.nytimes.com/2012/05/01/world/obamas-counterterrorism-aide-defends-drone-strikes.html?_r=0.

Shah, Murtaza A. 2012. "US Drone Attacks Not War Crimes; To End Soon: Nawaz." *News*, October 25. http://www.thenews.com.pk/Todays-News-13-26261-US-drone-attacks-not-war-crimes;-to-end-soon-Nawaz.

Sifton, John. 2012. "A Brief History of Drones: With the Invention of Drones, We Crossed into a New Frontier: Killing that's Risk-Free, Remote, and Detached from Human Cues." *Nation*, February 7. http://www.thenation.com/article/166124/brief-history-drones#ixzz2V4xA1E9V.

"Silent, Fearful Support for US Drones in Tribal Pakistan." 2013. Arabia. MSN.com, November 12. http://arabia.msn.com/news/world/2170794/silent-fearful-support-for-us-drones-in-tribal-pakistan/.

Singh, Ritika. 2012. "Lawfare Podcast Episode #20: Daniel Markey on U.S.-Pakistan Terrorism Cooperation and Pakistan's Extremist Groups." *Lawfare*, September 27. http://www.lawfareblog.com/2012/09/daniel-markey-on-u-s-pakistan-terrorism-cooperation-and-pakistans-extremist-groups.

Singh, Ritika. 2013. "Drone Strikes Kill Innocent People. Why Is It So Hard to Know How Many?" NewRepublic.com, October 25. http://www.newrepublic.com/article/115353/civilian-casualties-drone-strikes-why-we-know-so-little.

Spector, Paul. 2004. "Social Desirability Bias." In *Sage Encyclopedia of Social Science Research Methods*, edited by Michael S. Lewis-Beck, Alan Bryman, and Tim Futing Liao. Thousand Oaks, CA: Sage. http://srmo.sagepub.com/view/the-sage-encyclopedia-of-social-science-research-methods/n932.xml.

Stolzenberg, Stacia, and Kathy Pezdek. 2013. "Interviewing Child Witnesses: The Effect of Forced Confabulation on Event Memory." *Journal of Experimental Child Psychology* 114 (1): 77–88.

Swami, Praveen. 2007. *India, Pakistan and the Secret Jihad: The Covert War in Kashmir, 1947–2004*. London: Routledge.

Taj, Farhat. 2010. "Analysis: Dangerous Abyss of Perceptions." *Daily Times*, January 30.

Taj, Farhat. 2012. "Analysis: Drone Attacks: Challenging Some Fabrications." *Daily Times*, January 2. http://archives.dailytimes.com.pk/editorial/02-Jan-2010/analysis-drone-attacks-challenging-some-fabrications-farhat-taj.

Trimble, Stephen. 2011. "DUBAI: F-16s Powered Up Pakistan's Counter-Insurgency Strikes." FlightGlobal.com, November 13. http://www.flightglobal.com/news/articles/dubai-f-16s-powered-up-pakistan39s-counter-insurgency-364727/.

United Nations Office for the Coordination of Humanitarian Affairs (UNOCHA). 2014a. "Humanitarian Dashboard—Pakistan." Geneva, March. http://www.unocha.org/pakistan/reports-media/ocha-reports.

United Nations Office for the Coordination of Humanitarian Affairs (UNOCHA). 2014b. "Pakistan: North Waziristan Displacements Situation Report No. 3." Geneva. https://pak.humanitarianresponse.info/system/files/documents/files/OCHA%20Pakistan_NWA%20Displacements_Situation%20Report%20No.%203_ah.pdf.

U.S. Air Force. 2010a. "MQ-9 Reaper." Langley, VA: Air Combat Command, Public Affairs Office, August 18. http://www.af.mil/AboutUs/FactSheets/Display/tabid/224/Article/104470/mq-9-reaper.aspx.

U.S. Air Force. 2010b. "MQ-1B Predator." Langley, VA: Air Combat Command, Public Affairs Office, July 20. http://www.af.mil/AboutUs/FactSheets/Display/tabid/224/Article/104469/mq-1b-predator.aspx.

Wall, Andru E. 2011. "Demystifying the Title-10 Title 50 Debate: Distinguishing Military Operations, Intelligence Activities and Covert Action." *Harvard National Security Journal* 3 (1): 85–142.

Walsh, Declan. 2011. "A CIA Spy, a Hail of Bullets, Three Killed and a US-Pakistan Diplomatic Row." *Guardian*, February 20. http://www.theguardian.com/world/2011/feb/20/cia-agent-lahore-civilian-deaths.

Watson Institute for International and Public Affairs. 2014. "Costs of War—Pakistan." Providence, RI: Brown University, May. http://costsofwar.org/article/pakistani-civilians.

White, Joshua T. 2008. "The Shape of Frontier Rule: Governance and Transition, from the Raj to the Modern Pakistani Frontier." *Asian Security* 4 (3): 219–243.

Yousefzai, Zmarak. 2012. "Voice of a Native Son: Drones May Be a Necessary Evil." ForeignPolicy.com (South Asia Channel), October 15. http://southasia.foreignpolicy.com/posts/2012/10/15/voice_of_a_native_son_drones_may_be_a_necessary_evil.

Zalmay, Kahar. 2013. "In Praise of Drone Strikes." *LA Progressive*, June 8. http://www.laprogressive.com/praising-drone-strikes/.

Zenko, Micah. 2013. "Reforming U.S. Drone Strike Policies." Council Special Report No. 65. New York: Council on Foreign Relations, January. http://www.cfr.org/wars-and-warfare/reforming-us-drone-strike-policies/p29736?co=C009601.

PART III

The Future of Preventive Force

9

The Contemporary Practice of Self-Defense

Evolving Toward the Use of Preemptive or Preventive Force?

AVERY PLAW AND JOÃO FRANCO REIS

Preventive and preemptive uses of force by the United States over the last dozen years, including its current use of drone strikes in places like Pakistan and Yemen, are having a significant impact on the international legal regime and particularly the *jus ad bellum*. Yet this impact need not, as many commentators have worried, be all bad.[1] Indeed, we argue in this chapter that it may have some unintended positive effects on the law of self-defense and particularly on the interpretation of Article 51 of the UN Charter.

We begin by briefly reviewing three main schools of thought—strict construction, inherent right, and preventive force—which define the key debates over the meaning of self-defense under Article 51 in the era before 2001. Next, we outline two leading approaches to the interpretation of international customary law and apply these to the current character and trajectory of the law of self-defense. In doing so, we sketch an interpretation of how the United States' uses of force, and increasingly those of other states like Israel, Turkey, Russia, and Colombia, are reshaping those debates and creating a new practice permitting limited, episodic exercises of force in response to cross-border attacks by non-state actors. We then develop our argument that this practice has the potential to resolve some inconclusive and sterile debates over the meaning of Article 51—and might give states the necessary flexibility to confront immediate terrorist threats while avoiding the slippery slope to full-scale preventive war. We conclude with a discussion of three further claims and how they might inform future research.

Self-Defense: Three Schools of Thought

The first and perhaps central goal set out in the UN Charter (1945) is "to save succeeding generations from the scourge of war, which twice in our lifetime has brought untold sorrow to mankind." Central to that project is Article 2(4) of the charter, which declares that "All Members shall refrain in their international relations from the threat or use of force against the territorial integrity or political independence of any state, or in any other manner inconsistent with the Purposes of the United Nations." As previous chapters in this volume have already made clear, the charter also allows two exceptions to this general prohibition, specifically in Articles 42 and 51. Since only Article 51 is about the unilateral use of force, this will be our focus. To review, Article 51 states:

> Nothing in the present Charter shall impair the inherent right of individual or collective self-defence if an armed attack occurs against a Member of the United Nations, until the Security Council has taken measures necessary to maintain international peace and security.

This move reflects a laudable desire to minimize states' resort to force. But in narrowing the circumstances in which states could legitimately resort to force in this way, and particularly in light of the general reluctance of the Security Council to authorize it, enormous pressure is packed into the meaning of self-defense.

The exact parameters of self-defense as envisioned in Article 51 of the UN Charter have been the source of academic debate for almost seventy years, and this chapter addresses only one of many issues connected with them—that is, the question of whether or not preemptive or preventive self-defense against terrorist groups has any support in international law. In particular, we focus on how the debate over this question is being informed by contemporary practice. We will greatly simplify the complex debate around the meaning of this article by organizing it into three contending schools of thought, which we will term (1) strict construction, (2) inherent right, and (3) preventive force.

While these three "schools" are admittedly crude constructions, they do capture some essential contemporary points of division over self-defense. They are perhaps most quickly and clearly distinguished in terms of the attitude they take to the language of Article 51. The strict construction school focuses on the qualifying phrase "if an armed attack occurs," while the inherent right school concentrates on "Nothing . . . shall impair the inherent right," and the preventive force school seizes on the overriding goal of maintaining "peace and security." These differences of emphasis lead to rather different readings of the law.

The Strict Construction School

The most influential source of the "strict construction" view is the jurisprudence of the International Court of Justice (ICJ). The ICJ has held not only that any right of self-defense must be temporally preceded by an armed attack but also, at least in cases primarily concerned with attacks by non-state actors, that the attack must attain a certain scope and magnitude. As Christopher Greenwood aptly summarizes,

> In its judgment in the *Nicaragua Case*, the ICJ also stated that the use of force would not amount to an armed attack for the purposes of the right of self-defence unless it was of a particular scale and effect, and it contrasted an "armed attack" with a "mere frontier incident." . . . According to the Court, if a use of force did not rise above the level of a "mere frontier incident," then, even though it constituted a violation of Art. 2 (4) UN Charter, the victim of that violation was not entitled to respond by way of action in self-defence. (2011, ¶ 12)[2]

Commentators have put particular emphasis on this requirement. Antonio Cassese, for example, has insisted that in order to qualify as an armed attack, what terrorists commit must be "a very serious attack either on the territory of the injured state or on its agents or citizens while at home or abroad" (1989, 596). One obvious challenge that arises on this reading is the difficulty of defining what exactly constitutes a "very serious" (as opposed to merely "serious") attack, and why one merits a right of response and the other does not.

A second point stressed by the ICJ and the strict construction school is that acts of terrorism can count as armed attacks only if the terrorist group is under the "effective control" of a state (ICJ 1986, 65). This interpretation, established in the Nicaragua case, was reaffirmed by the ICJ again in its 2004 decision on the *Legal Consequences of the Construction of a Wall in the Occupied Palestinian Territory*,[3] in which it denied the Israeli claim to be acting in self-defense in part on the following grounds:

> Article 51 of the Charter . . . recognizes the existence of an inherent right of self-defense in the case of armed attack by one state against another state. However, Israel does not claim that the attacks against it are imputable to a foreign state. (2004, 56)

In essence, the court's majority regarded the invocation of the right of self-defense as inapplicable because the terrorist attacks in question were not attributable to a foreign state. This view has been criticized even from within the court itself. For example, in a 2005 separate opinion on the *Armed Activities on the Territory of the Congo* case, Judge Kooijmans asserted that "if armed attacks are carried out by irregular bands . . . against a neighboring state, they are still armed attacks even if they cannot be attributed to the territorial state. It would be unreasonable to deny the attacked state the right of self-defense merely because there is no attacker state, and the Charter does not so require" (2005, 314).

A third feature of strict construction implicit in the ICJ's jurisprudence is that the exercise of self-defense must follow fairly quickly after the armed attack that triggers it. As Jan Kittrick summarizes,

> Traditionally, self-defense responses must be an immediate reaction to the previous armed attack: the defensive measures must be exercised within a reasonable time frame after the attack. Should an unreasonable amount of time elapse between the original armed attack and the defensive response, the reaction could be rendered illegal. Timely response is therefore of the utmost importance. (2009, 159)

However, the requirement that an armed response be "timely" is vague and therefore invites controversy. It is also particularly difficult to apply in cases of attack by non-state actors. As Yoram Dinstein observes, "since 'the source of [terrorist] attacks may not be immediately obvious,' the process of gathering intelligence data and pinning the blame on a particular non-state group . . . may ineluctably stretch the interval between the armed attack and the forcible response" (2005, 250).

The Inherent Right School

A second, "inherent right" school of thought asserts that the right of self-defense existed prior to the UN Charter and that Article 51 even acknowledged this by saying that "nothing in the present Charter shall impair the inherent right of individual or collective self-defence." As Tom Ruys explains, some scholars believe that the reference to the word "inherent" and the *travaux préparatoires* (drafting history) "indicate that Article 51 was only intended to give particular emphasis in a declaratory manner for self-defence in the case of an armed attack" (2013, 256). Other scholars suggest a way of reconciling the seemingly contradictory commitments of Article 51 by reading the "if an armed attack occurs" as one possible trigger among others. For example, Myres McDougal and Florentino Feliciano have argued that "a proposition that 'if A, then B' is not equivalent to, and does not necessarily imply, the proposition that 'if, and only if, A, then B'" (quoted in Ruys 2013, 256). Therefore "if an armed attack occurs" would not be the same as "if, and only if, an armed attack occurs." Judge Stephen M. Schwebel, in his dissenting opinion to the ICJ Nicaragua judgment, similarly argued that Article 51 should not be interpreted as permitting self-defense "if, and only if, an armed attack occurs" (quoted in Moir 2011, 13).

From a temporal perspective, this group denies the need for an attack to occur before one can act in self-defense as long as the attack is imminent. They say, in Ruys's words, "that the Nuremberg and Tokyo Military Tribunal implicitly accepted the legality of anticipatory self-defense" (2013, 256) and ICJ judges like Judge Rosalyn Higgins seem to agree with Greenwood in saying that "the continued validity of this pre-charter law on anticipatory self-defence is consistent with the reference in Art. 51 to the right of self-defence being 'inherent'" (quoted in Lubell 2011, 58). They also believe that a non-state actor acting alone can commit "an armed attack" triggering the right of self-defense. Naturally, this means they allow the extraterritorial use of force in self-defense even if there is no link at all between the non-state actor perpetrating the armed attack and a supporting state.

This "inherent right" group usually links their legal arguments concerning the customary nature of the right of self-defense to pre-Charter law and "emphasis is often placed on what has come to be called the Caroline doctrine" (Quigley 2013, 152). As detailed in Chapter 1, in 1837, British troops attacked Canadian rebels and torched a boat named *Caroline* in U.S. territorial waters. In a diplomatic protest U.S. Secretary of State Daniel Webster wrote that "for self-defense to be legitimate, the British had to demonstrate a 'necessity of self-defence, instant, overwhelming, leaving no choice of means, and no moment for deliberation'" (Frank 2009, 98), adding that the defensive act could not be "unreasonable or excessive" (Jennings and Watts 1992, 420–27). Some authors like Michael N. Schmitt argue that "over time, this standard, and its implicit criteria, has become universally accepted as the keystone in the law of self-defense" (2003, 530).

Yet some authors are extremely critical of any attempt to use the Caroline doctrine as a precedent supporting anticipatory self-defense. Dinstein, for example, writes, "Reliance on that incident in the context of anticipatory self-defense is misplaced. There was nothing anticipatory about the British action against the *Caroline* steamboat on U.S. soil, inasmuch as use of the *Caroline* for transporting men and materials across the Niagara River—in support of an anti-British rebellion in Canada—had already been in progress" (2005, 184–85). For Dinstein, the *Caroline* incident is at best an example of the traditional use of self-defense, since the attacks to which the British were responding were already underway. Therefore this case cannot be used as a precedent for preemptive self-defense.

Some commentators also question whether the precedent would be compelling even if it did represent a case of preemptive force. Ruys, for instance, argues that "the 'episodic reference' to the 1837 Caroline incident is considered anachronistic and misguided. . . . [It] dates from an age where States were essentially free to resort to war against one another and lacking a legal regime of Self-defense" (2013, 258).

Other commentators like James Dever and John Dever have argued in a more practical vein that reliance on "customary approaches [like the *Caroline* precedent] are flawed in the current global environment" (2013, 165–66). In particular, they suggest that given the increased harm that terrorist attacks can do, and the difficulty of uncovering plots being hatched by secret, transnational networks, both the "necessity" and "imminence" requirements invoked in the Webster formulation must be relaxed.

The Preventive Force School

Proponents of the third, "preventive force" school believe that there is no need for an "armed attack to occur," nor does there even need to be an "imminent threat of attack," before action in self-defense can be taken. Just "a threat" is needed, whether from a state or non-state actor. On September 17, 2002, the U.S. government published a new National Security Strategy, which laid the base for the famous "Bush doctrine," the epitome of the preventive approach: "The greater the threat, the greater is the risk of inaction—and the more compelling the case for taking anticipatory action to defend ourselves, even if uncertainty remains as to the time and place of the enemy's attack." As President Bush put it himself, "I believe it is essential—I believe it is essential—that when we see a threat, we deal with those threats before they become imminent" (quoted in Shue 2005, 11).

In essence, the preventive force approach "attempts to break the link between first-use of force and aggression" (Shue 2005, 2). That is probably the main reason that, in Ruys's words, "on the whole, legal scholars have almost *unisono* denounced the doctrine of *preventive* self-defense" (2013, 322). More specifically, at least three important objections have been raised:

1. This theory "lacks clear legal footing under the text of Article 51" (Maggs 2006–2007, 479).
2. It invites a premature resort to armed force, increasing the likelihood of armed conflict (Svarc 2007).
3. It was exploited by some of the most reprehensible regimes to carry out deplorable acts, including the Nazis when they invaded the Soviet Union during the Second World War (Maggs 2006–2007).

The preventive force school has arguably the weakest legal arguments when it comes to legal precedents, *opinio juris*, and the interpretation of Article 51. However, it could still be supported by state practice, which could evolve into customary law. But is current practice and legal opinion moving in this direction, or is it consolidating around one of the other schools of thought or some new position? This question encapsulates the main focus of this chapter, but to answer it requires an interpretation of where the customary law of self-defense stands today and how it is changing, particularly in relation to the three schools outlined above. Unfortunately, the interpretation of customary law is itself a controversial area, with at least two main contending approaches. In the next section, we will focus our attention on outlining these two leading approaches to the interpretation of international customary law and then will apply them to the current character and trajectory of the law of self-defense.

Customary Law: Two Approaches

Where the content or meaning of conventions or treaties are ambiguous or contested, a typical recourse is to look to the second source of international law (after "international conventions")—namely, "international custom, as evidence of a general practice accepted as law," as classically stated in Article 38 of the Statute of the International Court of Justice. Of course, determining the current content of customary law is a notoriously difficult enterprise. Indeed, even the means and materials for formulating such an interpretation are the subjects of profound disagreement. Olivier Corten influentially distinguished two leading approaches to interpreting customary international law in a 2006 article. One leading school, which can be described as "realist," offers an "extensive" approach, embracing the view that "actions speak louder than words." This approach focuses primarily on what powerful states do, particularly their use of force, as the primary indication of what they think is permissible (2006, 804–12). Realists correspondingly accept the possibility of rapid and dramatic shifts in customary law, as the behavior of leading states changes, and sometimes swiftly.

In contrast, a second leading school, which can be described as "institutionalist," offers a more "restrictive" approach, focusing on the crucial importance of oral and written opinions of state officials (especially as formulated in formal documents and articulated in prominent public venues), which establish the *opinio juris* of states. After all, political actions are subject to a wide range of interpretations and justifications, and it is hard to determine what precedent is being set until one knows what the state taking action thinks it is doing and why, and what other states think of that justification for action (Corten 2006). In the extreme case, if a state declines to justify and/or acknowledge an action, this might well be thought to reflect the state's own doubts regarding the legality of the action. Consequently, far from establishing a new precedent, the state would in fact be reinforcing the very prohibition that it simultaneously violates. The key point here is that it is the express statements (or lack thereof) of all states that determines the legal significance of actions. Institutionalists correspondingly see customary law as changing very gradually, as states slowly realign in their express understanding of what is and is not permissible. Corten (2006) suggests that this approach is especially closely associated with the jurisprudence of the International Court of Justice (see also Ruys 2013).

The following section leans more toward the realist approach than the institutionalist for two main reasons. The first is that the realist approach is more straightforward and manageable within a relatively short chapter, as we can briefly refer to the recent actions of states and avoid a detailed examination of the stated opinion of every state regarding these actions. The second reason is that the authors lean toward the realist view anyway. Nevertheless, we will say at least a few words about expressed *opinio juris* in the following section and suggest that, in certain respects, it aligns with the evolving behavior of important states. Still, we recognize that our review and analysis are provisional and therefore are vulnerable to criticism from a more systematically institutionalist perspective.

For present purposes, we are going to focus on illustrating a common pattern, which has recently emerged among a number of important states. The essence of that pattern is a resort to lethal force against non-state actors that are engaged in a pattern of violent attacks on the population of the acting state, and especially (although not exclusively) against their leadership and key personnel. There is also a growing tendency to employ such force openly and to insist on its legality as self-defense.

We will argue that this pattern of behavior offers at least a partial resolution to the logjam of legal debate over the meaning and scope of the right of self-defense reflected in the diverse disputes among the three schools of thought outlined above. In essence, we suggest that the emerging practice of states described below represents a stylized synthesis of the three main schools of thought on self-defense, at least in relation to using counterforce against terrorist groups engaged in campaigns of attack from abroad. Moreover, we will argue that this synthesis is not without virtue, as illustrated by the possibility of answering many of the telling criticisms raised in the foregoing sections against the three main traditional interpretations of self-defense. Consider the following cases.

Case 1: Russian Targeted Killing, 2004

Russian agents killed Zelimkhan Yandarbiyev in Doha, Qatar, on February 13, 2004 by planting a car bomb in his Land Rover. Yandarbiyev had been the president of Chechnya from 1996 to 1997 and was an active leader of Chechen resistance groups. He had been the subject of three failed extradition attempts by the Russian government. Russian officials "accused the former writer and poet of complicity in the incursions by Chechen rebels into the neighbouring republic of Dagestan in August 1999" (BBC News 2004). They also accused him of having "helped orchestrate the October 2002 theater siege in which Chechen rebels raided a Moscow theater, taking hundreds hostage" (CNN News 2004). The United States and the United Nations both added him to "a list of people suspected of ties to Al Qaeda and other terrorist organizations" (CNN News 2004).

The two assassins were arrested in Qatar following Mr. Yandarbiyev's killing, and the judge presiding over their trial said they were acting on orders from Russian leadership. Given life sentences in Qatar, they were later extradited to Russia where they were welcomed as heroes and released from prison (BBC News 2005).

In July 2006, the Russian Parliament passed a law (No. 35-FZ) permitting the Russian security services to kill alleged terrorists overseas if authorized to do so by the president. The authorization appeared to include the use of intelligence agents as well as the military. According to press reports at the time, "Russian legislators stressed that the law was designed to target terrorists hiding in failed States and that in other situations the security services would work with foreign intelligence services to pursue their goals" (Alston 2010, ¶24). The law itself asserts that this authorization is consistent with both the Russian Constitution and International Law (see Articles 1 and 2 of the law).

Case 2: Israeli Counterterrorist Offensive in Lebanon, 2006

In the summer of 2006, Israel launched a military campaign in Lebanon. The campaign was precipitated by a Hizbullah incursion across the Israeli-Lebanese border on July 12, which resulted in the killing of three soldiers and the kidnapping of two others. The Israeli government reacted by sending IDF units over the border in an attempt to rescue the kidnapped soldiers. The operation failed to free the captured soldiers, but did result in the deaths of five more soldiers. The Israeli Prime Minister, Ehud Olmert, then declared that the Hizbullah attack was an "act of war" and launched a full scale military campaign, beginning with extensive air strikes across Lebanon and leading to a major ground campaign in the south of the country (Urquhart and McGreal 2006). The fighting continued until a UN ceasefire went into effect on August 14.

On July 16th, the Israeli Cabinet issued the following statement: "Israel is not fighting Lebanon but the terrorist element there, led by Nasrallah and his cohorts, who have made Lebanon a hostage and created Syrian- and Iranian-sponsored terrorist enclaves of murder" (Israeli Ministry of Foreign Affairs 2006). Israel's operations were directed mainly against suspected Hizbullah militants (although arguably with insufficient precision), and deliberately sought to avoid confrontation with the Lebanese army, which equally sought to avoid engagement. Hizbullah, for its part, flung defiance at the Israeli government, daring Israel to launch a ground assault, and launching over 4,000 missiles into Northern Israel. It was clear to most observers, then, that the fight was between Israel and Hizbullah with unfortunate Lebanese civilians too often trapped in between. The final list of fatalities included somewhere between 250 and 600 Hizbullah fighters, 119 Israeli soldiers, and somewhere in the vicinity of 850 and 1,191 Lebanese civilians (Amnesty International 2006).

A striking feature of this brief conflict was that there was widespread support in the international community for Israel's right to defend itself by pursuing a military campaign against Hizbullah under the laws of armed conflict. While it is true that the international community became increasingly critical of the manner in which Israel prosecuted the armed conflict, its criticism was almost entirely framed in terms of humanitarian law. In an address to the UN Security Council on July 30 Secretary General Kofi Annan affirmed that "no one disputes Israel's right to defend itself" but noted that there was growing concern about "its manner of doing so," and in particular about what appeared to be "grave breaches of humanitarian law" (UN News Centre 2006). Similarly, the G-8, meeting in St. Petersburg, recognized Israel's "right to defend itself" against Hizbullah's aggression but cautioned Israel "to be mindful of the strategic and humanitarian consequences of its actions" and in particular to avoid "casualties among innocent civilians" (G8 2006). At the national level, most countries focused their criticisms exclusively on what Turkey called "bombing civilians"—that is, on Israel's purported failure to fully comply with humanitarian principles of proportionality and distinction (Kuwait News Agency 2006). In essence, the international community accepted that a country attacked by a terrorist organization had a right to use military force against that organization in self-defense, even on the territory of another state, provided that it complied with the humanitarian law of armed conflict.

A second Israeli use of lethal force against alleged terrorists also requires mention, specifically its use of targeted killing outside its official borders. According to the human rights NGO B'Tselem (2014), Israel has killed 459 Palestinians in the course of targeted killing operations in the West Bank and Gaza Strip since the beginning of 2000. Of these, 273 were identified as intended targets and the remaining 186 as unintended targets. The Center for the Study of Targeted Killing (2014) at the University of Massachusetts has identified twenty-five Israeli drone strikes, most of these in the Gaza Strip, but also one each in Sudan, Egypt, Lebanon, and Syria. News reports suggest that these strikes have killed thirty militants and six civilians. While Israeli targeted killings were initially widely criticized when they started in 2000, in recent years other countries have largely ignored them. This reinforces the conclusion that states like Israel are regularly employing military force against terrorist groups under the aegis of self-defense, and that other states are increasingly tolerant of such uses of force.

Case 3: Turkey versus PKK in Iraq, 2007 to Present

On October 17, 2007, the Turkish Parliament voted overwhelmingly (507 to 19) to authorize Prime Minister Recep Tayyip Erdoğan to order strategic strikes or large-scale military operations against the Kurdistan

Workers' Party (PKK) in Iraq for a one-year period (*Washington Post* 2007). Two weeks prior, the PKK had stepped up attacks inside Turkey, killing thirty-one people, including a busload of civilians and thirteen military commandos. Following the decision, the Turkish military launched air strikes and conducted a ground invasion in northern Iraq. According to a report by CNN (2008a), as many as ten thousand Turkish troops were involved in the operation, which lasted from February 22 to 29, 2008.

The Iraqi government never consented to its neighbor's armed incursion into its territory, but seemed to tolerate it to some extent (although Iraq's Kurdish president, Massoud Barzani, warned that the regional government would not stand by "if the Turks struck civilians"). On July 10, 2008, Turkey and Iraq signed a joint political declaration establishing a "high-level strategic cooperation council" between the governments that would help forge a "long-term strategic partnership" (CNN 2008b). Turkey has continued to launch cross-border attacks against the PKK.

On July 25, 2012, Prime Minister Erdoğan asserted that Turkey also has an "undisputed right" to intervene if Kurdish terrorists in northern Syria pose a threat to Turkey:

> We will not let the terrorist group to set up camps [in northern Syria] and pose a threat to us. . . . No one should attempt to provoke us. We will not bow to provocation but rather take whatever steps are necessary against terrorism. (*Today's Zaman* 2012)

In particular, Erdoğan asserted that Turkey would take measures to prevent Syrian Kurdish groups affiliated with the PKK from controlling northern Syria. He mentioned the possibility of the military creating a "buffer zone" inside Syria to prevent terrorist camps.

Case 4: Colombian Attack on the FARC in Ecuador, 2008

On March 1, 2008, shortly after midnight, a Colombian air strike targeted Luis Edgar Devia Silva, better known by his *nom de guerre*, Raúl Reyes, inside Ecuadorian territory. The strike killed an estimated twenty-five people, including four Mexicans and one Ecuadorian (Podur 2008). Reyes's personal computer, which was recovered from the site, indicated a link between the Revolutionary Armed Forces of Colombia (FARC) and the governments of Ecuador and Venezuela. Ecuador and Venezuela immediately cut diplomatic ties and sent troops to the Colombian border. Within a week of the operation, an emergency Organization of American States session was held in which the majority of member states criticized the cross-border attack. Colombia was accused of violating Ecuador's territorial sovereignty, while Colombia asserted that its action was self-defense.

On March 7, a meeting of the Rio Group (an organization of Latin American and Caribbean states) was held in the Dominican Republic to defuse the threat of direct conflict between nations. At the summit the presidents of Colombia, Ecuador, Venezuela, and Nicaragua publicly shook hands "marking the end of a diplomatic crisis" (BBC News2008).

Case 5: U.S. Targeted Killings, 2002 to Present

President Obama has greatly accelerated an expansion of the U.S. drone campaign that began in July of 2008, during the last year of President Bush's tenure (Woodward 2010).[4] Specifically, after a grand total of only eleven strikes during the first four years of the campaign in Pakistan (2004–2007), the United States carried out 402 attacks from 2008 to 2014. Reported fatalities resulting from the strikes have correspondingly risen, from 184 during the first four years of the campaign, to 2,925 from 2008 to 2014, according to the Center for the Study of Targeted Killing (2014). At least 109 of the fatalities between 2008 and 2014 have widely been reported as civilians, and another 456 are contested. Strikes in Yemen were concentrated between 2010 and 2014 (with the exception of one in November 2002), and totaled 110 strikes with 608 killed, at least 57 of whom were widely reported as civilians, with another 12 contested. At least 7 strikes have also been carried out in Somalia, with another 20 fatalities. In all, the United States has carried out at least 530 drone strikes resulting in 3,737 dead, including at least 223 civilians and perhaps as many as 764 (Center for the Study of Targeted Killing 2014).

The United States has also carried out a number of other lethal counterterrorist operations, including the commando raid carried out into Osama bin Laden's compound in Abbottabad on May 1–2, 2011. Notably, although this last operation involved the insertion of troops on a lethal mission into the heart of Pakistan without the permission or even knowledge of the Pakistani government, there was very little public criticism of the operation outside of Pakistan itself.

This same muted reaction is perceptible in regard to the U.S. drone campaign, particularly in the early years. A number of commentators, both sympathetic and opposed to drone strikes, noted that U.S. terrorist targetings "have not yet provoked much controversy" (Byman 2006, 96; O'Connell 2005; Bennett 2012). Following the acceleration of the program in 2008–2010, there has been a good deal of domestic and international criticism, exemplified by the public protests organized by CODEPINK (a left-wing NGO also called "Women for Peace") as well as by a number of studies published by U.S. law schools (including the "Living under Drones" report by Stanford and New York University, co-authored and briefly described by Stephan Sonnenberg[5] in Chapter 5 of this volume), and on the international side by a UN investigation led by UN Special Rapporteur Ben Emmerson. The UN special rapporteurs for Summary, Arbitrary, and Extrajudicial Killings, Philip Alston and Christof Heyns, have also been outspoken in raising questions about the strikes, demanding greater transparency and condemning certain egregious practices like "double tap" signature strikes, in which drones pause after an initial strike and then launch additional strikes against those rushing to the scene.

What is perhaps most striking is the reluctance of states to openly condemn the U.S. practice, which has led some legal scholars like Mark Osiel (2009) and Jordan Paust (2010) to suggest that by tolerating this practice over a period of years, the great majority of states have accepted it as a legitimate practice. While that conclusion may still be premature, it does appear that there is vastly more tolerance of this practice than there was fourteen years ago, when the United States was itself actively condemning it.

The Case for Evolution in Customary Law

With regard to the cases examined above and their implications for the state of contemporary customary law, a number of caveats are in order. First, this short treatment has not provided a historical basis for comparison—for example, similar incidents might have occurred before 2000. Certainly there are at least some historical cases, most famously the Israeli "Wrath of God" operation following the massacre of Israeli athletes at the 1972 Munich Olympics. Still, there is good reason to believe that the scope of lethal counterterror operations has greatly expanded since 2000. We are not aware of any databases that track all cross-border counter-terror operations, but Bryan Price and Patrick Johnston have published important studies of "decapitation" operations intended to eliminate "top terrorists" or, in Johnston's case, "insurgent leaders" often involved with terrorism, based on datasets that each assembled. Price documents 204 decapitations of terrorist leaders between 1970 and 2008 and Johnston documents 118 decapitation attempts of insurgent groups between 1975 and 2003. Price's data (2012, 2014) indicates that the number of targeted killings of terrorist leaders recorded between 2000 and 2008 more than doubled the total of any previous decade. Johnston remarks that his data "indicates that both the rate of leadership decapitation attempts and of successes increased around 2000. This increase coincided with escalated high-value targeting efforts after the terrorist attacks of September 11, 2001" (2012, 58). Of course, these datasets focus on the killing of leaders, while this chapter considers lethal transborder counterterror operations more generally, so the increase in the first does not prove an increase in the second. But if one considers the very large proportion of low-level targeting in the U.S. drone campaign in Pakistan—Reuters, for example, reports 92.2 percent low-level targeting in the period between 2008 and 2010, near the height of the campaign—it seems very likely that terrorist targeting in general has also risen (Entous 2010).

A second caveat is that there are ways to explain these extraterritorial uses of force in ways that do not appeal (primarily) to self-defense. In some cases, there may be a non-international armed conflict (NIAC) and consent to intervene in it from the territorial state (see Chapter 7 in this volume for further discussion of NIAC). The existence of a non-international armed conflict would explain why military force could be employed without appeal to self-defense, while consent of the state on whose territory counterterror operations are conducted can resolve the apparent violation of sovereignty without appeal to self-defense. But it seems fairly clear that both of these conditions apply to few, if any, of the foregoing cases. The Russian, Turkish, Israeli, and Colombian cases clearly lack consent, and the Pakistani government at least publicly refuses consent to the United States, although the Yemeni and Somali presidents have explicitly consented. Also, although a NIAC clearly exists after 2009 in the Colombian case (and probably the Pakistani), it is doubtful it did in the others at the time of the events in question. So self-defense seems an essential justification in most operations. The only case where it seems likely that there is both consent and a NIAC is Pakistan after 2009; in fact, U.S. officials still habitually invoke self-defense (along with NIAC) in justification of their program, so it is difficult to deny its relevance.

Those caveats aside, the foregoing cases provide a suggestive set of materials for assessing customary law in line with the more realist, "extensive" approach. In the first place, these cases exhibit a common pattern of action: they are all cases of states both confronting terrorist attacks emanating from the territory of another state and responding by employing lethal force against individuals they perceive as responsible. None of the states received permission from the UN Security Council to take such action, and when they have defended it, they have appealed to self-defense. In no case does the terrorist group targeted appear to be under the control of another state and, in most, the initial attack(s) that precipitated the invocation of self-defense was relatively small in comparison to a military assault (the 9/11 attack obviously presents an exception, but it is worth remembering that the United States had already used lethal force against al-Qaeda in 1998). These states' common behavior constitutes a collective challenge to the narrow, conventional interpretation of self-defense as being justifiably triggered only by a major attack by another state or a by terrorist group under the "effective control" of another state.

It is true that in some cases—most notably the Russian—the actions were taken covertly, and have never officially been acknowledged. But here it is worth remembering that what is most salient for a realist analysis is what states actually do, rather than what they say about it (and Russia does, at any rate, explicitly assert the right to kill terrorists abroad). It is also significant from a realist perspective that some of the most important and influential states in the world are leading the way— including the United States, Russia, and Turkey—and that the states are widely distributed around the world. Most importantly, perhaps, the pattern seems to be accelerating, as tentative initial moves from the United States and Israel have evolved into full-fledged campaigns and an increasing number of other countries have begun to follow suit (see Fisk and Ramos 2014).

The upshot is to suggest one of those rapid shifts in customary law that adherents of the realist approach think possible, especially when led by major players. In essence, the following practice appears to be emerging:

1. When a non-state actor (even without state sponsorship) engages in a lethal attack from abroad, and explicitly or manifestly intends more, and
2. When the state from which the attack emanates is not willing or able to suppress the threat,
3. It has become permissible for a state to act in self-defense to prevent further attacks by using brief, preemptive, repeated[6] (or "episodic") military targetings of suspected terrorists, especially leaders.
4. Such actions are permissible even if the targeted state does not have detailed knowledge of specific imminent plots.

There is also some evidence of evolution in the law from the perspective of a more "restrictive" institutionalist analysis. Two developments stand out. In the first place, the states carrying out targeted killing and broader military campaigns against non-state actors on foreign territory have begun to recognize these actions officially and to defend them as legal.[7] The United States, Israel, Turkey, and Colombia all publicly owned and defended their operations in the name of self-defense. Moreover, the Israeli High Court of Justice determined in 2006 that some targeted killings of this sort are legal under both international and Israeli law. This marks a sharp shift from the bulk of pre-2000 state practice, which more typically followed the pattern of covert action and public denial.

The second key point here is that most other states seem to be willing to acquiesce in this reinterpretation of the customary law of self-defense. This is reflected in two main ways. The first is the eloquent silence with which reports of many targetings are met—most famously and notably the killing of Osama bin Laden. The second is the nature of the criticism that is offered.

The best illustrations of both patterns come from international reactions to the U.S. adoption of targeted killing. As previously noted, there was little vocal public opposition to U.S. drone policy until the massive acceleration of strikes that began in the summer of 2008. Even when vocal international opposition appeared, it has been led more by UN officials such as Philip Alston, Christof Heyns, and Ben Emmerson than by state governments.[8]

Moreover, the tenor of criticism tends to be particular rather than general. Alston (2009, 2010), for example, has generally limited himself to saying that some drone strikes *could* be illegal if undertaken outside the context of self-defense or armed conflict and, therefore, that states conducting them should be more transparent about the legal rationale, procedures, and safeguards for their use. Emmerson (2013) is more inclined to object to the idea of drone strikes but is compelled to acknowledge that there is legitimate disagreement over when international law permits lethal force to be used. And, like many others, Heyns (2013) primarily criticizes the most dubious species of drone strikes (such as double taps) on the grounds that these seem to violate the laws of war. In each case, the critics allow that at least some of the U.S. drone strikes outside Afghanistan may well be legal, and focus on gathering evidence and identifying the cases that clearly fall outside the bounds of the permissible. The principal debate over drone strikes is thus not over whether they are ever legal, but increasingly over when and against whom.

Still, the overall picture is far from one-sided, and those skeptical of the new self-defense norm can take comfort in the fact that at least some of the criticism of the U.S. drone program challenges the right of states to use military force against non-state actors outside of their own territory (see O'Connell 2009, 2010a, for example). Moreover, two important recent decisions of the International Court of Justice, *Legal Consequences of the Construction of a Wall in the Occupied Palestinian Territory* (2004) and *Democratic Republic of the Congo v. Uganda* (2005), emphasized the traditional view that a state has the right to use military force in self-defense against a non-state actor only if that non-state actor is acting on behalf of another state. Even in the ICJ decisions, however, justices offered forceful dissents on exactly the points at issue here, observing that the majorities on the court were out of touch with current state practice.

In summary, then, while courts and scholars remain divided on the breadth of the right of self-defense, there is an increasingly forceful case on realist (extensive) grounds, and even to some degree on institutionalist (restrictive) grounds, that the customary law of self-defense is evolving toward allowing the use of episodic military force against non-state actors engaged in a pattern of attacks against the citizens of a sovereign state.

Conclusion

Rather than offer a conventional summary in this final section, we will briefly outline three further claims building on the foregoing analysis that warrant further research: (1.) that the legal basis for this new, current practice of self-defense is dubious but improving; (2) that current practice synthesizes some key elements from the three traditional schools on self-defense; and (3) that it is not without some intuitive merits.

The first part of the first claim is that the legal foundation for current practice is dubious. In point of fact, such military operations look illegal from the perspective of the first two schools. Thus, on the one hand, for strict constructionists, there is no state in "effective control" of targeted terrorists; nor is there a tight temporal link between attack and response, nor even in some cases an initial "armed attack" of sufficient scope to trigger the right of self-defense. On the other hand, for advocates of "inherent right," there is no evidence of knowledge of a particular imminent attack, which creates a "necessity of self-defence, instant, overwhelming, leaving no choice of means, and no moment for deliberation." So, at first glance, it appears that the current practice can be justified only in terms of the preventive force school, which, as noted above, is rejected by the great majority of legal scholars.

There is, however, an alternative approach, which, while still legally weak, offers more promise. We call this approach "Caroline plus." In essence, scholars have followed the logic of the inherent right school back to the precedent of the *Caroline* and the idea of preemption, then abstracted the criteria of imminence, necessity, and proportionality from Webster's formulation, and finally applied them more flexibly to contemporary practice, particularly with regard to imminence. This also appears to be the strategy of the Department of Justice white paper on targeting American citizens, which invokes the ideas of preemption and imminence, but then defines imminence in strikingly broad terms:

> The condition that an operational leader present an "imminent" threat of violent attack against the United States does not require the United States to have clear and specific knowledge that a specific attack on U.S. persons and interests will take place in the immediate future.... The threat posed by al-Qa'ida and its associated forces demands a broader concept of imminence.... Imminence must incorporate considerations of the relevant window of opportunity, the possibility of reducing collateral damage to civilians, and the likelihood of heading off future disastrous attacks on Americans. (Department of Justice undated, 7)

This argument is inherently weak in that it both relies on the Caroline precedent (despite all the potential objections raised in the foregoing discussion of the "inherent right" school), but also undermines its relevance as a precedent by seeking to radically revise its terms. Nonetheless, its prospects are gradually improving because of the proliferation of contemporary precedents (some of which are described in the foregoing case studies), which rely on ideas of imminence and necessity similar to those outlined above.

The second claim is that the current practice offers an interesting synthesis of some key elements of each of the three schools. This is particularly clear with regard to the strict constructionist school, for the appeal to self-defense is triggered by an initial armed attack, albeit potentially smaller than might conventionally be allowed, and without a state sponsor. So the requirement of an initial attack is retained, but with a broader interpretation of "armed attack," which offers some answer to critics (like Judge Kooijmans) who condemn the strict constructionists for unreasonably denying states the right to defend themselves against unaffiliated non-state actors. Moreover, the current practice provides this additional leeway while retaining the great value in the strict constructionist approach and its requirement of a prior attack, which is to restrain legitimate state uses of force to cases of response to aggression and thus to prevent irresponsible adventurism.

The state practice we described in the case studies also retains the criteria of imminence and necessity from the inherent right school, while on the one hand broadening them to permit sustained attacks against a terrorist group conducting an ongoing campaign, and on the other marrying them to the requirement of a prior terrorist attack described in the last paragraph. In other words, a broader interpretation of imminence is allowed if and only if there was an initial attack in the past and there is a reasonable assumption of a threat in the near future. To put this in terms employed elsewhere in this book, where a standard "inherent right" view of imminence is well-expressed by Eviatar (Chapter 7)—"It is impossible to define the 'imminent' threat needed to justify lethal force in self-defense as anything other than a threat of an aggressive action that is 'about to happen"—those who embrace the "Caroline plus" view insist that context matters with regard to how imminence applies. In particular, in the case of a terrorist attack and an ongoing threat from abroad, the application that makes most sense, and is actually being used, is closer to the idea of "lagged imminence" proposed by John Emery and Daniel Brunstetter in Chapter 10, where they refer to the notion of "a real threat [that is] always on the horizon, albeit not immediately. Statesmen lack the ability to pinpoint the precise moment when such a threat will be actualized, but cannot simply ignore it."

This broader interpretation of imminence offers some answer to the criticism of Dever and Dever (2013), the U.S. Department of Justice (in its *White Paper*, undated), and others who insist that a strict Caroline standard would leave states in a permanent position of trying to stop terrorist attacks only once they are underway and it is already too late.

Finally, the new practice of self-defense also draws on the "preventive force" approach while also providing a basis for responding to some of the most telling criticism of it. On the one hand, it offers some room for states to act "preventively"—that is, as indicated above, against terrorist threats before they fully materialize in a concrete plan already in preparation or execution. On the other hand, the type of preventive action that it permits does not sever the link to initial aggression (for an initial attack must precede the use of force in self-defense) and thus avoids "unilateral adventurism."

Our concluding claim is that this new practice is not without merits, in that it offers a workable compromise in the long, stale debate among the three schools of self-defense interpretation outlined at the beginning of this chapter. While the legal foundations of this view remain tenuous, they may well become stronger if states come to appreciate the practicality of the compromise and adopt it, both in practice and in *opinio juris*. In this respect, it may be worth stressing the advantages suggested above, most notably that the new practice offers states significant flexibility in combating the real and sustained threats posed by transnational terrorist groups while continuing to permit the exercise of self-defense only against those who have already committed armed attacks and are intent on doing so again.[9]

That said, we acknowledge that the possible evolution of international law toward this compromise also carries the danger of not only eroding the prohibition of the use of force in Article 2(4) of the UN Charter, but also permitting future violations of sovereignty. However, as noted by Thomas Nichols (Chapter 4), sovereignty should not be treated as an unlimited right to ignore or violate international law, especially if this results in harm to citizens of other states. There must be limits to sovereignty. One such limit should be another state's right to defend its citizens against terrorist violence, including by force if necessary and if used justly.[10] International law must find a sustainable way to balance these considerations.

In suggesting that the emerging practice is not without merits, we do not mean to present it as constituting current law (*lex lata*) or even the law as it necessarily should be (*lex ferenda*). We cautiously suggest that this seems an increasingly likely direction for state practice—and, after it, customary law—to move, and we give some thought to what coherence it can be given, and how it might be shaped to maximize benefits and minimize costs.

NOTES

1 For a very critical approach to the U.S. policy of using drones see, for example, O'Connell 2010 and 2012.
2 Military and Paramilitary Activities in and against Nicaragua (*Nicaragua v. United States of America*), Jurisdiction and Admissibility. 1984. 392 ICJ REP. June 27, 1986. This case was decided by the International Court of Justice in 1986. The court ruled in favor of the Republic of Nicaragua, which claimed that the United States violated multiple norms of International Law, including the prohibition of the use of force (art. 2[4] of the UN Charter), by mining Nica-

rgua's harbors and supporting the Contras, a rebel group fighting Nicaragua's government.
3   For further references to ICJ cases, see O'Connell 2011.
4   This is also discussed in earlier chapters of this volume (Chapters 1, 4, 5–7).
5   For an interesting and different view on how drones might have an impact on *jus ad bellum* norms see Stephan Sonnenberg, Chapter 5. The authors agree that the misuse of drones raises serious legal issues but remain cautious regarding Sonnenberg's conclusions.
6   "Repeated" in this context should not be interpreted as allowing signature or double tap strikes, which may violate the humanitarian principle of distinction.
7   See President Obama speech at the National Defense University, May 23, 2013. For further discussion, see O'Connell 2011.
8   On February 27, 2014, the European Parliament also voted (534 to 49) in favor of a resolution that demanded that European Union member states "oppose and ban practices of extra judicial targeted killings" and made reference to drone strikes (*Express Tribune* 2014).
9   It is worth noting once more that the authors agree with David Glazier's (Chapter 6) statement that "any strikes conducted abroad without consent of the nation whose territory is involved violate international rules about sovereignty and prohibitions against the use of force unless the exacting *jus ad bellum* criteria for self-defense of necessity, proportionality, and imminence are satisfied." In other words, "Caroline Plus," as the name itself suggests, is not just about "imminence." The "necessity" and "proportionality" requirements are as relevant as ever.
10 The authors acknowledge that applying a Caroline plus standard justly may also involve some elevated constraints on the way force is used, although unfortunately these issues of *jus in bello* cannot be addressed here. Two interesting discussions of what elevated standards might look like can be found in Braun and Brunstetter (2014) and Plaw and Colon (2015).

BIBLIOGRAPHY

Alston, Philip. 2009. "UN Rights Expert Voices Concern over Use of Unmanned Drones by United States." UN News Centre, October 28. http://www.un.org/apps/news/story.asp?NewsID=32764&Cr=alston=#.VAaVDvldV8E.

Alston, Philip. 2010. "Report of the Special Rapporteur on Extrajudicial, Summary or Arbitrary Executions." New York: United Nations Human Rights Council, May 28. http://www2.ohchr.org/english/bodies/hrcouncil/docs/14session/A.HRC.14.24.Add6.pdf.

"Report of the Special Rapporteur for Extrajudicial, Summary and Arbitrary Executions." United Nations A/HRC/14/24/Add.6.

Amnesty International. 2006. "Israel/Lebanon Deliberate Destruction or 'Collateral Damage'?" August 23. AI Index: 18/007/2006. http://www.refworld.org/docid/4517a71c4.html.

BBC News. 2004. "Russia 'Behind Chechen Murder.'" June 30. http://news.bbc.co.uk/2/hi/middle_east/3852697.stm.

BBC News. 2005. "Convicted Russia Agents 'Missing.'" February 17. http://news.bbc.co.uk/2/hi/europe/4275147.stm.

BBC News. 2008. "Leaders Say Colombia Crisis Over." March 8. http://news.bbc.co.uk/2/hi/americas/7284597.stm.

Bennett, Jonathan. 2012. "Exploring the Legal and Moral Bases for Conducting Targeted Strikes Outside of the Defined Combat Zone." *Notre Dame Journal of Law, Ethics and Public Policy* 26 (2): 549–580.

B'Tselem. 2014. "Statistics." B'Tselem.org. http://www.btselem.org/statistics.

Braun, Megan, and Daniel R. Brunstetter. 2013. "Rethinking the Criterion for Assessing CIA-Targeted Killings: Drones, Proportionality, and *Jus ad Vim*." *Journal of Military Ethics* 12 (4): 304–324.

Byman, Daniel. 2006. "Do Targeted Killings Work?" *Foreign Affairs* 85 (March–April): 95–112.

Cassese, Antonio. 1989. "The International Community's 'Legal' Response to Terrorism." *International and Comparative Law Quarterly* 38 (July): 589–608.

Center for the Study of Targeted Killing. 2014. Database. July 1. http://targetedkilling.org/#/strikes.

CNN News. 2004. "Blast Killed Exiled Chechen Leader." February 13. http://edition.cnn.com/2004/WORLD/meast/02/13/qatar.chechen/index.html.

CNN News. 2008a. "Turkey Launches Major Iraq Incursion." February 23. http://edition.cnn.com/2008/WORLD/meast/02/22/turkey.iraq/index.html.

CNN News. 2008b. "Iraq, Turkey Sign Border Security Pact." July 10. http://www.cnn.com/2008/WORLD/meast/07/10/iraq.turkey/.

Corten, Olivier. 2006. "The Controversies over the Customary Prohibition on the Use of Force." *European Journal of International Law* 16 (5): 803–822.

Dever, James, and John Dever, Jr. 2013. "Making Waves: Refitting the Caroline Doctrine for the Twenty-First Century." *Quinnipac Law Review* 31:165–193.

Dinstein, Yoram. 2005. *War, Aggression and Self-Defence*, 5th ed. New York: Cambridge University Press.

Emmerson, Ben. 2013. "Promotion and Protection of Human Rights while Countering Terrorism." United Nations A/68/389.

Entous, Adam. 2010. "CIA Drones Hit a Wider Range of Targets in Pakistan." Reuters, May 5. http://www.reuters.com/article/2010/05/06/us-pakistan-usa-cia-idUSTRE6450KT20100506.

Express Tribune. 2014. "European Parliament Condemns Drone Strikes." February 28. http://tribune.com.pk/story/677202/european-parliament-condemns-drone-strikes/.

Fisk, Kerstin, and Jennifer M. Ramos. 2014. "Actions Speak Louder than Words: Preventive Self-Defense as a Cascading Norm." *International Studies Perspectives* 15 (2): 163–185.

Frank, Thomas M. 2009. *Recourse to Force*. New York: Cambridge University Press.
G-8. 2006. "St. Petersburg Declaration on the Middle East." St. Petersburg, July 16. http://en.g8russia.ru/documents/.
Greenwood, Christopher. 2011. "Self-Defense." *Max Planck Encyclopedia of Public International Law*, April. http://opil.ouplaw.com/view/10.1093/law:epil/9780199231690/law-9780199231690-e401?rskey=xMiUjV&result=271&q=&prd=EPIL.
Heyns, Christof. 2013. "Report of the Special Rapporteur on Extrajudicial, Summary or Arbitrary Executions." New York: United Nations General Assembly, September 13. http://justsecurity.org/wp-content/uploads/2013/10/UN-Special-Rapporteur-Extrajudicial-Christof-Heyns-Report-Drones.pdf.
International Court of Justice (ICJ). 1986. "Merits, Case Concerning Military and Paramilitary Activities in and against Nicaragua." The Hague, June 27. http://www.icj-cij.org/docket/files/70/6503.pdf.
International Court of Justice (ICJ). 2004. "Advisory Opinion Concerning Legal Consequences of the Construction of a Wall in the Occupied Palestinian Territory." The Hague, July 9. http://www.refworld.org/docid/414ad9a719.html.
International Court of Justice (ICJ). 2005. "Armed Activities on the Territory of the Congo." December 19. The Hague. http://www.icj-cij.org/docket/files/116/10455.pdf.
Israeli Ministry of Foreign Affairs. 2006. "The Second Lebanon War (2006)." http://www.mfa.gov.il/mfa/foreignpolicy/terrorism/hizbullah/pages/hizbullah%20attack%20in%20northern%20israel%20and%20israels%20response%2012-jul-2006.aspx.
Jennings, Robert, and Sir Arthur Watts. 1992. *Oppenheim's International Law*, 9th ed. Harlow, UK: Longman.
Johnston, Patrick. 2012. "Does Decapitation Work?" *International Security* 36 (Spring): 47–79.
Kittrich, Jan. 2009. "Can Self-Defense Serve as an Appropriate Tool Against International Terrorism?" *Maine Law Review* 61:133–169.
Kooijmans, Pieter. 2005. "Separate Opinion of Judge Kooijmans." The Hague: International Court of Justice. http://www.icj-cij.org/docket/files/116/10463.pdf.
Kuwait News Agency. 2006. "World Efforts Continue to Stop Escalating Israeli Violence in Middle East." July 14. http://www.kuna.net.kw/home/story.aspx?Language=en&DSNO=886234.
Lubell, Noam. 2011. *Extraterritorial Use of Force against Non-State Actors*. New York: Oxford University Press.
Maggs, Gregory E. 2006–2007. "How the United States Might Justify a Preemptive Strike on a Rogue Nation's Nuclear Weapon Development Facilities under the U.N. Charter." *Syracuse Law Review* 57:465–496.
Military and Paramilitary Activities in and against Nicaragua (*Nicaragua v. United States of America*), Jurisdiction and Admissibility. 1984. 392 ICJ REP. June 27, 1986.
Moir, Lindsay. 2011. *Reappraising the Resort to Force*. Oxford, UK: Hart Publishing.
Obama, Barack. 2013. "Remarks by the President at the National Defense University." Washington, DC: White House, Office of the Press Secre-

tary, May 23. https://www.whitehouse.gov/the-press-office/2013/05/23/remarks-president-national-defense-university.

O'Connell, Mary Ellen. 2005. "When Is a War Not a War?" *ILSA Journal of Comparative and International Law* 12 (2): 536–539.

O'Connell, Mary Ellen. 2009. "Combatants and the Combat Zone." *University of Richmond Law Review* 43:845–864.

O'Connell, Mary Ellen. 2010a. "Lawful Use of Combat Drones." Testimony to Congressional Subcommittee on National Security and Foreign Affairs, April 28. https://www.fas.org/irp/congress/2010_hr/042810oconnell.pdf.

O'Connell, Mary Ellen. 2010b. "The Choice of Law against Terrorism." *Journal of National Security Law and Policy* 4:343–368.

O'Connell, Mary Ellen. 2011. "Seductive Drones—Learning from a Decade of Lethal Operations." *Journal of Law, Information and Science*, special issue. DOI: 10.5778/JLIS.2011.21.Oconnell.1.

O'Connell, Mary Ellen. 2012. "Unlawful Killing with Combat Drones: A Case Study of Pakistan, 2004–2009." In *Shooting to Kill: Socio-Legal Perspectives on the Use of Lethal Force*, edited by Simon Bronitt, Miriam Gani, and Saskia Hufnagel, 263–292. Oxford, UK: Hart Publishing.

Osiel, Mark. 2009. *The End of Reciprocity*. Cambridge, UK: Cambridge University Press.

Paust, Jordan J. 2010. "Self-Defense Targetings of Non-State Actors and Permissibility of U.S. Use of Drones in Pakistan." *Journal of Transnational Law and Policy* 19 (2): 237–280.

Plaw, Avery, and Carlos Colon. 2015. "Correcting the Record: Civilians, Proportionality and the *Jus ad Vim*." In *Legitimacy and Drones*, edited by Steven J. Barela. Surrey, UK: Ashgate Publishing, 2015.

Podur, Justin. 2008. "Colombia Assassinates FARC Commander Raúl Reyes." *Colombia Journal*, March 3. http://colombiajournal.org/colombia-assassinates-farc-commander-raul-reyes.htm.

Price, Bryan. 2012. "Targeting Top Terrorists." *International Security* 36 (Spring): 9–46.

Price, Bryan. 2014. "Data." Bryan C. Price: Home Page: Data. http://bryancprice.com/uploads/Terrorist_Group_Leadership_Decapitation__1970-2008.xls.

Quigley, John. 2013. *The Six-Day War and Israeli Self-Defense*. New York: Cambridge University Press.

Ruys, Tom. 2013. *"Armed Attack" and Article 51 of the UN Charter*. New York: Cambridge University Press.

Schmitt, Michael. 2003. "Preemptive Strategies in International Law." *Michigan Journal of International Law* 24:513–548.

Shue, Henry. 2005. "Preemption, Prevention, and Predation." *Philosophic Exchange* 35:5–17.

Svarc, Dominika. 2007. "Anticipatory and Preventative Force under International Law." *Peace Review* 19:217–225.

Today's Zaman. 2012. "Don't Provoke Us, Erdoğan Says in Stern Warning to Syrian Kurds." TodaysZaman.com, July 26. http://www.todayszaman.com/news-287715-dont-provoke-us-erdogan-says-in-stern-warning-to-syrian-kurds.html.

UN News Centre. 2006. "Security Council Must Condemn Israeli Attack, Demand Cessation of Hostilities, Annan Says." *UN Daily News*, July 30. http://www.un.org/apps/ news/story.asp?NewsID=19345&#.Vkok5narSUk.

Urquhart, Conal, and Chris McGreal. 2006. "Israelis Invade Lebanon after Soldiers Are Seized." *Guardian*, July 12. http://www.theguardian.com/world/2006/jul/12/israelandthepalestinians.lebanon.

U.S. Department of Justice. Undated. *White Paper: Lawfulness of a Lethal Operation Directed against a U.S. Citizen Who Is a Senior Operational Leader of Al-Qa'ida or an Associated Force*. Washington, DC: Department of Justice. http://msnbcmedia.msn.com/i/msnbc/sections/news/020413_DOJ_White_Paper.pdf.

Washington Post. 2007. "Turkey Authorizes Iraq Incursion." October 18. http://www.washingtonpost.com/wp-dyn/content/article/2007/10/17/AR2007101700967.html.

Woodward, Bob. 2010. *Obama's Wars*. London: Simon and Schuster.

10

Restricting the Preventive Use of Force

Drones, the Struggle against Non-State Actors, and Jus ad Vim

JOHN EMERY AND DANIEL R. BRUNSTETTER

The idea of preventive force short of war has taken on greater relevance since President Obama was elected in 2008. While Obama rejected Bush's preventive war doctrine—altering the U.S. national security strategy to reflect a more robust understanding of the *jus ad bellum* principle of last resort—he has utilized limited preventive force on a much larger scale (Brunstetter and Braun 2013). His expansion of the drone program illustrates a willingness to use limited force in a preventive manner by adapting the blurred notion of imminence from the Bush doctrine to the targeted killing of individuals (that is, suspected terrorists).[1]

Obama's use of preventive drone strikes reflects the comparative advantages of drones (as opposed to more robust military options) to use force in a more precise manner (targeting terrorists with a reduced risk to civilians), without putting U.S. soldiers at risk. Moreover, as Kenneth Anderson argued during a congressional hearing on drones, the prominent strategic advantage of drones is their ability to provide a "limited, pinprick, covert strike" in order "to avoid a wider war" (2010a, 5).

In this chapter, we address one of the questions posed in the introduction to this volume: To what extent is the United States' use of preventive force a slippery slope to preventive war, and what are the legal and ethical implications of these actions? Our main claim is that drones employed outside the traditional battlefield are a form of limited preventive force aimed at *avoiding* a larger war, but the legal and moral justifications currently provided by the U.S. government (as outlined in the introduction of this volume) are considerably too permissive. Our argument makes two important assumptions. The first is that the preventive use of limited force by a state is justifiable under certain limited circumstances. Building off the work of Whitley Kaufman (2005), who argues that preventive force is permitted in the historical just war tradition, and Michael Doyle (2008), who sketches the conditions wherein a turn to preventive force would be legitimate under the UN Charter, we assert that a state's right to self-defense includes the right to employ limited preventive force.[2] The second assumption, explored in recent scholarly work, is that antiterrorist actions "cannot be well governed within either the law-enforcement paradigm or the war convention" (McMahan 2012, 155) and that the principles of just war do not neatly transfer to instances of limited force such as drone strikes (Brunstetter and Braun 2013). The legal ambiguity, muddled notion of imminence, and misplaced use of just war principles that have characterized the Obama administration's justification of drones point to the need for a hybrid moral framework. It should be calibrated specifically to limited force that lies somewhere between law enforcement and just war, and combines elements of both. We thus employ a hybrid law enforcement–war ethical framework termed *jus ad vim*—the just use of force short of war (Walzer 2006a; Ford 2014; Brunstetter and Braun 2011, 2013).[3]

Our chapter attempts to provide normative guidance for addressing the legal ambiguities surrounding drone use. While numerous scholars in this volume discuss the legal merit of international humanitarian law (the Laws of Armed Conflict) and international human rights law (law enforcement) in governing drone use, the contours of the drone debate point to clear disagreement as to which body of law is most applicable (Abizaid and Brooks 2014). Indeed, the most controversial drone strikes take place in what Michael Walzer (2007) has called the "in-between spaces," which are not zones of war where armies fight, nor zones of peace where police action can be undertaken (such as areas of Yemen, Pakistan, and Somalia).

In proposing a hybrid law enforcement–war ethic, our chapter can be read as being in conversation with Chapter 11, in which Ben Jones and John Parrish ultimately resist the call for a new ethical space between war and law enforcement. Jones and Parrish argue that admitting such a space exists would relax legal restraints on force, thus imposing wartime legal standards that diminish respect for human rights relative to law enforcement standards. While there is certainly merit to such arguments, Jones and Parrish give perhaps too much weight to the law enforcement ethic, given that the in-between spaces exist *because* law enforcement does not function adequately in these areas. When the existence of these in-between spaces is *not* acknowledged and when law enforcement standards are instead imposed where conditions are less than ideal, the populations living in these areas are condemned to suffer under the rule imposed by terrorists. Without adequate law enforcement, the very real threat of future terrorist attacks remains unresolved.

We concur that the laws governing a zone of war should not be imposed wherever terrorists operate, and therefore reject the more permissive laws of armed conflict subscribed to by the Obama administration. However, we argue that elements of the law enforcement ethic can be blended with the *jus ad bellum* principle of last resort to form an ethical framework that morally circumscribes the use of preventive force in these in-between spaces as well as providing for national security concerns.

Our argument unfolds in the following way. Many scholars who discuss targeted killing do not make a clear distinction between preemption and prevention; thus, the two are often blurred, even though they remain categorically different. In the first section, we therefore briefly define what we mean by a preventive drone strike, distinguishing between a strike against an imminent threat (which would be preemptive and very rare) and a strike against a more distant threat—what we term "lagged imminence"—in places where capture is unfeasible. In the second section, we delineate a notion of last resort for *jus ad vim* that draws from both the law enforcement and just war paradigms to circumscribe the scope of drones as a preventive use of force. Thus, we propose three necessary criteria of last resort: a transparent process to define who is targeted and why, with evidence presented against them (through indictment by grand jury or trial *in absentia*); clear evidence that the target poses a demonstrable ongoing threat; and affording those targeted an opportunity to surrender (which entails giving serious consideration to pursuing options that prioritize capture). This notion of last resort would rule out so-called "signature strikes"—killing unidentified men who appear to fit the characteristics of a militant/terrorist—and summary executions for past crimes committed, but would allow for exceptional cases where preventive lethal force could be used. Each of these sections contains illuminating case studies of drone strikes and law enforcement measures to illustrate the relevant points.

Preventive Force and Lagged Imminence

Much of the controversy surrounding drone strikes revolves around the question of imminence. In truth, there is little debate over whether a strike would be permitted if an enemy attack were truly imminent—that is, just about to take place (although, in reality, law enforcement mechanisms are often sufficient).[4] Even in the more restrictive law enforcement setting, lethal force would be justified under such circumstances.[5] Nevertheless, such uses of force would, by definition, be preemptive and not preventive, because they would be a response in self-defense to an *immediate and impending* threat to national security. The challenge of drone policy is that most strikes do not fit this scenario because they are not a response to an imminent threat, but rather are preventive in nature.

Traditionally, an "imminent threat" has been understood to mean "instant, overwhelming, and leaving no choice of means, and no moment for deliberation" (Abizaid and Brooks 2014, 35). But drone strikes against suspected terrorists are different. The United States claims *perpetual imminence* of terrorist threat—that is, it claims that there is *always* the threat of an attack about to be realized, even though one can never be sure of when it will occur. This claim would suggest that we have crossed the threshold of last resort, meaning lethal options would be legitimate. However, rarely do those targeted by drones present such an immediate threat; rather, most would be considered a potential terrorist threat that may emerge someday if he or she acquired the means and opportunity to do so. We term this threat one of *lagged imminence*, meaning that there is a real threat always on the horizon, albeit not immediately. Statesmen lack the ability to pinpoint the precise moment when such a threat will be actualized, but cannot simply ignore it.

Our conception of lagged imminence is far more restrictive than the perpetual imminence currently claimed by the United States, but less restrictive than the Caroline precedent discussed in other chapters in this volume. We understand why David Glazier and Daphne Eviatar (Chapters 6 and 7, respectively) would be hesitant to allow for the preventive use of force against a threat labeled as somewhat imminent, as current U.S. justifications of drone strikes have rendered the term meaningless. Indeed, Glazier views this as a slippery slope toward more killing that undermines global respect for human rights and the rule of law. Nevertheless, the notion of lagged imminence accounts for the extant threat that terrorists can pose, but also recognizes that the Caroline doctrine of 1842, which applies adequately to interstate conflict between opposing militaries, does not fit the current threat of loosely affiliated transnational terrorist networks. Thus, our conceptualization of lagged imminence would be most in line with the "Caroline plus" view discussed in Chapter 9 by Avery Plaw and João Franco Reis. Ultimately, viewing the threat of terrorism through the lens of lagged imminence, as opposed to perpetual imminence, places renewed emphasis on the just war principle of last resort. We will outline the parameters of *jus ad vim* below.

The Law Enforcement and Just War Divide

Scholars clearly are divided on the issue of targeted killings by drone strikes, which do not fit neatly into current legal and moral frameworks. The divide essentially lies between those who believe terrorism is best fought through law enforcement means and those who believe the principles of just war apply. Chapters 5, 6, and 7 in this volume discuss the differences between the legal distinctions of whether the law of armed conflict or international human rights law (IHRL) applies to U.S. drone strikes. While the United States claims the laws of armed conflict generally apply, the United Nations Human Rights Office of the High Commissioner and other NGOs (including Amnesty International, Human Rights Watch, and the International Committee of the Red Cross) claim IHRL is the most applicable body of law (see also Eviatar, Chapter 7). Given this legal disagreement, exploring the broader ethical concerns of drone use can provide insight into how legal norms could evolve to fit the in-between spaces where controversial drone strikes tend to occur.

The main critique of proponents of the law enforcement paradigm is that the individuals targeted are killed without due process or an opportunity to surrender. As Fernando Tesón argues:

> During peacetime, the state can use lethal force only in very limited circumstances, mostly in self-defense or to protect persons from deadly threats. Beyond that, a suspected criminal is entitled to due process and

may not be killed except in execution of a lawful sentence pronounced by a court of law after a finding of guilt. (2012, 45)

Law enforcement has indeed succeeded in fighting terrorism in certain contexts, as will be demonstrated in several of the cases we present below. Drones, however, have contributed to an expansion of targeted killing that goes well beyond the bounds of traditional law enforcement because they are seen as part of an ongoing war against a permanently imminent terrorist threat. If seen through the lens of war, drone use falls under the laws of armed conflict.

The view that current drone use is covered under the laws of armed conflict is echoed in the testimony of Kenneth Anderson, a professor of law at American University's Washington College of Law, during the 2010 congressional hearings on drones:

> The charge has been leveled that this [current drone use] is a violation of international human rights standards. It constitutes extra-judicial execution without having any charges, without having attempted to arrest the person. We respond by saying the person is a terrorist combatant and can be targeted at any point. We are not obligated to try and detain them or to capture them. (Anderson 2010b)

It is important to note that by adopting the just war paradigm, the question of preventive force is all but moot because in the context of an ongoing war, one has the right to use lethal force against most combatants. Although the right to use lethal force must be tempered by the principles of *jus in bello* (just conduct during war) for deciding whether to use force and what level of force to employ—this qualification does not, in our view, provide enough moral constraint on the use of force.

The Limitations of *Jus ad Bellum* and *Jus in Bello* in Governing Drone Strikes

In a 2013 speech defending American use of drones, President Obama referenced just war principles: "We are at war with an organization that right now would kill as many Americans as they could if we did not stop them first. So this is a just war—a war waged proportionally, in last resort, and in self-defense." Echoing the tenets of *jus ad bellum* (justice of going to war), Obama's main ethical claims are that the United States has just cause to use force in self-defense against those who attacked it as well as continue to threaten its citizens or interests *and* that the threshold of last resort has been crossed, meaning all realistic nonviolent options have been exhausted. What is significant about this claim is that it assumes the *jus ad bellum* criteria have been satisfied, and thus imposes the laws of armed conflict on the struggle against al-Qaeda. Doing so has substantial implications with respect to who can be targeted. The U.S. government's turn to just war principles is meant to provide ethical guidelines to curb the use of drones. However, the administration's instrumentalization of just war principles is problematic on multiple accounts. It broadens the use of lethal force in a way that undermines the restrictive power of the last resort criterion and distorts the principle of necessity, while adopting a controversial understanding of the principles of distinction and proportionality.

For one thing, the U.S. claim to use lethal force outside territorially defined battlefields such as Afghanistan, Iraq, or Libya is a controversial interpretation of international law. Imposing the laws of war paradigm onto the global struggle against al-Qaeda and associated forces gives exceptionally wide latitude for U.S. officials to use lethal force. As David Luban argues, this broadening of lethal force "depresses human rights from their peacetime standard to the war-time standard" (2002, 14). Imposing the war paradigm essentially undermines any notion of last resort by giving the United States almost free reign to target anyone deemed an enemy combatant, because almost anything can be justified by claiming military necessity. While the United States invokes the *jus ad bellum* notion of last resort by stating that it prefers to capture terrorists, the notion of imminence it employs is exceptionally vague. Rosa Brooks's account of the leaked 2011 Department of Justice white paper, which details President Obama's legal team's view on the targeting killing of American citizens, illustrates the problematic logic:

> Any person deemed to be an operational leader of al Qaeda or its "associated forces" inherently presents an imminent threat at all times—and as a result, the United States can lawfully target such persons at all times, even

in the absence of specific knowledge relating to planned future attacks. (2014, 94)

This understanding of imminence characterizes it as a permanent state, thus diluting the traditional meaning, namely, that there must be concrete and reasonable suspicion of an impending attack at a specific moment in order to justify a preemptive response.

Instead, the use of drones represents a preventive use of force wherein a "*lack* of knowledge of a future attack [is used] as the justification for using force." Since the United States "'may not be aware of all al-Qaeda plots . . . and thus cannot be confident that none is about to occur,' force is presumed to *always* be justified" against the kinds of people considered likely to engage in attacks if given the opportunity to do so (Brooks 2014, 94; emphases in original). In view of this perception of a permanent imminence, preventive force becomes a matter of military necessity, even though the situation involves what is arguably lagged imminence, which could afford the time for nonviolent measures to be tried first. This situation is where a robust notion of last resort is morally necessary.

In the past, the U.S. drone ethic linking permanent imminence to military necessity has justified the highly controversial policy of signature strikes, as well as the targeting of both key leaders and low-level operatives. This is a wide—too wide—range of preventive force. As Brunstetter and Braun assert:

> Drones forestall the threshold of last resort for larger military deployment, but the last resort criterion does not apply to drone strikes themselves because the targeted killing of (alleged) terrorists becomes the default tactic. Thus, the use of drones as a means to enhance a state's capacity to act on just cause proportionately and discriminately may lead to the propensity to do the opposite. (2011, 346)

Thus, when drones are a default tactic of preventive force, allowing states to act on what they feel are just causes, such an expansion ultimately undermines *jus in bello* principles.

Moreover, the notion of proportionality has been misinterpreted to the point of rendering it meaningless. Drone strikes have been called proportionate compared to the firebombing of Dresden in World War II, but as Braun and Brunstetter argue, this is not a relevant benchmark:

> To claim drones are proportionate because they are more accurate and discriminating than other weapons/tactics is to suffer from what we call *proportionality relativism*—the use of comparisons between drones and other historic eras, weapons or tactics as a basis for establishing proportionality. We contend that such comparisons are misleading. (2013, 305)

The erroneous belief in the inherent proportionality of drones has led to their expanded use, especially in Pakistan during the period of 2010–2011. As Sarah Kreps and John Kaag argue, such use distorts the meaning of proportionality: "Instead of the ends of a war determining the appropriate means of war (which is prescribed in the norm of proportionality), the means of modern warfare are determining objectives" (2012, 278). Hence, more drone strikes translate into elevated risk for collateral damage, which raises serious concerns about the extent to which drones satisfy the principle of distinction, given that it is often difficult to determine who is an appropriate target in nonconventional war. Walzer calls attention to these concerns: "But here is the difficulty: the technology is so good that the criteria for using it are likely to be steadily relaxed. That's what seems to have happened with the U.S. Army or with the CIA in Pakistan and Yemen" (2013). He, too, worries about the restraining mechanism applied to drones. Referencing the "Living under Drones" report, co-authored by Stephan Sonnenberg (Chapter 5), he also warns against the overuse of drones by highlighting the costs they impose upon civilian populations and on America's global image.

The purview of preventive force as understood by the United States sets a dangerous precedent, which, if adopted by other countries, would have grave repercussions for international peace and security (see Fisk and Ramos 2014; Chapters 5–7 of this volume). By assuming that last resort does not apply to addressing the terrorist threat and that drones inherently satisfy the requirements of both discrimination and proportionality, states with drone technology can more easily engage in conflicts they would otherwise have been hesitant to participate in. As articulated in a recent report on U.S. drone policy (Abizaid and Brooks 2014), the United States is traversing a slippery slope toward preventive war. It is worth quoting at length the report, authored by legal, military, and technology experts, concerning UAVs, or drones (as we refer to them throughout the chapter):

> The increasing use of lethal UAVs may create a slippery slope leading to continual or wider wars. The seemingly low-risk and low-cost missions enabled by UAV technologies may encourage the United States to fly such missions more often, pursuing targets with UAVs that would be deemed not worth pursuing if manned aircraft or special operation forces had to be put at risk. For similar reasons, however, adversarial states may be quicker to use force against American UAVs than against U.S. manned aircraft or military personnel. UAVs also create an escalation risk insofar as they may lower the bar to enter a conflict, without increasing the likelihood of a satisfactory outcome. The U.S. use of lethal UAVs for targeted strikes outside of hot battlefields is likely to be imitated by other states. Such potential future increase in the use of lethal UAV strikes by foreign states may cause or increase instability, and further increase the risk of widening conflicts in regions around the globe. (Abizaid and Brooks 2014, 11)

Thus, if the U.S. precedent holds, states will be able to engage more frequently in preventive force for national security reasons while maintaining the appearance of waging moral war. To avoid such a precedent being set, a more restrictive moral framework is needed.

Toward a Hybrid Law Enforcement–Just War Paradigm

The upshot of these concerns is that the Obama administration's turn toward just war principles imposes a paradigm that is too permissive when it comes to using drones for preventive lethal force. Their allure inspires the tendency to use them too often. In light of the concerns raised above, as well as issues raised in recent reports by the United Nations, Amnesty International, and Human Rights Watch (and discussed in detail in other chapters of this volume), it is clear that an alternative moral framework is needed. Clearly, both the law enforcement and just war paradigms have their limitations and setbacks when it comes to engaging the threat posed by non-state actors operating throughout the in-between spaces. However, while proponents of drone strikes argue the law enforcement paradigm is too restrictive, this does not mean that the war paradigm is necessarily the correct choice. What might a framework in between these two look like?

David Glazier explains an alternative logic in his response to drone proponents at the April 2010 congressional hearings on drones:

> One choice is to use this paradigm of the law of armed conflict. But another paradigm is . . . to choose to treat terrorists essentially as pirates, or as terrorists have been treated in the past, using the military under constraints that are much more akin to law enforcement than to law of war.[6]

Glazier's intuition points to an ethical space where traditional law enforcement is too strict to be effective, but the laws of war are too permissive. This is a space where drones are used preventively, and where a hybrid model between law enforcement and the laws of war is needed.

Bearing in mind the problematic use of drones, scholars have argued that the just war principles do not neatly fit this ethical space. Brunstetter and Braun (2013), for example, point out that the ethical principles informing the requirements of *jus ad bellum* are not transferable to all contexts within the international system as the meaning of the principles changes significantly in a context of limited force, and they conclude that a theory of the just use of limited force—*jus ad vim*—is needed. In their development of *jus ad vim*, they recalibrate some of the traditional *jus ad bellum* criteria to fit a limited force context. Thus, they contend that a new principle called the "probability of escalation" is required: "If engaging in limited strikes has a high probability of resulting in escalation, then such actions are not justifiable, even though there may be just cause" (Brunstetter 2013). This reconceptualization of the use of limited force deepens the link between *jus ad vim* and the *jus in bello* principles. It also has clear implications for circumscribing the preventive use of drones. As Rosa Brooks remarks in the conclusion of her study of U.S. drone use and international law, "The international community needs to develop a *jus ad vim* to occupy the space between war and peace: a law and ethics relating to ongoing but discrete smaller scale uses of force" (2014, 99).

Preventive Drone Strikes and *Jus Ad Vim*

In addition to lagged imminence, one of the keys to legitimating preventive drones strikes is to recognize the existence of geographical spaces—the in-between spaces we discussed above—where law enforcement can at times be extremely difficult, or perhaps even impossible. These spaces—parts of Yemen, Pakistan, and Somalia, and perhaps new areas in the future—are those where the respective governments have failed to maintain control over significant portions of their territories and/or lack the will to confront terrorists who thrive there.[7] The use of force in such geographical spaces, as Walzer explains, has a "different feel" than traditional war zones because it occurs in a kind of "darkness . . . outside the moral and legal conventions of ordinary warfare" (2007, 484). While war used to be easily defined as a zone of combat where lethal force was justified (to be distinguished from a zone of peace, where it was not), the struggle against terrorism has created these in-between spaces of moral uncertainty where force is used on a limited scale, but war is not declared.

Sympathetic to the limitations of both the law enforcement and just war paradigms cited above, Walzer asks the critical question regarding the use of preventive force against suspected terrorists: "What standards apply to the secret war against terror?" (2007, 481). The challenge here is in identifying the moral status of those against whom one is fighting. For Jeff McMahan, although terrorists have some elements of both criminals and combatants, terrorists "lack some of the defining characteristics of combatants and are considerably more dangerous than ordinary criminals" (2012, 155). Thus, he concludes that antiterrorist actions "cannot be well governed within either the law-enforcement paradigm or the war convention" (2012, 155). While this would suggest a need for a hybrid model that combines the elements of both paradigms to enable limited uses of force, previous attempts to develop one have selectively deployed elements of both paradigms to loosen the constraints on the use of force and thereby created a framework that is too permissive (see Luban 2002; Coady 2008). Refining the approach, Brunstetter and Braun (2013) argue that the implied permissiveness of such a framework can be circumscribed by clear restraining mechanisms that limit the way a state uses limited force, while also permitting the necessary force to respond to threats. In what follows, we combine elements of both paradigms—specifically, the permissive right of a state to kill members of a terrorist group as one can kill enemy soldiers under the laws of war and the right to due process and restrictive collateral damage policy inherent in the law enforcement paradigm—to establish a notion of last resort for preventive drone strikes in a context of lagged imminence.

Drones and Last Resort

In the previous section, we argued that the U.S. government's notion of imminence is too broad. If we were also to categorically reject the use of force in cases of lagged imminence and use drones only preemptively, then, as Mark Totten recognizes, we would "fail to provide states with the security they require" because "against the new threat of global terrorism the point of last resort may arrive prior to the point of imminence" as it is traditionally understood (2010, 184, 186). But even if Totten is correct in arguing that there is a *narrow* space for the just preventive use of force because the threshold of last resort is crossed before imminence, this still begs the question: What does last resort mean with regard to a preventive drone strike? Stated differently, given the notion of lagged imminence that characterizes the nature of the threat from terrorist organizations, what should be tried before resorting to a preventive drone strike? In this section, we sketch a distinct ethical theory of last resort that is calibrated to the act of targeted killing by drones.

In the just war paradigm, last resort places emphasis on considering all feasible nonviolent measures before resorting to war. The threshold of last resort does not mean that every imaginable resolution must be attempted before resorting to war, because there is always something more to try. Rather, last resort is a marker that all reasonable alternatives have been tried and failed "before you 'let loose the dogs of war'" (Walzer 2004, 155). For Walzer, political leaders must cross that threshold with "great reluctance and trepidation" (2004, 88). How does one capture this "reluctance and trepidation" for a preventive drone strike?[8]

The first part of the process of satisfying last resort for a drone strike is identifying a legitimate target. According to the just war paradigm, the *jus in bello* principles require a state to take due care to avoid civilian casualties. The principles of distinction, proportionality, and necessity regulate (and supposedly limit) the scope of lethal force that can be employed. Civilians cannot be directly targeted, the use of force must not exceed the expected military gain, and lethal force must not be used unless for the sake of military advantage. While the U.S. government claims to satisfy these principles, critics claim that those being killed are summarily executed without due process and that civilians are unjustly placed in the line of fire (Human Rights Watch 2013; Amnesty International 2013). Whether accidental or as a result of the U.S. government's policy of proportionality balancing (the conscious decision that anticipated military advantage outweighs collateral damage), these consequences "on the ground" point to significant moral concerns regarding the American administration's ethical stance.

While no framework can entirely eliminate mistakes or control the reaction of international actors to a specific drone strike, we contend that a more robust, hybrid notion of last resort could satisfy the call for greater protection of civilian human rights across the globe while also satisfying concerns of national security. This would set a more rigorous, human rights-oriented precedent for the United States and for states that may acquire lethal drones in the future. To this end, we propose a notion of last resort that would require the following from states proposing to carry out preventive drone strikes: (1) that the person targeted is known by name; (2) that the level of ongoing threat he or she poses be explicated publicly (the first two are accomplished by public indictment); and (3) that he or she be afforded the opportunity to surrender and capture before a preventive strike is justified.

To begin explaining our rationale, we must answer the question: Why would knowing the identity of the target be so important? This criterion is linked to the *jus in bello* principle of distinction, and a direct reaction to the U.S. policy of signature strikes. Take, for example, one such "signature" drone strike made on August 29, 2012, in the village of Kashamir, Yemen. According to Human Rights Watch (2013), this strike appears to have killed three al-Qaeda in the Arabian Peninsula (AQAP) militants and two civilians, described as "pillars of the community." These two men were Salim Jaber, a cleric who preached against violent Islamist militancy, and his cousin Walid Jaber, one of the village's police officers. The scenario appears to have been that the three AQAP militants had traveled to the village to seek out the cleric and challenge his preaching against them. Perhaps they intended to do him harm, but the drone killed all of them when they met together. The result of this strike was three dead militants whose actual threat to the United States remains unknown (which raises concerns of necessity), and two civilians whose deaths cause immeasurable damage to the fabric of the local community.

Killing a local cleric who both preached against terrorism and was an ally in the fight against AQAP has arguably done more harm to American security than the threat those three unknown militants may have posed. This case thus illustrates the lack of discrimination and necessity that exists with signature strikes. One could find many problems with such a strike—it is difficult for a drone operator in the Nevada desert to determine who is civilian and who is a militant posing an imminent threat, for example—but what is clear is that if a more restrictive notion of last resort that rejects targeting individuals based on a suspicious "signature" been in place, this strike would never have occurred.

Assuming, then, the just use of preventive force is contingent on identifying a legitimate target, what conditions need to be satisfied to determine a legitimate target? One could imagine post-hoc justifications. In 2002, Qaed Senyan al-Harthi, an alleged al-Qaeda leader and mastermind of the U.S.S. *Cole* bombing, was killed in the first purported CIA drone strike outside a declared war zone in Yemen. Kenneth Roth, executive director of Human Rights Watch, did not criticize the killing in his official report because it appeared that al-Harthi was a combatant and reasonable attempts at capture had failed (indeed, eighteen Yemeni police had been killed trying to capture him). The report was nevertheless critical of the U.S. government because the government "made no public attempt" to justify this use of force (Roth 2003). While some scholars suggest that after-the-fact justifications are the best way to assuage the tension between national security and transparency (Walzer 2007; Tesón 2012), we argue that justification must happen before preventive use of force becomes a viable alternative. While we accept portions of the just war paradigm, namely that preventive drone strikes may be just under certain, restricted circumstances, the moral significance of the hybrid-law model rests on adapting key elements of the law enforcement paradigm as well—namely the presumption of innocence and due process.

Our view of last resort thus draws heavily upon the law enforcement paradigm. Before lethal preventive force can be justified, evidence must be presented in a transparent manner that communicates to the public at large—domestically and internationally—the identity of a specific individual and the crime with which he or she is charged. Despite the necessities of national security, the U.S government's turn to "secret kill-lists" and/or public most-wanted lists without publicly disclosed evidence has cast doubt over the CIA drone program, especially where the lack of transparency and sufficient moral checks has opened the way for ever-expanding target lists. Rather than adopting a notion of lagged imminence, which brings the essence of last resort to the fore by recognizing there is time for nonlethal measures to be undertaken, the government has viewed terrorism through the lens of perpetual imminence and thereby legitimated expanding kill-lists to include targets of opportunity, punitive strikes, signature strikes, and strikes to deny safe haven. As discussed in other chapters of this volume, what initially began as the selective and sparse targeted killings of alleged terrorists in 2002 became an expansive preventive drone campaign that had, by 2010, massive human rights consequences and a significant impact on international norms regarding the use of force. While the number of strikes has diminished since 2012, there is no guarantee an increase would not occur should another 9/11-type attack occur.

To avoid the impetus toward target expansion and its negative consequences, which the perceived technological advantages of drones could, in fact, augment, we propose that preventive drone strikes require a formal indictment or trial in absentia of prospective targets. At such public hearings, evidence must be presented before a grand jury.[9] Although this might appear to be an overly restrictive process that would appear too onerous in a counterterrorism setting, such restrictions have been used in combatting terrorism in the past.

For example, after the 1998 embassy bombings in Kenya and Tanzania and following a comprehensive law enforcement investigation, twenty-one men were indicted by grand jury in New York City.[10] Nine of these men were captured and tried in U.S. civilian courts and are either serving life sentences or still awaiting trial. Only three of the men remain fugitives (most prominently Ayman al Zawahiri); the remaining were killed in various incidents (some were killed in drone strikes; others died in Afghanistan, Somalia, or Pakistan). In addition, in the wake of the U.S.S. *Cole* bombing in 2000, the U.S. Department of Justice indicted two of the alleged masterminds on May 15, 2003—Jamal Ahmad Mohammad Al Badawi and Fahd al-Quso (Ashcroft 2003). Al Badawi was captured by Yemeni forces and was eventually released for unknown reasons, yet he remains on the FBI's most wanted list with a $5 million bounty (BBC News 2007). Al-Quso was killed in a CIA drone strike in Yemen on May 6, 2012 (*Guardian News* 2012).

Admittedly, both indictments came in the wake of terrorist attacks against the United States, before the advent of armed drones, and prior to the heightened perceptions of threat that followed 9/11. However, this should not diminish the importance of the indictment process, which can curtail the facile use of preventive force by drones. Moreover, the perception of perpetual imminence that followed 9/11 and imposes the war paradigm on the in-between spaces should not be permanent. It is important to note that an indictment need not be post-facto in nature (for example, after a major attack). Intelligence gathering and cooperation with allies can help identify individuals who might pose a current threat, while members of al-Qaeda who come into leadership positions might also be indicted. Note that even though the indictment process falls short of what is required by the law enforcement paradigm (because an indictment does not warrant execution), it nevertheless reduces the level of killing permitted under the laws of war paradigm by restricting drone strikes to specific individuals who are indicted. But who are these individuals? And what makes them liable to consideration for a preventive drone strike?

The purpose of an indictment process is to disaggregate the image of the enemy into those who are part of the terrorist organization but do not pose an immediate threat (such as low-level militants against whom law enforcement mechanisms must be employed) and those who pose an *ongoing demonstrable threat* and against whom limited preventive force can be used. There is an obvious difference between a driver and a bomb-maker, or between a foot soldier and a prominent leader of al-Qaeda. But what exactly does posing a demonstrable ongoing threat mean? Drawing from the law enforcement paradigm, we can say that the fact that an individual has previously participated in a criminal act does not justify killing him or her. This would be summary execution. Rather, threat is measured by an individual's active role in planning and seeking to carry out future attacks. Evidence of such intentions must be gained by ongoing intelligence gathering, which is enhanced by cooperation among concerned states. We return below to how indictments lay the groundwork for a just drone strike, but before we do so, it is important to state that indictment alone is not sufficient to justify a preventive targeted killing.

Claire Finkelstein (2012) argues that the permission to use preventive force is linked to the legitimacy of issuing a threat to use force that can be acted on *only* if the suspect does not surrender or cannot be captured. Stated differently, when engaged in a nontraditional combat operation in between law enforcement and war, states have a duty to pursue capture, but retain the right to use force if capture is impossible. This stance reflects a blending of the law enforcement and just war paradigms: a state does not have the right to kill any combatant, but only specific individuals who are threats, while these individuals have the right, as per law enforcement, to due process. It is, of course, unlikely that individuals will simply turn themselves in, meaning law enforcement agents or Special Forces units may need to be employed (or ruled out). As Finkelstein concludes, "When the threat is ignored, and the duty to capture cannot be met in other ways, it is then permissible in some cases to follow through on the threat" (2012, 182).

The United States has often stated a preference for capture, as Obama articulated in his 2013 speech at National Defense University: "America does not take strikes when we have the ability to capture individual terrorists; our preference is always to detain, interrogate, and prosecute." However, the perception of perpetual imminence reduces the commitment to capture. While there are clear limitations on capture due to rugged terrain and risk to civilians or law enforcement agents, the belief in perpetual imminence emphasizes a problematic framing of the terrorist threat, wherein a state must take advantage of "windows of opportunity" to eliminate a perceived threat before it materializes (Brennan 2012). The belief that if one does not strike now, the window of opportunity will be closed indefinitely involves an opportunistic view of preventive force that undermines the commitment to capture and the very notion of last resort.

The importance of the duty to capture and the challenges of viewing drone strikes through an elusive notion of "windows of opportunity" are highlighted in the Human Rights Watch (2013) account of the April

17, 2013, drone strike in Yemen that killed an alleged local AQAP leader, Hamid al-Radmi. Residents and security officials said that al-Radmi could have been arrested at any time after he returned to Wessab, Yemen, in 2011, upon his original release from prison. "He was in my office all the time and I could even have gone to his house to arrest him," said one ranking security officer in Wessab who knew al-Radmi. Yet there was never a call for al-Radmi's arrest from the Yemeni or American government. While some government officials argued that his capture would have been impossible, al-Radmi's cousin was critical of this view. Had authorities sought the family's help, he claims, they would have tried to turn al-Radmi over to local authorities. Because relatives play an important role in administering justice in Yemen's tightly knit family and tribal system, such an option is not out of the realm of possibility.

Indeed, failing to pursue such options had negative consequences. One cousin, an elderly farmer named Muhammad Ali Saleh, claims that the killing turned al-Radmi into a martyr:

> They should have taken him to court, brother. Charge him and keep him in prison and even hang him there up and down every day but not kill him like that if he committed a crime. Now people are crying about him everywhere. What does that accomplish? (Human Rights Watch 2013)

The killing of al-Ramdi by a U.S. drone thus begs two distinct questions: When did the window of opportunity for a targeted strike close? And when did the window of opportunity for capture close? There is no easy answer to either question. However, by viewing the threat of terrorism through the lens of lagged imminence, it is reasonable to think that the window of opportunity to kill is much larger than the United States perceives it to be in most cases.

The Hamid al-Radmi case showcases several steps that could have be taken to demonstrate a genuine attempt to capture. First, the case highlights the importance of a public indictment and of providing the possibility for surrender as criteria for last resort. An indictment would have opened a clear window of opportunity for capture by alerting local authorities to the legal justification for arresting the al-Radmi, while also affording him the opportunity to surrender (or convincing his relatives to turn him in). Second, it demonstrates the significance of cooperation with local authorities. When a suspected terrorist's wanted status is made known through indictment, channels of cooperation with local authorities are opened and can be exploited to attempt to capture these individuals. Finally, public indictment broadens the window of opportunity for capture because the legal process makes an immediate drone strike in response to perceived perpetual imminence impossible.

Two Cases: For and Against Preventive Drone Strikes

In order to demonstrate how the principle of last resort might function, we turn to two cases in which the individuals targeted had been indicted for their roles in the 1998 U.S. embassy bombings in Kenya and Tanzania and were therefore known to have ties with al-Qaeda. However, the level of ongoing threat and the feasibility of capture differed in each case.

The first case involves a situation in which the threshold of last resort was not crossed. In October 2013, on the streets of Tripoli, Libya, American troops, assisted by FBI and CIA agents, seized Abu Anas al-Libi, who had been indicted for his alleged role in the embassy bombing in Kenya and was a suspected leader of al-Qaeda. In 2000, New York prosecutors had charged him for his role in conducting "visual and photographic surveillance" of the United States Embassy in Nairobi in 1993 and again in 1995.[11] Prosecutors said in the indictment that "Abu Anas had discussed with another senior Qaeda figure the idea of attacking an American target in retaliation for the United States peacekeeping operation in Somalia" (Kirkpatrick, Kulish, and Schmitt 2013). Despite al-Libi's indictment for a past crime, it was not immediately justifiable to kill him because it was disputed whether he was continuing operations with al-Qaeda or had renounced his membership. Insofar as he was not considered to pose an ongoing threat, the preventive use of lethal force was not permitted. Thus, the window of opportunity for capture was prolonged, and in 2013, Abu Anas al-Libi was captured on the streets of Tripoli, Libya and brought to the United States for trial. While awaiting trial in New York City, he passed away from complications due to liver disease caused by hepatitis C (Deinst and Windrem 2015).

In the second case, the threshold of last resort was crossed, in that the targets posed an ongoing threat and capture was deemed unfeasible. On January 1, 2009, a predator drone strike killed two al-Qaeda militants in Pakistan's tribal areas. The two men killed were Kenyan citizens: Usama al-Kini, described as al-Qaeda's chief of operations in Pakistan, and his lieutenant, identified as Sheik Ahmed Salim Swedan. These two men were indicted in 2000 along with Osama bin Laden for their roles in the 1998 embassy bombings. While the two were on the run for nearly a decade, the United States gathered reliable intelligence that detailed their continuing role in future terrorist plots. Even if we were to grant that an al-Libi–type extraction would have been unlikely to occur because the risk this would pose to civilians, law enforcement agents and/or Special Forces was deemed too great, the opportunity to surrender was still present given a public indictment. These suspects knew that they were wanted and targeted (as did local authorities who were, for whatever reason, unable to pursue them over a substantial period of time), yet they continued to engage in activities that posed an ongoing threat. They were "suspected of overseeing" a September 2008 "deadly suicide bombing at a Marriot hotel in Pakistan's capital," and it could also be inferred, based on their continued allegiance to al-Qaeda, that they were plotting additional terrorist activities (Schmitt 2009). This particular drone strike crosses the threshold of last resort as we have delineated. These men were known to be wanted because evidence had been presented against them through an indictment by grand jury. They were deemed to pose a demonstrable, ongoing threat given their patterns of activity and through updated, reliable intelligence. By virtue of being publicly indicted, they were afforded an opportunity to surrender, while the ability to capture them was severely compromised by the geographical location in which they resided. Although one could imagine prolonging the window of opportunity for capture, last resort, as we see it, does not require doing so.

Conclusion

In order to address the moral obscurity that permeates the in-between spaces where drone strikes take place, we have laid the framework for a theory of last resort for *jus ad vim*—a hybrid law—just war paradigm calibrated for the use of limited preventive force. Current paradigms for preventive force against non-state actors are either too restrictive (law enforcement) or too permissive (international humanitarian law, or the Laws of Armed Conflict). Our discussion of last resort draws from both the law enforcement and just war paradigms and represents an attempt to delimit the ethical guidelines for using preventive force on a limited scale. As argued elsewhere in this volume, the legal application of such a principle poses other challenges. However, the principle of last resort for *jus ad vim* is a key element of a more restrictive moral paradigm calibrated to the use of limited force and justification for drone strikes.

Throughout our chapter we have focused on the use of limited preventive force short of war outside the traditional battlefield. We have asserted that preventive drone use aims to *avoid* a larger war. On the one hand, despite being highly controversial, drone strikes in the in-between spaces of Yemen and Pakistan, have succeed in helping to avoid a larger preventive war. On the other hand, their preventive use has resulted in a sustained—perhaps even perpetual—low-level conflict that has serious human rights implications for civilian population living under drones, as illustrated by Sonnenberg in Chapter 5 of this volume. While the current use of drones by the United States in these contexts does not represent a slippery slope toward preventive war—no one believes that Yemen or Pakistan would go to war with the United States over drone strikes—it does raise questions about whether preventive drone use may lead to an escalation of violence by setting a dangerous precedent for the preventive use of limited force as other countries develop and start to employ similar technologies. This is why developing a clear notion of last resort, in our view, is essential.

The responsibility for deciding when the threshold is crossed ultimately rests in the hands of statesmen and military leaders, but if last resort is to have meaning, then the threshold may be crossed only with "reluctance and trepidation," as Walzer (2004) writes. The three criteria of last resort we delimit—public indictment of a specific individual, demonstrable ongoing threat, and duty to capture—provide the vocabulary for critical reflection on preventive drone strikes on a case-by-case basis. While not ruling out preventive strikes entirely, the principle of last resort challenges the Obama administration's morally problematic default assumption that targeted killings are just, simply by nature of the enemy the United States faces and the presumed accuracy of drones. Our view of last resort marks an attempt to circumscribe the dangerous precedent regarding preventive force set by the United States discussed elsewhere in this volume and represents a critical point of departure for a discussion about how other states contemplating the use of preventive force on a limited scale morally evaluate their options. Ultimately, the United States should seek to reach a state of *jus post vim* by closing these in-between spaces through a continued emphasis on strengthening local law enforcement capabilities rather than risk resorting to preventive force.

NOTES

1. We use the term "drone" to reflect the terminology now commonly used in the media and in public speeches by U.S. officials, although other terminology, including "unmanned aerial vehicle" (UAV), is frequently employed in military circles.
2. For a more robust discussion on the debate of the legitimacy of preventive measures, see Buchanon and Keohane (2004) and Reichberg (2007).
3. The ethical framework of *jus ad vim* is not without its critics. Prominent philosopher C. A. J. Coady's (2008) main critique of *jus ad vim* is that it dangerously lowers the standards of last resort from the traditional *ad bellum* framework, which is why in this chapter we seek to establish a clear standard of last resort *ad vim*. Thus, we examine how the principles of *jus ad vim*, in particular a revised notion of last resort, can shed light on the question of a just preventive drone strike.
4. Take, for example, a case in 2012 in Yemen where a combination of cooperation with local authorities, increased security conditions after unusual "chatter" on the airways, and good old-fashioned double agents infiltrating the ranks of al Qaeda led to the unmasking of a terror cell planning attacks on Western targets and the dismantling of a local terrorist cell. What is significant about this case is that an effective law enforcement system was capable of apprehending the suspected terrorists. This may not always be feasible for the in-between spaces where the reach of law enforcement is limited. See Harris 2012.
5. For example, the two al-Shabaab members who were killed by Kenyan police as they attempted to bomb a power station in the town of Mandera in May of 2014. See Wabala and Otsialo 2014.
6. A transcript of the congressional hearing entitled "Rise of the Drones II: Examining the Legality of Unmanned Targeting" can be found at http://www.fas.org/irp/congress/2010_hr/drones2.pdf.
7. The United States has pointed to these characteristics as a justification for using drones (Brennan 2012). Also, Avery Plaw and João Franco Reis in Chapter 9 of this volume discuss countries deemed unwilling or unable to stop terrorism within their borders in relation to the evolution of customary law. David Glazier in Chapter 6 believes the unwilling or unable distinction to be an over-simplistic formulation that significantly misstates the law.
8. The development of our ideas was deeply influenced by Henry Shue's (2007) discussion of preventive force. While not discussing drones explicitly, Shue reaches four necessary conditions, the first three of which have important implications for preventive drone strikes. The attack must be: limited to the direct elimination of

the threat, undertaken only when military action is urgent, based on solid intelligence, and be substantively multilateral.
9 Due to space constraints, we leave aside questions regarding who should preside over such deliberations. Ideally, responsibility should fall with international organizations; individual state claims uncorroborated with clear evidence should be met with suspicion.
10 *United States v. Usama bin Laden et al.*, S(9) 98 Cr. 1023 (LBS) (New York, 2000), http://cns.miis.edu/reports/pdfs/binladen/indict.pdf.
11 A full copy of the indictment of Abu Anas al-Libi can be found under "Grand Jury Indictment of Abu Anas al-Libi," *New York Times*, October 5, 2013, http://www.nytimes.com/interactive/2013/10/05/world/africa/06libya-document.html?_r=0.

BIBLIOGRAPHY

Abizaid, General John P., and Rosa Brooks. 2014 "Recommendations and Report of the Task Force on U.S. Drone Policy." Washington, DC: Stimson Center, June. http://www.stimson.org/images/uploads/ task_force_report_final_web_062414.pdf.
Amnesty International. 2013. "Will I Be Next? US Drone Strikes in Pakistan." London: Amnesty International, October. https://www.amnestyusa.org/sites/default/files/asa330132013en.pdf.
Anderson, Kenneth. 2010a. "Rise of the Drones: Unmanned Systems and the Future of War." Written Testimony Submitted to Subcommittee on National Security and Foreign Affairs, Committee on Oversight and Government Reform, US House of Representatives. Subcommittee Hearing. March 23, 2010. 111th Cong., 2nd sess. Digital Commons, American University, Washington College of Law. http://digitalcommons.wcl.american.edu/cgi/viewcontent.cgi?article=1002&context=pub_disc_cong.
Anderson, Kenneth. 2010b. U.S. Congress, House of Representatives, Subcommittee on National Security and Foreign Affairs. "Rise of the Drones: Unmanned Systems and the Future of War." Hearing before the Subcommittee on National Security and Foreign Affairs, 111th Cong., 1st sess., March 23. http://www.gpo.gov/fdsys/pkg/CHRG-111hhrg64921/html/CHRG-111hhrg64921.htm.
Ashcroft, John. 2003. "Remarks of Attorney General John Ashcroft: Indictment for the Bombing of the *U.S.S. Cole*." Washington, DC: Department of Justice, May 15. http://www.justice.gov/archive/ag/speeches /2003/051503agremarksusscole.htm.
BBC News. 2007. "*U.S.S. Cole* Plotter Freed by Yemen." October 27. http://news.bbc.co.uk/2/hi/middle_east/7065074.stm.
Braun, Megan, and Daniel R. Brunstetter. 2013. "Rethinking the Criterion for Assessing CIA-Targeted Killings: Drones, Proportionality, and *Jus ad Vim*." *Journal of Military Ethics* 12 (4): 304–324.
Brennan, John O. 2012. "The Ethics and Efficacy of the President's Counterterrorism Strategy." Washington, DC: Wilson Center, April 30. http://www.wilsoncenter.org/event/the-efficacy-and-ethics-us-counterterrorism-strategy.
Brooks, Rosa. 2014. "Drones and the International Rule of Law." *Ethics & International Affairs* 28 (1): 83–103.

Brunstetter, Daniel. 2013. "Syria and the Just Use of Force Short of War." *Ethics & International Affairs* blog. New York: Carnegie Council, September 24. http://www.ethicsandinternationalaffairs.org/2013/syria-and-the-just-use-of-force-short-of-war/.

Brunstetter, Daniel, and Megan Braun. 2011. "The Implications of Drones on the Just War Tradition." *Ethics& International Affairs* 25 (3): 337–358.

Brunstetter, Daniel, and Megan Braun. 2013. "From *Jus ad Bellum* to *Jus ad Vim*: Recalibrating Our Understanding of the Moral Use of Force." *Ethics & International Affairs* 27 (1): 87–106.

Buchanan, Allen, and Robert O. Keohane. 2004. "The Preventive Use of Force: A Cosmopolitan Institutional Proposal." *Ethics & International Affairs* 18 (1): 1–22.

Coady, C. A. J. 2008. *Morality and Political Violence*. New York: Cambridge University Press.

Crawford, Neta C. 2003. "The Slippery Slope to Preventive War." *Ethics & International Affairs* 17 (1): 30–36.

Dienst, Jonathan, and Robert Windrem. 2015. "Suspected Plotter of U.S. Embassy Attacks Abu Anas al-Libi Dies in New York." *NBC News*, January 2. http://www.nbcnews.com/news/world/suspected-plotter-u-s-embassy-attacks-abu-anas-al-libi-n278866.

Doyle, Michael W. 2008. *Striking First: Preemption and Prevention in International Conflict*. Edited, with an introduction by Stephen Macedo. Princeton, NJ: Princeton University Press.

Finkelstein, Claire. 2012. "Targeted Killing as Preemptive Action." In *Targeted Killings: Law and Morality in an Asymmetrical World*, edited by Claire Finkelstein, Jens David Ohlin, and Andrew Altman, 156–183. Oxford, UK: Oxford University Press.

Fisk, Kerstin, and Jennifer M. Ramos. 2014. "Actions Speak Louder than Words: Preventive Self-Defense as a Cascading Norm." *International Studies Perspectives* 15 (2): 163–185.

Ford, S. Brandt. 2013 "*Jus ad Vim* and the Just Use of Lethal Force Short of War." In *Routledge Handbook of Ethics and War: Just War Theory in the 21st Century*, edited by Fritz Allhof, Nicholas Evans, and Adam Henschke, 63–75. New York: Routledge.

Guardian News. 2012. "U.S. Airstrike Kills Top al-Qaeda Leader in Yemen." May 7. http://www.theguardian.com/world/2012/may/07/us-airstrike-kills-al-Qaeda-leader-yemen.

Harris, Paul. 2012. "FBI Urges Renewal of Surveillance Measures after Foiled al-Qaeda Plot." *Guardian*, May 9. http://theguardian.com/world/2012/may/09/fbi-surveillance-measures-al-Qaeda.

Human Rights Watch. 2013. "'Between a Drone and al-Qaeda': The Civilian Cost of U.S. Targeted Killings in Yemen." New York, October 22. http://www.hrw.org/node/119909/section/7.

Kaufman, Whitley. 2005. "What's Wrong with Preventive War? The Moral and Legal Basis for the Preventive Use of Force." *Ethics & International Affairs* 19 (3): 23-38.

Kirkpatrick, David D., Nicholas Kulish, and Eric Schmitt. 2013. "U.S. Raids in Libya and Somalia Strike Terror Targets." *New York Times*, October 5. http://www.

nytimes.com/2013/ 10/06/world/africa/Al-Qaeda-Suspect-Wanted-in-US-Said-to-Be-Taken-in-Libya.html?pagewanted=all&_r=0.

Kreps, Sarah, and John Kaag. 2012 "The Use of Unmanned Aerial Vehicles in Contemporary Conflict: A Legal and Ethical Analysis." *Polity* 44 (2): 260–285.

Luban, David. 2002. "The War on Terrorism and the End of Human Rights." *Philosophy and Public Policy Quarterly* 22 (3): 9–14.

McMahan, Jeff. 2012. "Targeted Killing: Murder, Combat or Law Enforcement?" In *Targeted Killings: Law and Morality in an Asymmetrical World*, edited by Claire Finkelstein, Jens David Ohlin, and Andrew Altman, 135–156. Oxford, UK: Oxford University Press.

Obama, Barack. 2013. "Remarks by the President at the National Defense University." Washington, DC: White House Office of the Press Secretary, May 23. https://www.whitehouse.gov/the-press-office/2013/05/23/remarks-president-national-defense-university.

Reichberg, Gregory M. 2007. "Preventive War in Classical Just War Theory." *Journal of the History of International Law* 9 (1): 3–33.

Roth, Kenneth. 2003. "Justice and the 'War' Against Terrorism." Human Rights Watch, January 6. http://www.hrw.org/fr/news/2003/01/05/justice-and-war-against-terrorism.

Schmitt, Eric. 2009. "2 Qaeda Leaders Killed in U.S. Strike in Pakistan," *New York Times*, January 8. http://www.nytimes.com/2009/01/09/world/asia/09iht-09pstan.19209938.html.

Shue, Henry. 2007 "What Would a Justified Preventive Military Attack Look Like?" In *Preemption: Military Action and Moral Justification*, edited by Henry Shue and David Rodin, 222–247. New York: Oxford University Press.

Tesón, Fernando R. 2012. "Targeted Killing in War and Peace: A Philosophical Analysis." In *Targeted Killings: Law and Morality in an Asymmetrical World*, edited by Claire Finkelstein, Jens David Ohlin, and Andrew Altman, 403–434. Oxford, UK: Oxford University Press.

Totten, Mark. 2010 *First Strike: America, Terrorism, and Moral Tradition*. New Haven, CT: Yale University Press.

Wabala, Dominic, and Manase Otsialo. 2014. "Kenya: Police Kill Two Al-Shabaab Members in Manala." *All Africa*, May 3. http://allafrica.com/stories/201405050038.html.

Walzer, Michael. 2004. *Arguing about War*. New Haven, CT: Yale University Press.

Walzer, Michael, 2006a. *Just and Unjust Wars: A Moral Argument with Historical Illustrations*, 4th ed. New York: Basic Books.

Walzer, Michael. 2006b. "Regime Change and Just War." *Dissent* 53 (3): 103–108.

Walzer, Michael. 2007. "On Fighting Terrorism Justly." *International Relations* 21 (4): 480–484. doi:10.1177/0047117807083073.

Walzer, Michael. 2013. "Targeted Killing and Drone Warfare." *Dissent Magazine*, January 11. http://www.dissentmagazine.org/online_articles/targeted-killing-and-drone-warfare.

11

Drones and Dirty Hands

BEN JONES AND JOHN M. PARRISH

The period known as the "War on Terror" has prompted a revival of interest in the problem of "dirty hands": the idea that political actors frequently confront choices where they cannot fulfill their responsibilities to the public welfare without violating ordinary moral obligations to avoid violence, deception, and similar forms of wrongdoing (Walzer 1973). This renewed interest is hardly surprising given that the September 11 attacks led policymakers anxious to prevent another attack to reconsider the morally troubling tactics the literature about dirty hands addresses. Several motifs from the dirty hands literature have migrated into public discourse—in particular, the image of the anguished public leader committed to protecting the public's safety, but seemingly forced into distasteful acts of violence to achieve this noble end. Popular news stories regularly characterize President Obama as personally wrestling with the insoluble moral dilemmas inherent in an "attempt to apply the 'just war' theories of Christian philosophers to a brutal modern conflict," yet ultimately resolving these conflicts in favor of the more aggressive course of action (Becker and Shane 2012).

It is easy to understand why the notion of dirty hands appeals to political actors concerned with polishing the president's image. More surprising, however, is that some political theorists have begun to adopt a similar interpretation of contemporary targeted killing policies. Notable in this regard is political theorist Stephen de Wijze (2009), who contends that a policy of using preventive (lethal) force against terrorist actors is (under specified but not uncommon circumstances) an instance of a dirty-handed moral dilemma (see also Kaag and Kreps 2014). The identification of the practice of targeted killing with the problem of dirty hands carries far-reaching implications that are morally problematic and, we argue, reveal the dangers of such an approach.

In the first two sections of this chapter, we argue that, while dirty hands situations exist as a persistent problem of political life, it is a mistake to classify current policies of targeted killing as an example of dirty hands. Instead, as we argue in the third section, if these policies are to be justified at all, they must meet the more exacting standards of either the just war ethic or the law enforcement ethic, in particular the requirement (with limited and defined exceptions) that noncombatants be immune from violence. In the fourth section, we review a proposal by Michael Walzer (and by John Emery and David Brunstetter in Chapter 10 of this volume) that a third, "in-between" ethic is needed to accommodate the gap in circumstances under which the just war and law enforcement ethics apply. We conclude, however, that such an ethic is unnecessary and that it potentially undermines existing traditions and law in place to restrain force. The chapter's fifth section argues that the concept of dirty hands can prove insightful for moral analysis of some targeted killing scenarios, but that applying it to ongoing, established policies, rather than specific emergencies, opens the door to domains that the concept was never intended to address. In the sixth section, we examine the ethical problems of accountability associated with using drones, or unmanned aerial vehicles, for targeted killing (we opt for the more commonly used term "drone"). In the conclusion, we suggest some implications from the case of targeted killing for assessing the relationship of the dirty hands literature to the ethics of war more generally.

Dirty Hands

In his article "Targeted Killing: A 'Dirty Hands' Analysis," Stephen de Wijze argues that a policy of targeted killing "can indeed be justified" under certain circumstances, but nevertheless "also remains morally wrong and leaves a moral remainder that pollutes those who authorize, plan and execute it." This is what he calls a "dirty hands action" (2009, 308). De Wijze's underlying motivation in advancing this argument appears to be laudable. By appropriating the language of dirty hands, with its implication of a residue of moral pollution, he denies that those participating in targeted killings can wholly escape moral blame. His adoption of the dirty hands framework also permits him to argue that some dimension of targeted killing cannot be other than wrong, and consequently that "a policy of [targeted killing] must be adopted only with the greatest reluctance and as rarely as possible" (de Wijze 2009, 318). Important aspects of de Wijze's use of the dirty hands framework do seem valid. Some features of the moral quandary faced by government officials result from "the immoral and evil acts (or projects)" of the terrorist organizations themselves, including their unwillingness to engage in conventional struggle (precluding their easy identification as combatants) and their targeting of innocent civilians in peaceful contexts (making the imminence of the threat posed difficult to ascertain) (de Wijze 2005, 2009). Yet his central claim, that the idea of dirty hands is an appropriate description for the ongoing *practice* of targeted killing, is problematic.

De Wijze's argument that the practice of targeted killing qualifies as an example of dirty hands does not explain the *specific* moral principles violated by the practice. Instead, his analysis focuses on the regret and moral anguish that those engaged in targeted killings should experience. While these factors are often by-products of dirty-handed choices, they are not by themselves constitutive of what gives such choices their distinctive moral ambiguity. If regret and anguish were the only distinguishing aspects of dirty hands, the morally legitimate use of lethal force for self-defense in war or law enforcement would qualify as a dirty-handed action whenever the individual carrying out the killing felt moral anguish afterward. There may be good reason to desire that soldiers and law enforcement officers experience anguish over the use of lethal force, but classifying all who feel anguish as actors with dirty hands renders the concept overly broad and robs it of its specific explanatory power.

What makes the concept of dirty hands distinctive is its identification of a specific moral conflict between two real yet incompatible moral values or obligations, where no available action enables the agent to avoid violating a deeply held moral principle. In such circumstances, the necessity of choosing one of the available actions, and the correlative necessity of violating a contrary but still valid moral principle, generates what Bernard Williams (1973) termed a "remainder" of moral wrong incommensurable with, and therefore not wholly made up for by, the good vindicated by the choice. Not all instances of moral conflict rise to this level of complexity. In some cases, moral conflict may exist without yielding a particular ethical remainder: the violation of the moral principle is present, but is internally justified by moral values commensurable with those being compromised. For instance, in seeking to pass the Thirteenth Amendment to the Constitution abolishing slavery, Abraham Lincoln arranged for wavering Democrats to be promised government jobs and other favors in exchange for their votes and concealed this fact from public view on the grounds that it might scuttle the amendment. Despite the dubious nature of these actions, Lincoln could argue that the exclusionary impact of slavery (among its other harms) undermined democracy, and that the degree of prima facie corruption involved in his side deals were *internally justified*—fully made up for—by the same values they compromised (Parrish 2010). In other words, such forms of moral conflict really constitute cases of prima facie obligations being overridden by competing moral considerations of greater force.

Dirty-handed actions, however, go further than this. They signify a real difference in kind between the competing moral values at stake—an inability to resolve the conflict by appeal to a common moral standard. In the true dirty hands situation, whatever one does, there is a morally significant remainder to be accounted for. That is, the moral principle violated is incommensurable with the competing moral principle that motivates the violation, such that the wrong caused by the one cannot be made up for by the good of the other. This degree of moral conflict may not rise to the level of a truly tragic dilemma involving such brutal alternatives that there exists no discernable best action (as we arguably find in Sartre's [2007] dilemma of the soldier who must either forsake the resistance or abandon his ailing mother). But even when we can say with reasonable assurance that our choice is the best possible one given the circumstances, there remains, where dirty hands exist, a real wrong for which we are accountable.

The problem with de Wijze's analysis of targeted killing is vagueness: he does not specify precisely which aspects of moral conflict make it an instance of dirty hands. He distinguishes targeted killing from forms of political assassination (aimed at civilian leaders to promote regime change or other political gains), and instead defines targeted killing as the killing of individuals who (1) pose an "imminent threat"; (2) have "a proven record of actively planning and/or executing terrorist attacks against civilians" as well as a perceptible intention to continue to do so; and (3) present officials with "no realistic possibility of preventing such attacks by nonlethal methods and bringing the perpetrators before a proper court of law" (2009, 307–8). These broad criteria encompass a host of actions that seem to be distinct from each other in morally significant ways. For example, a law enforcement officer who kills a terrorist driving a vehicle laden with explosives toward a crowded building of civilians meets each of de Wijze's separate criteria: it is the killing of an individual posing an imminent threat, with a clear intent to harm civilians, where nonlethal means are not feasible for stopping the threat. Yet we recognize this case as a paradigm example of the legitimate use of force within a domestic law enforcement context. This case differs importantly from the dirty hands example offered in de Wijze's 2009 article, which details the Israeli military's targeted killing of Hamas leader Salah Shehada, resulting in fourteen civilian fatalities. According to de Wijze, any use of force meeting his broad criteria for a targeted killing should count as an instance of dirty hands. But as these two examples show, key morally relevant features of a targeted killing—such as whether it results in noncombatant casualties—offer a basis for differentiating among alternative categories of targeted killings, with some being legitimate, some morally questionable, and some morally blameworthy. By ignoring key features for evaluating these distinct cases, de Wijze mistakenly collapses many varieties of this practice under the general category of dirty-handed dilemmas.

Emergency Ethics

In contrast to de Wijze's approach, we believe the best way to understand the concept of dirty hands is as a dilemma that characteristically arises in emergency situations that are difficult to foresee and plan for. A promising place to explore this connection is the work of Michael Walzer, since he repeatedly addresses (with varying degrees of directness) the relationship between dirty hands and emergency ethics. In his influential treatment of just war theory, *Just and Unjust Wars*, Walzer does not appeal directly to the idea of dirty hands, though he does approach closely to it in his chapter on "Supreme Emergency." Walzer argues that situations combining an imminent threat with enormously consequential stakes—no mere danger, but catastrophe "of an unusual and horrifying kind"—may create a situation of "supreme emergency" in which "one might well be required to override the rights of innocent people and shatter the war convention" (2000, 259). Walzer illustrates the kind of scenario he has in mind by referencing Britain's decision to bomb German cities in 1940, when Britain teetered on the edge of defeat at the hands of the Nazis—a situation that fulfilled the criteria of both imminence (defeat was at hand) and outsized consequences (not just defeat, but surrendering to the inhumanity of the Nazi system). He contrasts this situation, where the emergency does seem genuinely supreme, with Britain's decisions to bomb German cities later in the war (when British defeat was no longer a realistic possibility) and the American decision to bomb Hiroshima and Nagasaki (where a negotiated peace might have been had if the Allies had not demanded unconditional surrender by Japan). In his later essay, "Emergency Ethics," Walzer (2004) explicitly identifies a supreme emergency in war as a specific instance of the problem of dirty hands.

Some commentators, notably C. A. J. Coady, have puzzled over an apparent shift in Walzer's thinking on dirty hands, noting that his original characterization in "Political Action" included such seemingly mundane cases as campaign finance corruption, which could never approach the degree of justificatory burden he establishes for a supreme emergency. Coady notes that in "Political Action" Walzer's argument "seemed much closer to a utilitarianism of extremity with the extreme being nowhere near the limit set by supreme emergency," since even in his most dramatic illustration, the ticking-bomb case, the stakes, while high, amounted to the lives of at most a few hundred innocents, not the survival of a free way of life (2008, 84–85; see also Coady 2014). In "Emergency Ethics," however, Walzer repeatedly invokes the image of dirty hands in describing the moral status of the person who authorizes mass murder of noncombatants in a supreme emergency. Coady suggests this represents a change in Walzer's view, one that substantially narrows the scope and raises the stakes for when dirty hands justifications are valid.

We believe this alleged shift is better explained by interpreting Walzer's supreme emergency as an instance of dirty hands, but not the only instance. Bombing cities is one possible kind of dirty-handed action, presumably at the farthest extreme of justificatory burden, with bribing a local boss to win an election being another instance closer to our regular experience. Walzer does suggest different thresholds for invoking dirty hands in each of these cases, but not because his view about the broad category has shifted. It is more likely that the differing thresholds reflect a divergence in the harm contemplated (one needs a better justification for bombing a city than for a bribe). Admittedly, Walzer contributes to the ambiguity, in particular by his statement that "dirty hands aren't permissible (or necessary) when anything less than the ongoingness of the community is at stake, or when the danger that we face is anything less than communal death" (2004, 46). On closer examination, however, Walzer's actual point appears to be that the form of the problem of dirty hands *that typically arises in cases of war*—namely, violating the just war protections for noncombatants—requires a higher threshold to be reached. This claim is fully consistent with the possibility (indeed, likelihood) that a lower standard is required for the commission of lesser evils.

Thus, Walzer's considered view appears to be that the deliberate killing of noncombatants as a direct, instrumental means to one's end (as opposed to a foreseen, but not directly intended, side-effect) requires the highest conceivable standard of justification—a supreme emergency. This standard is higher than other varieties of dirty hands, such as bribery or torture (though Walzer still requires a compelling justification in these cases, too). For the remaining cases, we have what Walzer calls "the war convention"—that is, the traditional rules of *jus in bello*,[1] including the requirement of proportional force and the principle of noncombatant immunity. Drawing this distinction between dirty hands broadly understood and emergency ethics more narrowly construed allows us to note the crucial point of contrast between Walzer's view of dirty hands and de Wijze's: Walzer explicitly rules out the idea that the dramatic action demanded by an emergency could ever become the normal state of affairs or that a dirty-handed action could ever become an ongoing *policy*.

The very nature of Walzer's concept of supreme emergency implicitly confines its applicability to a temporally limited context. "Even in wars where the stakes are very high," Walzer cautions, "they may not be so high at every moment in the course of the war as to bring the supreme-emergency argument into play. Each moment is a moment-in-itself; we make judgments again and again, not once for each war" (2004, 46). Most importantly, this caveat means that emergency ethics (the form of dirty hands associated with taking human life) cannot ever be permitted to become rationalized, bureaucratized, made a matter of habit: "We must resist the routinization of emergency, reminding ourselves again and again that the threats we force others to live with, and live with ourselves, are immoral threats. . . . This is the essential feature of emergency ethics: that we recognize at the same time the evil we oppose and the evil we do, and that we set ourselves, so far as possible, against both" (2004, 49). Dirty hands is thus the wrong terminology to use to describe an ongoing policy that provides for the intentional taking of life under specified and predictably recurring conditions. Rather, that is what the principles governing the use of force in law enforcement and just war are for: to mark out the limits within which the necessary evil of killing can and cannot be *justifiably* pursued.

Walzer clearly intends his conception of dirty hands (and the related topic of emergency ethics) as a distinct category from the ordinary actions a soldier or domestic law enforcement officer takes in accordance with the conventions governing force in the spheres of war and peace, respectively. This distinction does not imply that there is no relation between dirty hands and the violence associated with war or law enforcement. War in particular, writ large, may at bottom constitute a kind of dirty-handed dilemma, an awful necessity posed by the nature of violence and public order.[2] Similarly, the overall choice to enter into a specific war, particularly a non-defensive war, might frequently meet many of the criteria for a dirty-handed dilemma. But within the context of a specific war, if the rules of conduct associated with war have any meaning at all, they claim to mark out territory within which soldiers may follow their directives with moral safety, knowing that actions taken in accord with both the rules' spirit and their letter will be internally justifiable moral actions (see Parrish 2010).

Just War and Law Enforcement

The previous sections highlight the problem with applying the concept of dirty hands too readily to war, for which there is already a well-developed body of ethical theory to assess its practices. The concept of dirty hands describes moral conflict at the margins, where we experience friction between spheres of value in unexpected and incommensurate ways. Just war theory, however, expects moral conflict in its midst. It *begins* with a recurring experience of moral conflict, the kind arising in war, and purports to map out the terrain so that soldiers and commanders may each follow a path with relative ethical safety. Perhaps just war theory is wrong to assume that such a mapping is possible (McMahan 2009), but this is undoubtedly what the theory supposes itself to be doing.

In this sense, just war theory describes the ethical dimension of a *practice*—soldiering and commanding in war—that is of ongoing relevance to a regrettable sphere of human life. In the domestic context, a similar practice exists—law enforcement—and it too is guided by an ethical theory. We will call these theories the *just war ethic* (JWE) and the *law enforcement ethic* (LEE), respectively. For both these areas, the ethical theories in question offer a basis for guiding the practices they circumscribe along morally permissible paths. If the practice itself is morally justified and the ethic properly describes the normative permissions and prohibitions that apply *systematically* within the practice, then the practitioner—the soldier, the police officer—may refer directly to the practice's ethic with confidence, without needing in ordinary circumstances to go beyond it. As John Rawls (1955) argues in his essay "Two Concepts of Rules," prescriptions that arise *within* a morally justified practice can claim a kind of day-to-day insulation from the broader, all-things-considered judgments that grander ethical theories—such as consequentialism and deontology—invite. We can critique the practice itself from whatever theoretical vantage point we find most compelling. We can also question which practice applies, or ought to apply, to our present circumstance. What we cannot do, in Rawls's view, is critique the applicability of the rules from a standpoint located within the practice itself. We may debate what just war theory should prescribe in circumstances of war and whether this really is a circumstance of war. But *as* soldiers, we do what the code of soldiers prescribes.

In the case of targeted killing via drones, the two ethics, the JWE and LEE, present potential guides. The JWE, as enshrined in international humanitarian law, requires that military actions respect the following principles: (1) only combatants are legitimate intended targets in military operations; (2) there must be reasonable certainty in distinguishing between combatants and noncombatants when carrying out attacks against combatants; (3) the force used against combatants must be proportional to the threat; and (4) the military advantage gained from an attack must outweigh any unintended harm that the attack inflicts on civilians (Melzer 2008a, 2008b; Walzer 2000; also see Chapters 5–7 and 8–10, this volume). This ethic does not place an outright prohibition on knowingly causing noncombatant casualties; instead, it emphasizes basic protections for noncombatants and seeks to minimize harm to them.

The LEE, as reflected in international human rights law, serves as an alternative moral framework for evaluating drone strikes outside traditional combat contexts. In contrast to the JWE, the LEE puts forward the following more stringent criteria governing the use of force: (1) only imminent threats to life permit the use of lethal force; (2) there must be certain identification of a threat before using force against him or her; (3) lethal force is justified only when nonlethal measures are not feasible for stopping a threat; and (4) any use of force must avoid the foreseeable risk of civilian casualties.[3] The LEE prioritizes guaranteeing due process and the presumption of innocence to those suspected of wrongdoing over swift action against them. Therefore this ethic greatly restrains the use of lethal force, permitting it only in instances where it is absolutely necessary to stop an imminent threat to life (Melzer 2008b; also see Eviatar, Chapter 7 in this volume).

Some drone strikes constitute relatively easy cases, where it is clear which ethic applies. Contemporary militaries now employ armed drones in conventional warfare. In a conventional war, a military is justified in using drones against targets that do not present an imminent threat, as long as these operations adhere to the principles of the JWE. Drones may be a new technology, but there is nothing inherently unjust about the way they kill enemy combatants. The same criteria used to evaluate other military operations—such as a cruise missile strike—apply to drone strikes in conflict zones (Alston 2010; Strawser 2010).

There is greater moral ambiguity, however, when drone strikes against suspected terrorists occur in areas that are not conventional conflict zones—such as Pakistan, Yemen, and Somalia—since the JWE's more permissive rules for armed attacks would not ordinarily apply to these areas. The United States has authorized over five hundred strikes in these regions since 2002, when the targeted killing of suspected terrorists by drones started becoming an entrenched practice of U.S. foreign policy (Bureau of Investigative Journalism 2015). How we morally assess the U.S. program of targeted drone strikes hinges on whether the JWE or LEE should guide these operations.

The United States has made its position clear: the policy of targeted killings falls under the JWE and is consistent with international humanitarian law. According to both the Bush and Obama administrations, the United States is engaged in an ongoing non-international armed conflict[4] against al-Qaeda and its affiliates. Congress's 2001 Authorization for the Use of Military Force provides legal legitimacy for these ongoing strikes against suspected terrorists and extends beyond the battlefields of recognized war zones (Iraq and Afghanistan) to the more ambiguous cases of zones experiencing periodic conflict (Pakistan, Yemen, and Somalia) (Brennan 2011; U.S. Department of Justice undated; Holder 2012; Koh 2010; Obama 2013). Many in the international community, however, express skepticism toward the administration's legal rationale for its program of targeted killings. In particular, critics find the U.S. government's expansive interpretation of its current conflict with al-Qaeda—a conflict without geographic or obvious temporal limits—to be unprecedented and dangerous (Brumfield and Morgenstein 2013).

If the LEE should govern targeted killings by the United States outside of traditional conflict zones, these strikes clearly lack legitimacy. Under the LEE, targeted killing is justified only in those rare cases when lethal force is necessary to incapacitate an imminent threat to life, and never as a punitive response (Alston 2010; Melzer 2008b; Eviatar, Chapter 7). U.S. targeted killings by drones often fall short of the LEE's imminent threat criterion. As Chapter 1 in this volume highlights, the United States has sought to circumvent this limitation by adopting a definition of imminence that undermines its basic meaning (also see Fisk and Ramos 2014). A leaked Department of Justice legal memo shows that, under U.S. policy, "the condition that an operational leader present an 'imminent' threat of violent attack against the United States does not require the United States to have clear evidence that a specific attack on U.S. persons and interests will take place in the immediate future" (U.S. Department of Justice undated, 7). A pattern of planning attacks—not active participation in a specific future plot—is sufficient to qualify an individual as an imminent threat.

There are similar shortcomings in the application of the LEE's certainty standard. In addition to "personality strikes"—drone strikes against named terrorist suspects—the CIA carries out "signature strikes" against individuals based on their pattern of behavior, though the targets' identity is unknown (Klaidman 2012). Since U.S. officials do not even know the identities of the targets of signature strikes, this practice clearly violates the LEE's requirement that officials identify a target with high certainty before authorizing a strike.

In addition, the very nature of drone strikes—killing at a distance by machine—often puts in jeopardy any hope of meeting the LEE's requirement to use nonlethal measures in stopping a threat and to opt for lethal measures only as a last resort. General James Cartwright, former vice-chair of the Joint Chiefs of Staff, points out that a suspected terrorist facing a drone strike has little opportunity to surrender when there is no one on the ground in conjunction with a drone (McKelvey 2012). Defenders of U.S. drone policy argue that strikes occur only in areas where it is impossible for troops to seize suspected terrorists. Yet the risk associated with putting boots on the ground, along with the controversy inherent in incarcerating and trying suspected terrorists on U.S. soil, create strong political incentives to choose drone strikes as a first rather than last resort (Klaidman 2012).

Finally and importantly, the CIA drone program violates the LEE's—and arguably also the JWE's—noncombatant protection standard by authorizing strikes that entail likely civilian casualties. One example occurred in North Waziristan in Pakistan on March 17, 2011, when missiles hit a large gathering for a *jirga*, a tribal assembly for resolving disputes. Under the LEE, the presence of militants at this gathering in no way justifies the strike, given the large number—between nineteen and forty-one—of foreseeable civilian casualties. Also, so called "double tap" strikes—a follow up strike in the same location as an initial strike—often kill civilians, especially first responders and mourners who rush to the scene of a strike (International Human Rights and Conflict Resolution Clinic and Global Justice Clinic 2012; Sonnenberg, Chapter 5, this volume). Thus evidence suggests that the U.S. drone policy regularly violates all four of the LEE's criteria, as well as the less stringent JWE standards on occasion.

Walzer's "In-Between Zones"

Michael Walzer in recent writings offers an alternative approach to the problem of targeted killing outside traditional combat zones that avoids some of the difficulties of de Wijze's position, but raises other concerns. For Walzer, targeted killings are justified in some instances, but at the same time he finds the lack of restraints on the current U.S. drone program troubling. In particular, Walzer (2013) points out that the U.S. drone program as currently practiced regularly violates the just war principle of proportionality—a concern obscured by the U.S. practice of counting all military-age males killed in a strike as militants unless explicit evidence proves their innocence (Becker and Shane 2012). But Walzer's criticism goes beyond concerns that current U.S. drone policy is not living up to the principles of just war theory, suggesting instead that, for targeted killings outside of conflict zones, the JWE is not the proper perspective for evaluating these strikes. He does not object to the term "war on terror," but argues that in such a war governments normally should follow the rules governing law enforcement:

> Though the risks are larger in the "war" against terror than they are in the "war" against crime, I believe that the first of these can be conducted—certainly we should try to conduct it—within the moral and constitutional constraints that [hold in a zone of peace]. The details of the constraints have to be negotiated, of course, and they are negotiated through ordinary democratic processes. And it is entirely legitimate that sometimes they will be less restrictive with regard to what the police can do and sometimes more so. But the basic principles of morality and constitutionalism should be defended, even in hard times. (2007, 481)

Here Walzer gestures at a legal and moral framework for evaluating antiterrorism activities such as targeted killings that closely follows the LEE in many respects but could deviate from it in particular instances.

When Walzer specifically discusses the practice of targeted killings of suspected terrorists (2006, 2007, 2013), he never sets out a systematic framework for evaluating targeted killings outside of a war zone. Walzer at times expresses wariness about the potential hazards of developing a new framework distinct from either the LEE or JWE (2007). But together his recommendations appear to suggest that conditions in areas such as Pakistan and Yemen do require applying a novel moral and legal framework that synthesizes these two existing ethics governing the legitimate use of force.

To reconstruct Walzer's proposal, we begin by examining what he believes is an instance of a justified targeted killing. He approvingly cites a strike against al-Qaeda militants in the desert of Yemen in 2002.[5] This strike, according to Walzer, occurred in a context that does not fit neatly into either a war zone or a zone of peace: "Yemen lies somewhere between Afghanistan and Philadelphia. It is not a zone of war where armies fight, and it is not a zone of peace where the police do their work. The state's writ does not run in the desert of South Yemen" (2007, 481). Walzer appeals to the fact that certain presumptive features underlying the LEE do not apply to "in-between zones" such as Yemen. For instance, in the strike Walzer cites, the Yemeni government had previously tried to capture the militants, and it was only when these efforts failed that the United States resorted to a targeted killing.[6] When police work fails to stop those actively engaged in terrorist activities, Walzer implies, Special Forces are justified in adopting measures more in line with just war principles.

The example from Yemen shows that, under Walzer's proposed ethic for in-between zones, targets need not constitute an imminent threat to justify a strike. Walzer cites no evidence that the targets in Yemen posed an imminent threat in the sense that we normally understand "imminent"—"close at hand in its incidence; coming on shortly" (*Oxford English Dictionary*). The targets' suspected involvement in terrorist activities could qualify them as imminent threats according to the Department of Justice's understanding of the term, which explicitly does not "require . . . clear evidence that a specific attack . . . will take place in the immediate future" (undated, 7). But this definition, by dropping the requirement of immediacy, takes us far afield from traditional understandings of imminence. If Walzer were to retain the requirement that threats must be truly imminent to justify strikes in the in-between zones, he would have to point to a specific immediate attack disrupted by the strike in Yemen. Instead, he justifies the strike on the grounds that attempts to capture the targets proved unfeasible. Walzer also suggests relaxing, within these in-between zones, the LEE's strict requirement for nonlethal force whenever feasible—especially in cases where pursuing nonlethal options would put Special Forces in greater danger.

Though Walzer suggests some targeting criteria for the in-between zones similar to those governing the use of force in war, he emphasizes that protections for noncombatants in these zones remain similar to those applying in a zone of peace. Specifically, Walzer proposes two limits on the practice of targeted killing for in-between zones that are more demanding than what the JWE requires. First, he states, we must attain a higher standard of certainty than required under the JWE: to perform a targeted killing in an in-between zone, we must "be as sure as we can be, without judge or jury, that the people we are aiming at are really Al Qaeda militants or, more generally, that they are engaged in planning and carrying out terrorist attacks." Second, Walzer claims, the noncombatant protection requirement must also be strengthened: "We have to be as sure as we can be that we are able to hit the targeted person without killing innocent people in his (or her) vicinity" (2006, 11). To stop an imminent threat in an in-between zone, Walzer reluctantly concedes that noncombatant casualties may be permissible, but cautions that crossing this line easily can lead to brutality that cannot be justified (2007).

Thus, Walzer sets forth targeting criteria for in-between zones that combine principles from the JWE and LEE. With respect to (1) legitimate targets, only individuals actively engaged in terror activities count as legitimate targets in the in-between zones (a standard closer to the JWE requirements since the target does not need to be an *imminent* threat). With respect to (2) certainty, there must be certain identification of a threat before using force against him or her (corresponding to the LEE requirement). Regarding (3) the permissible degree of force, the force used against suspected terrorists must be proportional to the threat (the requirement under the JWE). Finally, in terms of (4) risk to noncombatants, any use of force must avoid the foreseeable risk of civilian casualties, except in emergencies (a slightly relaxed variation on the LEE standard). Table 11.1 shows how Walzer's in-between zone proposal compares with the requirements of the JWE and LEE.

TABLE 11.1 Comparison of the Criteria Governing the Use of Force

	Just War (War Zone)	Law Enforcement (Zone of Peace)	Walzer's Proposal (In-Between Zone)
Legitimate Targets	Combatants	Imminent threats only	Individuals actively engaged in terrorist activities but not necessarily imminent threats
Certainty	Reasonable certainty in distinguishing combatants from noncombatants	Certain identification of threat	Certain identification of threat
Degree of Force	Proportionality	Always nonlethal measures, except when not feasible	Proportionality
Risk to Noncombatants	Principle of double effect limits the acceptable risk of noncombatant casualties	Avoid the foreseeable risk of noncombatant casualties	Avoid the foreseeable risk of noncombatant casualties (except in emergencies)

Note: Shading denotes which criteria in Walzer's proposal correspond with the just war ethic and which correspond with the law enforcement ethic.

On its surface, Walzer's proposal appears to offer a sensible compromise. For in-between zones where U.S. drone strikes are common, Walzer recommends that we "maneuver between our conception of combat and our conception of police work, between international conflict and domestic crime, between zones of war and peace" (2006, 12). This compromise, however, proves to be more far-reaching than it seems. Through it, Walzer does not merely critique U.S. officials for incorrectly applying existing principles governing the use of force. He further suggests that the principles of the JWE and LEE are inadequate to provide guidance on the proper use of force against suspected terrorists. In other words, concern over the U.S. practice of targeted killings stems partly from the need to define a new practice altogether—the use of force in the in-between zones—with its own distinct rules. There is no category in international humanitarian law or international human rights law corresponding to the in-between zones that Walzer discusses. International law prohibits the targeted killing of a non-imminent threat outside of a war zone, for example, whereas Walzer's proposal permits such action in certain instances. Walzer's compromise, then, requires carving out a new theoretical category.

Admittedly, the two ethical traditions governing the use of force are not set in stone: like any practices, their rules and norms have changed over time.[7] Indeed, Avery Plaw and João Franco Reis in this volume make the case that international norms regarding targeting killing currently may be undergoing a transformation. The temptation for creating a third category is understandable, since dis-analogies between the domestic context and the more conflict-ridden territories Walzer refers to are not negligible. Domestic terrorism occurs in a context where the rule of law is strong, which fosters an environment favorable for law enforcement working to stop terrorist activities. The planning, preparation, and execution of global terrorism, however, often take place in distant locations where the reach of domestic law enforcement is weak. Success in countering international terrorism therefore depends on the ability of domestic law enforcement agencies to cooperate with their counterparts overseas, who may be ineffective or unwilling to assist. When law enforcement finds itself powerless to act in such instances, this challenge might be thought to justify relaxing the restrictions imposed by the LEE and adopting the alternate criteria suggested by Walzer for in-between zones.

Yet it is important to balance this perceived need for a new "in-between ethic" with a sober understanding of the implications of adopting it. Walzer's proposal would weaken the (theoretical) protections against force enjoyed by civilians and suspects in areas that, under the current dichotomous framework, generally are characterized as non-conflict zones. Perhaps a partial step toward the JWE is better than holding onto the LEE when it proves impotent to deliver its promises of security. Daniel Brunstetter and John Emery in this volume propose such a partial step, which places specific restraints on the use of force that go beyond Walzer's account of in-between zones. But it is important to remember that, even as the JWE places constraints on the devastating force brought to bear by war, within these constraints its effects on human beings remain no less devastating. For this reason, any conceptual step toward expanding the zone of war is to be accepted only with considerable caution.

The key step in Walzer's argument comes in drawing the conclusion that when law enforcement's efforts are stymied, the LEE's moral force no longer holds, opening the door to systematically permitting lethal force as a response to foreign terrorism in contexts where it would ordinarily be forbidden. The problems with this conclusion can be seen by considering an analogy to domestic terrorism. In the domestic context, where the LEE clearly applies, there is a strict prohibition against police exercising lethal force as a punitive response to someone suspected of terrorism. In the absence of an identifiable imminent threat, the use of lethal force by the police strikes us as unwarranted. Even if police have an opportunity to exact punitive measures directly against a suspect, we instead require them to follow the law, take the suspect into custody, and let the judicial process take its course. This conclusion still holds in cases where we have doubts that the justice system can be trusted to succeed in its objectives. For instance, suppose all the evidence points to the guilt of a suspected domestic terrorist, yet it is known that she is likely to get off on a technicality or never be brought to trial because certain witnesses are too fearful to testify. This scenario is not far-fetched, especially given the history of domestic terrorist groups like the Ku Klux Klan, which used violence and intimidation to operate with impunity in the Jim Crow era, when acts of domestic terrorism (such as lynching) occurred with far greater frequency than today (Equal Justice Initiative 2015). Despite such failures in the justice system, we remain wary of discarding the LEE to allow lethal force to punish past actions, due to its risk of further escalating violence.

For these reasons, we resist Walzer's implication that in-between zones require a new ethic distinct from the LEE or JWE (see Daphne Eviatar's chapter in this volume, which takes a similar position). Certainly, there is undeniable value in Walzer's analysis of these cases, for it brings into focus a set of challenges that arises outside of traditional war zones and may pose peculiar moral quandaries for those committed to following the prescriptions of the LEE and JWE. Where terrorism makes peace precarious, the impetus is greatest to resort to lethal force in situations that fall short of satisfying the criteria demanded by the LEE. The danger in choosing this path, however, lies in its broader impact on communities and noncombatants. Systematically relaxing the restraints on the use of force against suspected terrorists puts not only these suspects at greater risk, but also the communities of innocent men, women, and children where the suspects reside. In fact, such an approach to in-between zones has the effect of empowering terrorists to bring conditions approaching war wherever they go, even areas where, though the rule of law may be less than ideal, many of the characteristics of domestic stability and peace still obtain.

It is precisely this consequence of the United States' choice to define its counterterrorism efforts against al-Qaeda and its affiliates as a war without geographic limits that raises concerns. As Steven Ratner (2007) argues, by understanding this conflict as a *global* war that extends to wherever al-Qaeda is found, the United States pursues a policy that makes in-between zones both more vulnerable and more prevalent. Indeed, these areas have suffered the most under the CIA covert drone program and, unsurprisingly, strikes in these areas have proven the most controversial (International Human Rights and Conflict Resolution Clinic and Global Justice Clinic 2012). Walzer's proposal does seek to place limits that are currently absent from the CIA program. But the implications of his proposal still tend to weaken noncombatant protections in ways that are difficult to justify categorically, given the strong resemblance that in-between zones continue to share with non-war zones.

Across a variety of contexts, there should be a strong presumption to hold onto the protections of the LEE. The mere presence of a dangerous individual domestically does not compel us to abandon the LEE, nor should this dynamic alone lead to a different conclusion for a faraway land. The principles of *jus ad bellum* articulate demanding criteria that must be met before ever bringing war into a community (Johnson 1999; Walzer 2000; Yoder 1996). In particular, the principle of proportionality demands that any analysis of the expected harms and benefits of going to war cannot give privileged consideration to one's fellow civilians, but must give equal weight to foreign civilians whom the war could harm.[8] Due to the often-disastrous effects of war on civilians and the tendency of violence to escalate, the proportionality requirement sets a high bar to justify war. Furthermore, even if this bar is met, it is a mistake to claim that for entities as diverse and amorphous as terrorist organizations, a nation can make a broad proclamation of war against them once and then pursue war wherever these organizations are found. This perspective casts away important restraints on war and invites an expansive conflict. Decisions regarding war must be sensitive to conditions on the ground and made separately for distinct locales, on a case-by-case basis. Any attempt to systematically discard the LEE wherever the challenge of terrorism arises misses this critical point.

The LEE places more stringent restrictions on targeted killings and the use of force generally than the JWE, but it is not impotent in the face of terror and possesses resources for addressing it. Within Walzer's in-between zones, we observe that it is more likely that the third condition for a targeted killing—infeasibility of nonlethal measures—will be met than in standard domestic contexts because law enforcement is weak in such areas. Yet this observation by itself does not imply a weakening of the LEE's normative force. On the contrary, the LEE still holds for in-between zones, and officials retain a prima facie obligation to pursue nonlethal responses whenever such measures have a reasonable prospect of success. The implication of the purported distinction between zones of peace and in-between zones is in reality much more modest: if justified targeted killings of suspected terrorists become somewhat more plausible for in-between zones, that is only because the lack of effective law enforcement in these areas may tend to make capture and other nonlethal responses to *imminent* terrorist threats less feasible.

Potential Dirty Hands Dilemmas Raised by Drones and Terrorism

Our approach, therefore, begins by adopting the standpoint of the relevant ethic (JWE or LEE, depending on the context) as a starting point, and only then engages in adaptations of its strictures based on specific features of the situation at hand. Here the traditional concept of dirty hands again becomes relevant, particularly in emergencies where the morally obligatory features of the applicable ethic might potentially be overridden by competing consequentialist considerations. In this section, we focus on examples of what we take to be actual dirty-handed dilemmas, in order to highlight the truly hard cases that could arise when considering the targeted killing of a suspected terrorist. These examples offer a sharp contrast to the systematic abuses that often result from categorical exemptions to the LEE or JWE.

The complexity of terrorism makes it impossible to put forth a comprehensive list of potential dirty-handed actions related to the practice of targeted killings outside traditional combat zones. Nevertheless, below we review four brief hypothetical scenarios set in one of Walzer's in-between zones. In each example, one of the standard criteria for a targeted killing under what would normally be the default ethic outside traditional combat zones (the LEE) is not met. By eliminating one of the ordinarily required conditions of the LEE while leaving the other three intact, we can observe the moral stakes in each scenario more systematically.[9]

Example 1: Condition Requiring Presence of an Imminent Threat Is Not Met

A certain terrorist suspect is a charismatic leader with a history of enlisting new recruits and inspiring attacks against civilians—trends expected to continue as long as he remains in his leadership role. Nevertheless, there is no evidence that the terrorist suspect is currently participating in a specific plot endangering civilian lives. Capturing the leader through traditional law enforcement methods has proven impossible, which tempts officials to authorize a targeted strike against the leader based on his past actions and likelihood of similar future actions. Clearly, a strike in such a situation runs afoul of the criteria set forth by the LEE, which requires an imminent threat, not past actions, to justify a targeted killing. But because of the terrorist suspect's unique authority, some officials argue that they need to put aside the imminent threat requirement in this instance. Killing this individual will significantly damage the terrorist organization's capabilities and limit its ability to carry out future attacks.

Example 2: Condition Requiring Certainty of Target's Identity Is Not Met

An obscure terrorist group succeeds in carrying out several devastating attacks on civilians, and fear grips the country that more attacks are coming. The attacks catch intelligence officials off guard. Because officials do not have a long history of tracking this group, they lack in-depth knowledge of its members and operations. Officials succeed in piecing together parts of this terrorist network, but significant gaps in their knowledge remain. Based on chatter they have heard, intelligence officials believe with a high degree of certainty that more attacks are planned. Yet intelligence officials are less certain about the identity of the network's leaders and who should be priority targets in efforts to stop another attack. Given this imperfect information, any strikes would carry the significant risk of killing individuals unaffiliated with the terrorist network or who occupy relatively minor roles within it. Officials must decide whether the imminence and scale of the threat warrants relaxing the certainty requirement in this instance.

Example 3: Condition Requiring the Necessity of Lethal Force Is Not Met

Intelligence officials learn that a terrorist cell in a remote region is close to making operational a potentially devastating plot directed at civilians in a large city. The country where this terrorist cell resides wants to stop them just as much as the United States. In fact, the United States has participated with this country on joint operations to capture other suspected terrorists in its network. But the two countries have carried out these operations with varying degrees of success. Some operations have resulted in the successful capture of terrorist suspects. Other operations, however, have been bloody and cost both countries lives. Moreover, the success of this terrorist group in repelling external forces has helped its legend grow and brought in more recruits. Attempting to capture members of this terrorist cell through nonlethal means is an option, but based on past experiences some officials argue that targeted strikes are necessary to ensure that the plot is stopped, as well as to reduce projected military casualties and to preclude costly prestige and recruitment gains by the terrorist group.

Example 4: Condition Prohibiting Foreseen Civilian Casualties Is Not Met

For weeks, officials have been carrying out surveillance on a terrorist suspect. Officials have established that he is planning a chemical weapons attack in a populated area. Due to various factors, capturing the suspect proves impossible in his current location. The attack is imminent, so officials want to move forward with a strike. Members of the suspect's family are always near him, however. Those observing the suspect have tried to find an opportunity to strike when other family members are not in the immediate vicinity, but such an opportunity has yet to arise. Officials have to make a decision: carry out the targeted killing, which almost certainly will cause the death of noncombatant family members of the terrorist suspect, or allow the terrorist plot to proceed, which could result in high numbers of civilian casualties.

Whether a targeted killing in any of the above scenarios would be justified depends on the precise nature of the threat and the harm caused by forgoing a strike when not all the necessary conditions are met. Arguably the list of plausible cases of dirty hands could include examples when two or more conditions of the controlling ethic do not obtain, but as more conditions are not met, the required justification for permitting the legitimate use of lethal force becomes increasingly more demanding. For all dilemmas of dirty hands, the moral burden of proof rests with those employing violence to show that emergency circumstances compelled them to break the rules that normally bind them. The public has good reason to be skeptical of a dirty hands justification for a targeted killing, given how often governments offer dubious post hoc justifications for violations of moral rules. In any evaluation of dirty hands situations, it is critical to remember that by their nature such dilemmas ordinarily arise in emergency contexts. If advocates of a policy repeatedly have to resort to dirty hands arguments to defend implementation of it, that is a strong prima facie reason to suspect the policy. Dirty hands justifications may sometimes succeed in preserving the legitimacy of actions that normally would constitute ethical violations, but such justifications never can redeem an ongoing policy that routinely disregards basic moral principles.

Drones and Accountability

In his original essay on dirty hands, Walzer (1973) emphasizes the dangers of political actors stepping outside established moral frameworks to justify their actions. When a political actor breaks a moral rule, the worry is that she or he will become increasingly accustomed to wrongdoing. If nothing checks this behavior, the political actor may embrace the mindset that in service of the greater good, she or he can break moral rules with impunity. Because of this concern, Walzer suggests that political leaders who dirty their hands should face punishment. Walzer himself recognizes the challenges of implementing this proposal, but it conveys the idea that we must hold accountable leaders who dirty their hands for the public good. Ethical leaders embrace this aspect of the dirty hands situation: they recognize the legitimacy and binding nature of their moral and legal responsibilities, even as outside factors force them to regrettably forsake those responsibilities. By contrast, a political system that lacks specific mechanisms for accountability invites political actors to permanently claim and abuse emergency powers. That worry is particularly acute in the case of targeted killings. It is a short journey between the practice of targeted killings and government assuming the power to assassinate at will. It therefore is wise to be wary of attempts to legitimize an expansive program of targeted killings.

Political incentives, however, encourage expanding rather than limiting the U.S. practice of targeting killings. Current U.S. drone policy plays well in the court of domestic public opinion—though unsurprisingly, U.S. drone strikes are less popular abroad (Pew Research Center 2014). Equally important, this military technology allows the United States to wage war against suspected terrorists while avoiding what usually is an inevitable consequence of war—casualties among one's own soldiers. The actual human costs of drone strikes are easier for the media and public to ignore. From the perspective of electoral politics, then, there is little reason for politicians to curb drone strikes: successful strikes advance politicians' ambitions, while inaccurate strikes carry little risk of backlash.[10]

Is it possible to restrict targeted killings outside of war only to cases that pose an imminent threat to human life? Admittedly, we are only a few years into the experiment of using drone technology for targeted killings, but thus far the evidence casts strong doubts on government's ability to use this power in a way that avoids the excesses to which it is prone. Protocols for the current U.S. drone program are not even public. For the foreseeable future, no plausible scenario exists whereby the strikes themselves might be subject to any form of retrospective public review.[11] This lack of accountability has led to an expanded U.S. policy of targeted strikes that frequently kills and maims noncombatants and instills an atmosphere of fear in areas where strikes are most common. One Pakistani man, who lost his legs in a drone strike, poignantly describes the human toll of current U.S. policy: "Everyone is scared all the time. When we're sitting together to have a meeting, we're scared there might be a strike. When you can hear the drone circling in the sky, you think it might strike you" (International Human Rights and Conflict Resolution Clinic and Global Justice Clinic 2012, 81; see also Al-Muslimi 2013). The severe trauma experienced in communities impacted by drone strikes makes clear the urgent need to significantly limit the scope of this policy. In the absence of dramatic reforms, the inevitable conclusion seems to be that the capabilities drones provide are too tempting to insulate from systematic abuse. If that is the case, the most prudent step, suggested by Sonnenberg (Chapter 5), may be a broad policy restricting the use of drones for targeted killings to genuine combat zones.

Conclusion

Ethical analysis invites us to use our imaginations. It tends to draw our attention away from ordinary experience and toward the frontier of the unknown, to fix our gaze on situational outliers and eccentric hypothetical scenarios. In doing so, ethical analysis attunes us to recognize complexities in our moral experience that we might otherwise overlook. The politician dirtying his or her hands is such a case. But it is important when exploring these frontiers that we do not forget the distance we have traveled to arrive at them and that we do not import the ethic of the frontier to a context where it does not apply. As we have seen, targeted killing in the context of counterterrorism may present genuine dirty hands dilemmas. But this possibility in particular instances is insufficient to justify a *policy* that regularly violates fundamental moral and legal principles. It is hard not to reach the conclusion that dirty hands justifications for the U.S. drone program often serve as post hoc rationalizations for policies of dubious ethical standing.

Terrorism poses a special problem for ethical analysis because it seeks to import some of the conditions of guerilla or concealed warfare into otherwise peaceful contexts. The appropriate ethic for responding to terrorism where we encounter it directly is the ethic of war; but we rarely confront terrorism directly, for it conceals itself. Its obscene purpose is to subdue us to a life of flinching at the mere shadows of its terrors. If it induces us to lash out indiscriminately in response, to transpose the ethic of war to times and places of peace, all the better: that will admirably suit its end. And all the while, if it ever catches us unprepared, it gains the chance to start the whole dreadful cycle once more.

How to respond to terrorism is a matter of genuine ethical difficulty, and the notion of dirty hands captures its family resemblance with other familiar moral challenges that pervade public life. But we must not let these genuine challenges eradicate the ethical progress we have made. The codes of conduct restricting violence in the enforcement of law and in war reflect an ethical frontier never too far away, especially in counterterrorism efforts that often blur the lines between zones of war and peace. But even if terrorism is war made more insidious through concealment, it is still no *more* than war. If we know the ethic appropriate to war, terrorism cannot give us reasons to enact a policy that systematically goes *beyond* that frontier ethic. Indeed, it will frequently require us to take one or more steps back from that ethic of extremes, to rely on the ethic of law enforcement, until such time as war itself is indisputably at hand.

NOTES

1 *Jus in bello* refers to the system of ethical requirements pertaining to the conduct of war (such as proportionality and noncombatant immunity); it is to be distinguished from the ethical requirements of *jus ad bellum* pertaining to the decision to go to war (such as just cause and last resort). See Walzer 2000.
2 A view along these lines seems to be present in Augustine's just war theory, and more recent echoes are perceptible in Max Weber's political theory. See Parrish 2007 and Weber 1958.
3 One exception to the prohibition on foreseeable civilian casualties under the LEE is those rare cases when civilians find themselves trapped by a criminal's threat—as in a hijacked airplane headed toward a populated area—where stopping the threat regrettably entails the deaths of civilians within it.
4 The United States uses the term "non-international armed conflict" to specify that it is not at war with another state (what would be an *international* conflict), but rather at war with a non-state actor.
5 In separate articles, Walzer (2006, 2007) gives slightly different details to describe what appears to be the same strike. Walzer likely is referencing the first U.S. targeted killing by a drone outside a declared war zone, in which a CIA drone killed six people traveling by car in the desert in Yemen on November 3, 2002. Among the dead was al-Qaeda leader Ali Qaed Senyan al-Harithi (Melzer 2008b).
6 For the purpose of reconstructing Walzer's position, we do not question his account of the facts of the 2002 CIA strike in Yemen. It is important to point out, however, that the example is messier than Walzer implies. This strike, carried

out without consulting domestic law enforcement, killed a U.S. citizen, Kemal Darwish (Woods 2012).
7 For example, originally the rules of warfare required submarines to surface before they could attack, but these rules have since been abandoned (Yoder 1996).
8 A spokesman for the U.S. Conference of Catholic Bishops makes exactly this point when questioning the legitimacy of the current U.S. program of targeted killing (Pates 2013).
9 Some counterterrorism actions share features of the hypothetical scenarios outlined, and in particular there are potential parallels between the first scenario and the killing of Osama bin Laden. Nevertheless, we deliberately use hypothetical dirty hands dilemmas, rather than real-life examples, due to the secrecy surrounding most targeted killings by drones. Since key, morally relevant details often are missing in these cases, using them as examples presents difficulties for illustrating the dirty hands dilemma. We thus rely on hypothetical examples to show with greater precision different manifestations of the dirty hands dilemma that targeted killings can take.
10 Illustrating this point is the reported obsession with drone strikes of Rahm Emanuel as Obama's chief of staff. Anxious to learn of and publicize successful strikes, Emanuel recognized the political upsides of the drone program for the president (Klaidman 2012).
11 The U.S. Senate rejected a modest proposal to require the president to make public each year the number of people killed and injured in U.S. targeted killing operations (Mazzetti 2014).

BIBLIOGRAPHY

Al-Muslimi, Farea. 2013. "Drone Wars: The Constitutional and Counterterrorism Implications of Targeted Killing." Testimony before the United States Senate Judiciary Committee—Subcommittee on the Constitution, Civil Rights and Human Rights. Washington, DC, April 23.

Alston, Philip. 2010. "Report of the Special Rapporteur on Extrajudicial, Summary or Arbitrary Executions." New York: United Nations Human Rights Council, May 28. http://www2.ohchr.org/english/bodies/hrcouncil/docs/14session/A.HRC.14.24.Add6.pdf.

Becker, Jo, and Scott Shane. 2012. "Secret 'Kill List' Proves a Test of Obama's Principles and Will." *New York Times*, May 29. http://www.nytimes.com/2012/05/29/world/obamas-leadership-in-war-on-al-qaeda.html.

Brennan, John. 2011. "Obama Administration Counterterrorism Strategy." Speech at John Hopkins School of Advanced International Studies, Baltimore, June 29. http://www.c-span.org/video/?300266-1/obama-administration-counterterrorism-strategy.

Brumfield, Ben, and Mark Morgenstein. 2013. "Drones Killing Innocent Pakistanis U.N. Official Says." CNN.com, March 15. http://www.cnn.com/2013/03/15/world/asia/u-n-drone-objections.

Bureau of Investigative Journalism. 2015. "Covert Drone War." London. http://www.thebureauinvestigates.com/category/projects/drones/.
Coady, C. A. J. 2008. *Morality and Political Violence*. New York: Cambridge University Press.
———. 2014. "The Problem of Dirty Hands." In *The Stanford Encyclopedia of Philosophy*, edited by Edward Zalta. Stanford, CA. http://plato.stanford.edu/entries/dirty-hands/.
De Wijze, Stephen. 2005. "Tragic-Remorse—The Anguish of Dirty Hands." *Ethical Theory and Moral Practice* 7 (5): 453–471.
———. 2009. "Targeted Killing: A 'Dirty Hands' Analysis." *Contemporary Politics* 15 (3): 305–320.
Equal Justice Initiative. 2015. "Lynching in America: Confronting the Legacy of Racial Terror." Montgomery, AL, February.
Fisk, Kerstin, and Jennifer Ramos. 2014. "Actions Speak Louder than Words: Preventive Self-Defense as a Cascading Norm." *International Studies Perspective* 15(2): 163–185.
Holder, Eric. 2012. Speech at Northwestern University School of Law. Chicago, March 5. http://www.justice.gov/opa/speech/attorney-general-eric-holder-speaks-northwestern-university-school-law.
International Human Rights and Conflict Resolution Clinic at Stanford Law School and Global Justice Clinic at NYU School of Law. 2012. "Living under Drones: Death, Injury, and Trauma to Civilians from U.S. Drone Practices in Pakistan." September. http://law.stanford.edu/wp-content/uploads/sites/default/files/publication/313671/doc/slspublic/Stanford_NYU_LIVING_UNDER_DRONES.pdf.
Johnson, James Turner. 1999. *Morality and Contemporary Warfare*. New Haven, CT: Yale University Press.
Kaag, John, and Sarah Kreps. 2014. *Drone Warfare*. Cambridge, UK: Polity Press.
Klaidman, Daniel. 2012. *Kill or Capture: The War on Terror and the Soul of the Obama Presidency*. Boston: Houghton Mifflin Harcourt.
Koh, Harold Hongju. 2010. "The Obama Administration and International Law." Speech at the Annual Meeting of the American Society of International Law. Washington, DC, March 25. http://www.lawfareblog.com/wp-content/uploads/2013/01/Speech-by-Harold-Hongju-Koh-State-Department-Legal-Adviser-at-the-Annual-Meeting-of-the-American-Society-of-International-Law-Mar-25-2010.pdf.
Mazzetti, Mark. 2014. "Senate Drops Bid to Report on Drone Use." *New York Times*, April 28. http://www.nytimes.com/2014/04/29/world/senate-drops-plan-to-require-disclosure-on-drone-killings.html?_r=0.
McKelvey, Tara. 2012. "Interview with Harold Koh, Obama's Defender of Drone Strikes." *Daily Beast*, April 8. http://www.thedailybeast.com/articles/2012/04/08/interview-with-harold-koh-obama-s-defender-of-drone-strikes.html.
McMahan, Jeff. 2009. *Killing in War*. New York: Oxford University Press.
Melzer, Nils. 2008a. "Interpretive Guidance on the Notion of Direct Participation in Hostilities under International Humanitarian Law." *International Review of the Red Cross* 90 (872): 991–1047.
———. 2008b. *Targeted Killings in International Law*. New York: Oxford University Press.

Obama, Barack. 2013. "Remarks by the President at the National Defense University." Washington, DC: White House Office of the Press Secretary, May 23. https://www.whitehouse.gov/the-press-office/2013/05/23/remarks-president-national-defense-university.
Parrish, John. 2007. *Paradoxes of Political Ethics: From Dirty Hands to the Invisible Hand*. New York: Cambridge University Press.
———. 2010. "Benevolent Skulduggery." In *Corruption and American Politics*, edited by Michael Genovese and Victoria Farrar-Myers, 65–98. Amherst, NY: Cambria Press.
Pates, Richard. 2013. "Letter to National Security Advisor Thomas Donilon." Washington, DC: United States Conference of Catholic Bishops, May 13. http://www.usccb.org/issues-and-action/human-life-and-dignity/war-and-peace/arms-trade/upload/letter-to-administration-congress-on-drones-2013-05-17.pdf.
Pew Research Center. 2014. "Global Opposition to U.S. Surveillance and Drones, but Limited Harm to America's Image." Washington, DC, July 14. http://www.pewglobal.org/2014/07/14/global-opposition-to-u-s-surveillance-and-drones-but-limited-harm-to-americas-image/.
Ratner, Steven. 2007. "Predator and Prey: Seizing and Killing Suspected Terrorists Abroad." *Journal of Political Philosophy* 15 (3): 251–275.
Rawls, John. 1955. "Two Concepts of Rules." *Philosophical Review* 64 (1): 3–32.
Sartre, Jean-Paul. 2007. *Existentialism Is a Humanism*, translated by Carol Macomber and edited by John Kulka. New Haven, CT: Yale University Press.
Strawser, Bradley. 2010. "Moral Predators: The Duty to Employ Uninhabited Aerial Vehicles." *Journal of Military Ethics* 9 (4): 342–368.
U.S. Department of Justice. Undated. *White Paper: Lawfulness of a Lethal Operation Directed against a U.S. Citizen Who Is a Senior Operational Leader of Al-Qa'ida or an Associated Force*. Washington, DC: Department of Justice. http://msnbcmedia.msn.com/i/msnbc/sections/news/020413_DOJ_White_Paper.pdf.
Walzer, Michael. 1973. "Political Action: The Problem of Dirty Hands." *Philosophy and Public Affairs* 2 (2): 62–82.
———. 2000. *Just and Unjust Wars: A Moral Argument with Historical Illustrations*, 3rd ed. New York: Basic Books.
———. 2004. "Emergency Ethics." In *Arguing About War*. New Haven, CT: Yale University Press, 33–50.
———. 2006. "Terrorism and Just War." *Philosophia* 34 (1): 3–12.
———. 2007. "On Fighting Terrorism Justly." *International Relations* 21 (4): 480–484.
———. 2013. "Targeted Killing and Drone Warfare." *Dissent*, January 11. http://www.dissentmagazine.org/online_articles/targeted-killing-and-drone-warfare.
Weber, Max. 1958. "Politics as a Vocation." In *Max Weber: Essays in Sociology*, translated and edited by H. H. Gerth and C. Wright Mills, 77–128. New York: Oxford University Press.
Williams, Bernard. 1973. "Ethical Consistency." In *Problems of the Self: Philosophical Papers, 1956–72*. Cambridge, UK: Cambridge University Press.

Woods, Chris. 2012. "'Ok, Fine. Shoot Him.' Four Words that Heralded a Decade of Secret U.S. Drone Killings." London: Bureau of Investigative Journalism, November 3. http://www.thebureauinvestigates.com/2012/11/03/ok-fine-shoot-him-four-words-that-heralded-a-decade-of-secret-us-drone-killings/.

Yoder, John Howard. 1996. *When War Is Unjust: Being Honest in Just-War Thinking*, rev. ed. Maryknoll, NY: Orbis Books.

12

Beyond Preventive Force

Just Peace as Preventive Non-Intervention

DEEN CHATTERJEE

Commenting on the need for reconsidering U.S. engagement with the world, former U.S. ambassador to Saudi Arabia Chas Freeman said recently: "We are in the habit of thinking of the military first and only then about other things." He also said: "We have spent almost nothing looking at non-coercive methods of accomplishing things. So, we have become a very militaristic society in terms of how we approach the world. We don't even notice it anymore."[1] Nonetheless, people may say that the war-ready mindset is a prudent path toward securing the prospects of peace and security—anything less is naive and impractical in a dangerous world. For instance, General Philip Breedlove, commander of U.S. European Command and NATO Supreme Allied Commander, said in a recent interview: "The way to have no war is to be ready for war."[2]

However, Freeman is not a lofty idealist either. He would reject "an ideologically based foreign policy." In addressing the American mindset of perpetual war, he talks about "realism-based restraint" that prioritizes diplomacy and constructive engagement over war. In fact, he is not alone in this quest. Given that America's massive build-up for war and continual military engagements over the years have taken us nowhere near security and peace, many people have been progressively vocal in suggesting a need for a radical change in current thinking and policies. They realize that keeping war a live option in our search for peace has given us a climate (and politics) of pervasive fear at home and resentment and hostility abroad. Even the use of force that is short of full-fledged war, such as the deployment of killer drones or coercive sanctions, in the hope of extending the line of last resort to war has made us live under the perpetual shadow of war. This has made America into a "security nation," which is not conducive to peace or national health.

Accordingly, many scholars and policy makers have called for a shift in our national thinking that would prioritize proactive policies of peace over a reactive path to war. Tony Coady puts this idea well: "We can admit the grain of truth in the cynical slogan, 'If you want peace, prepare for war,' but the surer path to a more tranquil world is to prepare for peace directly" (2003, 293). This approach—which I call the just-peace approach—questions the just-war permissibility of preventive war. Indeed, the major thesis of this chapter is to show that to seek guidance in the just-war doctrine for working out the path toward peace and justice carries the risk of looking for ways to sanction preventive intervention, which could lead to more war and less peace. Especially because the U.S. war on terror is a new kind of military challenge, we need to look at the deeper issues of peace and justice, rather than simply wondering whether the new challenge would require that some of the criteria of the just-war doctrine be understood less stringently.

Though the long-standing just-war doctrine sanctions military self-defense, and international law endorses it, preventive wars in the name of self-defense when the danger is not actual or imminent raise moral and legal conundrums and lead to problematic outcomes. Especially with the changing nature of modern warfare, along with its blurring of the just-war distinction between preemption and prevention, matters of morality and legality related to preventive war are far from settled. This is so despite the fact that stipulated measures suggestive of preventive intervention are permissible with the UN Security Council's authorization, thus apparently making preventive war legal under current international law. However, given the reality of the sovereign politics of powerful states, the permissibility of duly constrained preventive force could morph into a broader mandate of intervention in the guise of just war.[3]

To counter this prospect, I propose a just-peace strategy that construes the idea of prevention through the lens of non-interventionist global justice. Instead of seeking legitimacy for prevention by invoking just-war for intervention, this non-interventionist platform would make the case for preventive war rare, if not redundant. I claim that this prospect is far more achievable and less costly in lives and resources than the current practice of perpetual war for perpetual peace. It offers a new vision with a pragmatic blend. It is worth keeping in mind Kant's (1983) critique of Grotius's and Vattel's doctrines of just war, which were prevalent during his time. In spite of the Grotian legacy of caution and moderation in international law, Kant had observed that the doctrines provided the pretext for justifying war.

In this chapter I briefly describe the moral, legal, and policy quandaries of legitimizing preventive intervention as a military option and explain how the new technologies of targeted killing compound these problems. My proposed alternative to the policy of preventive intervention for securing peace, which I call "preventive non-intervention," is to move away from the just-war provision and focus on what I call a just-peace approach. Though the just-war doctrine has been around for many centuries, the concept of just peace is a relatively recent one, indicating an emerging awareness that the broader issues of global justice transcend the confines of just-war concerns in promoting peace. My just-peace approach is part of the evolving global norm of justice, contextualized in the broader discourse of the normative and institutional challenges of globalization, with a focus on human development and wellbeing. I claim that it is high time that we shift our discourse from finding security in resorting to a just war to building security via a just peace.

Just War and Prevention

The legal, moral, and political quandaries surrounding preventive military intervention have become a major focus of controversy in response to the changing nature of warfare in the twenty-first century. The situation is made especially complicated by today's scenario of drone warfare in which a powerful nation using advanced military technology and assuming a global mandate for its military can undertake "riskless" warfare to thwart a perceived danger in another country and call it preventive intervention in the name of self-defense. This is the case with the targeted killing operations that have become a signature of the Obama administration's approach to terrorism.

The problems with claiming a global mandate for a nation's military, along with other trends of globalization, have raised serious questions about the relevance of the just-war doctrine in determining when a war is morally justified. This has prompted scholars to reexamine the theory and come to its defense. George Lucas, Jr., captures the nuanced just-war sentiment in this poignant question: "How can law-abiding peoples and nations avoid recourse to the destruction of war, while yet responsibly acting to protect themselves against legitimate threats to their security, welfare, and even to the rule of law itself?" (2013, 58).[4] The just-war doctrine attempts to respond to these legitimate concerns with an integrative approach that exhibits a balanced mix of realistic and normative components. No single criterion can define the entire doctrine or lead to an ideal resolution. Instead, the doctrine represents a comparative, comprehensive, and practical approach that has a limited prescription of what constitutes justified warfare and has long been the guiding principle of international law.

The late Jean Bethke Elshtain, a noted just-war theorist, gave us perhaps the most eloquent summation of the virtues of the just-war doctrine:

> Just-war thinking has many strengths. It is normative without generating a set of unyielding Kantian legalism. It is prudent without falling into crude realpolitik. It speaks to a general yearning for a world in which people are treated with minimal decency. It recognizes both necessity and responsibility. Thus, [use of reasonable force], as refracted through just-war thinking, will display a greater measure of gravity, being the result of serious moral reasoning, than conclusions arrived at using various shortcuts of either a narrowly legalistic or crudely reductionist sort. (2013, 26)

Nonetheless, the just-war theory has some inherent weaknesses that are quite formidable. One standard challenge it has faced is that the doctrine's criterion of just-cause legitimacy, along with its other criteria such as proportionality, necessity, and last resort, can be a matter of genuine disputation. This in turn can leave the doctrine open to self-serving interpretation by the contesting parties. Indeed, given the sovereign politics of powerful states vying to safeguard their regional or global hegemonic interests, the ambiguities of the just-war doctrine can be used, in the name of national security, as an open-ended license to respond militarily to any emerging or anticipatory threat at the discretion of the states themselves. This leads to a crafty blurring of the just-war distinction between preemption and prevention and allows powerful states to wage anticipatory wars in the name of dire necessity.

Preemptive war in self-defense is sanctified in the just-war tradition as well as in international law in response to an imminent threat,[5] but anticipatory use of force in response to a perceived threat that is not imminent—the so-called preventive war—has been a matter of dispute in the tradition. For scholars in the broadly legalistic Augustinian fold, preemption can be justified but a rigorous argument is required. No anticipatory war is justified in this thinking.[6] However, the later theorists in the Grotius-Hobbes tradition endorsed a permissive interpretation of judicious prevention as part of reasonable self-defense (Doyle 2008). Scholars also point out that a similar permissive interpretation of prevention, duly constrained, can also be found in Chapter 7 of the UN Charter pertaining to the Security Council in matters of international peace and security, thus making preventive war legal under current international law (Doyle 2008).

Nonetheless, the bar for preventive use of force is much higher than it is for preemption. To muster legitimacy, preventive war is predicated on multilateralism and other constraints to prevent the frequency of such wars being waged on the basis of mistaken assumptions or false pretenses. However, even a limited provision of preventive war, construed as a rare exception, can lead to a rather open-ended advocacy in the hands of a powerful state. We see this in the case of the so-called Bush doctrine—the Bush administration's National Security Strategy of 2002. Though couched in the language of preemption in the guise of vital national security concerns to make room for unilateralism, the doctrine embraces far-reaching preventive measures in its open-ended war on terror.

The doctrine is still largely in effect in current U.S. military policy, especially in its advocacy and practice of drone warfare in Pakistan, Yemen, Iraq, and Afghanistan. Hence, the issue is the moral permissibility of preventive war, regardless of its scope and the circumstances. The provision of preventive war in the guise of self-defense and humanitarian ventures can get unduly interventionist, making the world less secure. So, with the changing nature of warfare in the twenty-first century, the permissibility of preventive war with a broader mandate of intervention has become a major focus of controversy.

The growing reliance in modern warfare on advanced technologies such as weaponized drones makes the blurring of preemption and prevention all too likely and the slippery transition from the one to the other all too easy. President Obama's increasingly escalating use of drone attacks under the guise of just war is one such example. Targeted killings by drones where the victims are posing no imminent threat are (at best) cases of preventive military operations that would not merit unqualified just-war endorsement. Likewise, calling killer drones "moral predators," (Strawser 2012; see also Shane 2012) thus making obligatory their unlimited use whenever needed, also muddies the water. This trend has the likelihood of misuse of military options in the name of preventive intervention.

Though drones are unmanned military robots and exemplify the advanced sophistication of military technology, they are still a step away from the specter of cyber war. Even then, deployment of drones alters the reciprocal vulnerability of a conventional war and makes the asymmetries of power more pronounced by rendering military operations risk-free for the side using drones. In commenting on the frequent use of unmanned drones in today's U.S. combat overseas, Peter W. Singer writes:

> And now we possess a technology that removes the last political barriers to war. The strongest appeal of unmanned systems is that we don't have to send someone's son or daughter into harm's way. But when politicians can avoid . . . the impact that military casualties have on voters and on the news media—they no longer treat the previously weighty matters of war and peace the same way. (2012a)

In other words, risk-free combat technology can increase the likelihood of their use (Mazzetti 2014a; Enemark 2013). In fact, one can make the more general claim that the prospect of preventive use of force, if made easy, can make war all too tempting and frequent (Crawford 2003). The certainty factor of an imminent danger, already compromised in the need for expanded preemption due to the presence of WMD in today's unconventional warfare, is now put to severe test in view of the new challenges of drone warfare, making the claims of moral mandate in the slippery transition from preemption to prevention that much easier.

But this drift is making the world progressively less secure. It is not only that drones are being deployed in responding to the changing nature of today's war; it is that the very deployment of this new technology for lethal use is changing the nature of warfare and raising grave concerns over the laws of war and the future of warfare.

Deploying Lethal Drones: Legal and Moral Conundrums

International law lacks specific guidelines regarding the rules for covert operations. With regard to America's increasing reliance on drones for targeted killings in foreign countries, the CIA rather than the Pentagon is largely in charge. Shedding its more traditional role of carrying out espionage, intelligence collection, and data analysis, the CIA has become largely a paramilitary organization since the 9/11 attacks, but being a covert spy agency, it stays beyond the jurisdiction of international laws of military operations.[7]

The growing reliance of modern warfare on advanced technology is fraught with unresolved moral issues as well. In the case of combat drones, scholars have noted that the reliance on technology in targeted killing does not necessarily improve the *jus in bello* compliance of proportionality and discrimination. But the undue confidence in technology for precision killing in the murky arena of war, along with the perceived risk-free military operations for the side using such technology, especially if the combat is couched in the rhetoric of a humanitarian mission, can lower the threshold for the use of force (Singer 2012a; Sandvik and Lohne 2013).

This trend raises difficult moral issues. First of all, it calls into question the aptness of calling weaponized drones "moral predators". Robot ethics will not clean up combat, as Singer observes: "We shouldn't be seduced into thinking [that] advanced technology can make war something it isn't. Whether it's being fought with sticks and stones or Predator drones, war is still a story of humans: our causes, our decisions, our losses, and our ethics" (2012b; see also Berkowitz 2014).

Traditional just-war doctrine is grounded in the norm of moral equality of combatants, which holds that soldiers on both sides in a war have an equal right to kill even when they fight in an unjust war, as long as they adhere to the rules governing the conduct of war. One corollary to this norm is that soldiers are liable to attack only if there is reciprocity of risks—that is, if they stand in a position of mutual risk with their opponents. Accordingly, non-mobilized soldiers or those immobilized on the war front, who thus pose no immediate threat, should not be targeted for attack. A riskless warfare via robotic technology, due to its pronounced asymmetries of power, makes the combatants on the other side not an immediate threat, which makes them not liable to attack. This is a formidable roadblock to justifying a drone war as a just war. Moreover, if the combatants do not stand in a position of mutual risk in a drone encounter, then, paradoxically, they are not combatants anymore due to the loss of the distinction between combatants and non-combatants. In that case, the encounter again loses its just-war legitimacy, making it more like a police operation. This is one of the paradoxes of riskless warfare (see, for example, Kahn 2003).

But policing has its own legal guidelines, such as capturing a suspect for trial in a court of law, not killing a suspect unless faced with dire and immediate threat, nor attacking a suspected target in such a way that may endanger the lives of an entire family or of the people in the neighboring areas. These are strict guidelines, but they do not dictate the logistics of drone attacks. Thus drone warfare remains in a legal and moral limbo.[8]

One could argue that the moral and legal conundrums of riskless warfare are false dilemmas arising from an effort to make war moral. In reality, war is for safeguarding a nation's vital interests by vanquishing a sworn enemy by the least risky and most efficient means. We find an echo of this sentiment expressed by Captain Theodore Van Kirk, the navigator and, until his recent death, last surviving crew member of the *Enola Gay*, the B-29 Superfortress that dropped the atomic bomb on Hiroshima in the last days of World War II. Van Kirk joined his fellow crewmen in unwavering defense of the atomic raids, saying: "It's really hard to talk about morality and war in the same sentence. . . . I believe that when you're in a war, a nation must have the courage to do what it must to win the war with a minimum loss of lives" (quoted in Goldstein 2014).

Of course, this is a rather extreme view, and it stands in sharp contrast to the just-war approach. The just-war doctrine is known for its limited prescription of what constitutes justified warfare, with a balanced mix of facts and norms. As F. M. Kamm (2014) observes, both *jus ad bellum* and *jus in bello* conditions need to be evaluated in the context of the objectives of a specific combat, and the objectives themselves need to be moderated in view of these conditions and the broader issues of justice and politics. Accordingly, when we assess current U.S. drone operations abroad, undertaken in the name of combating terrorism, we may well conclude that the enhanced use of covert targeted killings, though riskless and efficient, not only is not meeting the long-term objectives of the United States, but may even be counterproductive. For instance, escalated drone attacks create widespread resentment and distrust of the United States in precisely those countries where combating terrorism requires winning the hearts and minds of the people via proactive nonmilitary means.[9]

Making war risk-free and easy increases the likelihood of the slippery transition from preemption to prevention. Preventive use of force short of a full-fledged war is still war—as it is, for instance, in drone warfare—and it could slip to a larger preventive war. This is where the moral conundrum of drone warfare is especially pronounced. As noted above, prevention is a disputed issue in just-war thinking, so just-war doctrine does not offer moral clarity in this debate, especially when such interventions are as open-ended, unilateral, and secretive, as is the case in the U.S. conduct of drone warfare. Yet the just-war criteria are used to tout predator drones for their moral and efficient ways of doing a job in a way that underscores the imperative of prevention in the U.S. military operations abroad.[10] Here we see how the global mandate of the U.S. military allows the Obama administration to justify the use of drones for preventive measures in faraway lands in the name of self-defense and national security, often couched in humanitarian jargon.

Just-War Dilemmas

My claim that the moral permissibility of preventive war, even with due limits, could gain undue legitimacy, leading to more war, not less, derives from two considerations: one is conceptual—namely, the slippery transition from preemption to prevention; the other is empirical—that is, even a limited provision of preventive war for justified self-defense, construed as a rare exception, can lead to a rather open-ended advocacy and use of it in the hands of a powerful state.

A major turn in rethinking just-war principles occurred during the Second World War, when the distinction between combatants and non-combatants blurred, making the war the first truly "total war." It compelled the Allied forces to navigate across a moral divide in deciding whether to undertake massive bombing of German civilian targets for both military and strategic reasons, which they eventually did. "I see this idea of just killing civilians and targeting civilians as being unethical—though the most unethical act in World War II for the Allies would have been allowing themselves to lose," says military historian Conrad Crane in the 2010 PBS Television's *American Experience* segment titled "The Bombing of Germany." We find an echo of Crane's words in Michael Walzer's restatement of the just-war doctrine in his classic *Just and Unjust Wars*, in which, despite his initial unequivocal stand against the violation of non-combatant immunity, he writes that in cases of "supreme emergency," "when we are face-to-face not merely with defeat but with a defeat likely to bring disaster to a political community" (2006, 268), rules of war can be breached.

The just-war dilemma of the Allied leaders in the bombing of the German civilians was prompted by German bombers' attacking London for fifty-seven consecutive nights, which meant that the response was directed to a "face-to-face" situation of a dire catastrophe. However, the quandary facing today's political theorists who draw from the just-war tradition is provoked by a new set of challenges unique to the new century. The question now has to do not only with justifying a first strike, but also with justifying how much in advance the strike is made, given the potential for catastrophic consequences. The understanding of a "face-to-face" danger in today's world could take a whole new meaning in view of the unconventional nature of warfare and the specter of WMD. Indeed, it could come eerily close to former U.S. Vice President Dick Cheney's "1 percent doctrine," which proclaimed that even a 1 percent possibility of the nation being attacked with WMD poses an unacceptable risk that would warrant massive "preemptive" measures. Here we see how the prospect of a catastrophic danger can radically stretch the certainty factor of an immediate danger. The slippery slide from expanded preemption to open-ended prevention due to this new challenge can push a nation to the extreme, paving the way for unilateralism, first strike, and massive killing of civilians in the name of self-defense.

The scenario of twenty-first century unconventional warfare has raised anew the question of how large of a threat justifies the massive killing of civilians. In finding ways to respond to this new challenge to peace and security, today's just-war scholars espouse an expanded version of self-defense that construes the idea of imminent danger not necessarily as immediate but in terms of high probability of a sufficiently serious risk, leaving the standard of imminence sufficiently open (see, for example, Luban 2004). This expanded reading gives them some latitude in justifying a first strike that they think is consistent with the spirit of just-war norms. Evidently, their moral high ground gives in to the demands of prudence at the midpoint of the spectrum of anticipation—one that ranges from the strict standard of "moral necessity of rejecting any attack that is merely preventive in character" (Walzer 2006, 80) to the other extreme of Dick Cheney's "1 percent doctrine."

Ironically, given recent trends of warfare and international law, the Bush doctrine's seemingly brazen unilateralism may not be that radical after all. Although the UN Security Council prohibits unilateralism in humanitarian military interventions, the slippery transition from multilateralism to unilateralism is implicitly left open in the practice, if not in the mandate, of the Security Council. Chapter 7 of the UN Charter permits multilateral use of force in response to threats to international peace and security, given the Council's authorization. Breach of human rights is not mentioned in the directive, but because humanitarian emergencies involving egregious violations of human rights pose challenges to international peace and security, coalitions of nations have undertaken rescue wars in such cases, even at times without the sanction of the Security Council.

But while collective self-defense can be taken to be an important part of international peace and security, providing distinct sets of guidelines in practice by separating self-defense and humanitarian intervention can be difficult. In fact, as Michael Doyle has noted, "Security Council practice . . . has [practically] merged the two" (2008, 136). The broad mandate of the Bush doctrine, which effectively makes the idea of "global safekeeping" an important part of national security strategy, gives the United States an open-ended and unilateral license to respond militarily, on the basis of its own internal perceptions and in the name of the "war on terror," to any acts or events in the world it deems threatening. Thus, it should not come as a surprise that the Security Council's implicit inclusion of the provision of (collective) self-defense in its authorization of multilateral military operations for peace and security has given the United States a convenient opening for claiming a global mandate in the name of national security.[11]

Predictably, most of the just-war theorists find the Bush doctrine too broad. Their concern in offensive wars is to devise means of effective accountability that would stay within the spirit of international law (Doyle 2008; see also Buchanan 2006, 2007; Buchanan and Keohane 2004; Luban 2004, 2007). Dubbed both substantive and procedural, their approach is a mix of the just-war criteria and legal propriety, putting emphasis on collaboration whenever possible and citing the UN Security Council as the venue for open arbitration and debate for procedural legitimacy. Their guidelines for assessing the gravity of the situation requiring permissible prevention display a judicious consideration of such factors as severity of threat, the likelihood of its occurrence, just-war criteria of legitimacy, and the legality of the proposed response. However, though mindful of the carefully set standards and procedures of multilateralism stipulated in the UN Charter, these theorists leave room for multilateral preventive response without the approval of the Security Council or even unilateral action by individual states in rare cases of extreme emergency, as long as these moves meet their carefully construed guidelines of prevention.

Nonetheless, these guidelines can also be open-ended and misused, especially if a unilateral provision is included in them. Legitimizing principled preventive war, however constrained, can give a powerful nation the moral license to expand the principle by pushing it in the direction of its own convenience. Even with *ex post* standards of accountability suggested by some just-war theorists, the damage is already done. And no specter of a "contingent contract" of *ex ante* accountability would deter a powerful state from trying to manipulate legitimacy in its own favor so long as the option of preventive intervention is in the vocabulary of legitimacy.[12] Moreover, the question of legality and procedural propriety seems to be moot in the context of claims of existential threat. Indeed, as Walzer has famously stated, "Necessity knows no rules" (2006, 254). Thus, the criteria proposed by the just-war scholars for the provision of a duly constrained and morally permissible preventive option can leave open the possibility that a powerful nation with a global mandate can construe the guidelines as an open-ended license to intervene unilaterally in the name of self-defense.

Indeed, the just-war permissibility of preventive war can get unduly interventionist, leading to more war. In addition, the deployment of unconventional drone technology in the war on terror creates its own set of moral and legal conundrums, which add to the list of challenges for just-war theorists seeking to defend the provision of preventive intervention. As noted above, the glaring asymmetry of power in today's riskless drone warfare makes the combatants on the other side not an immediate threat and thereby, paradoxically, renders them no longer combatants according to the norms of just war.

This blurring of the combatant/non-combatant distinction makes the war on terror a "total war," posing a renewed challenge to the just-war doctrine, as it did in the Second World War, when the combatant/non-combatant distinction also blurred. In addition, as the United States tries to minimize risks of a conventional war with the deployment of unconventional war machines in inept and failed states, the incentive for using unconventional tactics such as terrorism by the other side, in the absence of anything comparable, would be that much greater. But how could only one side—say, the terrorists—be blamed in a total war in which both sides have the objective of mitigating their existential threat by whatever means necessary? Under the circumstances, one side's claiming to follow the norms of just war would seem to be hollow. As Paul Kahn notes, "For the asymmetrically powerful to insist on the maintenance of the combatant/non-combatant distinction has the appearance of self-serving moralizing" (2003, 45).

Richard Falk (2014) has observed that drones are more dangerous than nuclear weapons. While this may be debatable, it is conceivable that given the introduction of killer drones to an unending total war in which the combatant/non-combatant distinction is blurred, the world could well enter into another dangerous arms race when the other side also gained rough parity in drone technology.

In view of these roadblocks to the applicability of the just-war doctrine in deciding on the justness of war, I propose that we need to shift our discourse from just war to just peace. I claim that the policy of perpetual war for perpetual peace has been a failed one. It has given us only war, not peace and justice. My just-peace approach blends the idea of justice with proactive policies that prioritize development and democracy over reactive military intervention for fostering lasting peace. It offers a new vision of liberalism with a pragmatic blend. It is different from the failed policies of liberal internationalism.

Liberal Internationalism

With its continuing involvement in Afghanistan, Iraq, and elsewhere in the name of spreading democracy and self-governance, the United States has become a de facto imperial democracy, even though the United States denies any imperial ambitions. Nonetheless, based on its role as the sole superpower with unprecedented political, economic, military, and cultural clout all over the world, the nation has been characterized as being "unmistakably an empire" (Ignatieff 2003). However, due to a proclaimed liberal commitment to promoting "sustainable models of national success" through democracy and free-market policies, the globally dominant United States is also termed a "liberal" empire (Ignatieff 2003). Although the old motif of civilizing the illiberal other is unmistakably present in today's imperial democracy, this "liberal imperialism" is garbed in the rhetoric of freeing people from the tyranny of their own rulers. Meanwhile, it is also being billed as part of the broader strategy of fighting terror and defending American interests. The liberating project of imperial democracy, thus, is also a "defensive imperialism," claiming to be consistent with human rights objectives and other cosmopolitan liberal values (Cooper 2002; see also Ignatieff 2003).

Perceived this way, the vision of liberal imperialism seems to embody "the fallacy of liberal internationalism" (Hoffman 1995),[13] according to which all good things can be delivered together without any internal tension. They usually cannot, and such efforts often turn out to be anything but good. The irony of promoting the democratic values of self-determination, human rights, and respect for international law through military intervention—unilaterally if needed and without the mandate of the United Nations if necessary—is palpably evident. Creating a dilemma for liberals, this situation also raises grave concerns about just distribution of power. Given the great asymmetry of power and resources among nations, powerful countries could more easily be able to keep weaker countries from gaining military parity and political equality in the name of war on terror, which is open-ended and "risk free" due to the advanced technology of weaponized drones.

In a *New Republic* cover article, "Superpowers Don't Get to Retire," Robert Kagan (2014) claims that recent events in Iraq have opened the door for interventionist revival and calls for the resumption of a more muscular approach to the world, going back to the days of liberal interventionism but avoiding the flaws that led to mistakes. This position, however, is full of unwarranted optimism. With liberal internationalism's latent "statist, geopolitical agenda" (Burke 2005, 86), it is unlikely that a more muscular approach would be able to avoid the past mistakes. The Libya operation that caused the ouster of Colonel Gaddafi, hailed by liberal internationalists like Anne-Marie Slaughter (2011) as a model for successful intervention, proved to be the exact opposite. As Anthony Burke observes, "My concern is that liberal internationalism has been inexorably drawn toward the norm of war and the instrumental images of the human Kant believed war would endanger" (2005, 86). His subsequent work thus takes up the question: "When security is grounded in exclusion and alienation, ethics licenses killing and war, and freedom is a mask for imperial violence, how should we act?" (2007, i). My normative vision of just peace as preventive non-intervention, in contrast to the just-war permissibility of preventive war, responds to this concern. In the section below, I present a philosophical roadmap of my proposal.

Just Peace as Preventive Non-Intervention

My idea of just peace eschews liberal interventionism, but it is squarely based on the project of a realistic and reformed liberal mission. Preventive non-intervention is an integral part of it. Liberal interventionism is a policy of traditional liberal internationalism, which is different from my proposal.

Pushing liberal reforms in non-liberal parts of the world is an exceedingly delicate project. Pursuing preventive military intervention to promote human rights initiatives and foster liberal institutions is not only counterproductive, but also against the liberal ethos itself. Situating the arguments for liberalism in the real world of diversity, need, and vulnerabilities is a bottom-up project, not a top-down strategy. Yet for far too long, liberalism has been making normative pronouncements that were either far too demanding or far too removed from the actual lives of people. Rather than viewing the world according to a facile thesis about a clash of civilizations, in which the so-called other is demonized, the focus instead should be on locating sources of prejudice and intolerance, which form a clash within. One need not be unduly deferential toward cultural demands by making liberalism a vacuous doctrine, nor should liberalism's pursuit of fairness and impartiality ignore cultural identities and variations (see, for example, Chatterjee 2012).

Liberalism needs to be more flexible and inclusive, showing more empathy toward cultural diversity and more sensitivity toward faith-based doctrines, without compromising its substantive principles. In other words, liberalism needs to be both substantive and negotiable as it faces the increasingly vocal challenges of a diverse world. My sketch here calls for a new vision of liberalism that is appropriate for a globalized "post-liberal" word where liberal emphasis on human rights would be cast in the context of human values that all cultures can understand and appreciate. In effect, in the face of tension, distrust, and militancy, the issue is not how liberal society should respond to illiberalism, but how liberalism can restructure itself as a political ideology, shun its image as a Western import, and confront the issues of justice and injustice in the real world (see, for example, Chatterjee 2013a).

Poverty, deprivation, and humiliation in our world of radical inequity are the chief reasons for the tensions and resentments that lead to hostilities. Confronting poverty and deprivation through proactive and sustained economic development and responding to humiliation through political recognition and accommodation would go a long way toward mitigating conflicts (see, for example, Dumas 2003). For this to occur, an equitable global structure is needed, along with domestic democratic reforms in individual countries. The latter is mostly ineffective and often hard to achieve without the former. Pervasive state failure to respond to citizens' broader human needs is linked to the inequity in the global political and economic order itself.

The gradual emergence of the global human rights culture in the last sixty years has achieved a certain level of international recognition for justice when there is an egregious violation of human rights, but severe poverty and radical inequity in the socioeconomic arena are still not properly recognized as urgent human rights concerns.[14] This "holocaust of neglect" (Shue 1996, 42) perpetuates deprivation, destabilization, and violence, creating the presumed need for the preventive intervention that often makes things even worse, whereas preventive non-intervention, if practiced as a systemic antidote to the inequity and neglect in the world, can take us beyond the need for preventive use of force.

In drawing attention to the priorities of global justice in responding to the problems of global inequity, Thomas Pogge writes:

> One-third of all human deaths are due to poverty-related causes, to malnutrition and to diseases that can be prevented or cured cheaply. Yet our politicians, academics, and mass media show little concern for how such poverty might be reduced. They are more interested in possible military interventions to stop human rights violations in developing countries, even though such interventions—at best—produce smaller benefits at greater cost. This Western priority may be rooted in self-interest. But it engenders, and is sustained by, a deeply flawed moral presentation of global economic cooperation. The new global economic order we impose aggravates global inequality and reproduces severe poverty on a massive scale. On any plausible understanding of our moral values, the prevention of such poverty is our foremost responsibility. (2001, 6)

Studies have shown that nonmilitary measures for mitigating the structural violence of poverty and inequality can be more effective in confronting terrorism than military actions, which may even prove to be counterproductive.[15] Former UN Secretary General Kofi Annan echoed this sentiment in a report released on May 1, 2005, when he famously observed that the ingredients of enduring global security lie not necessarily in deploying a nation's military force for global safekeeping but, more importantly, in promoting just development and comprehensive human rights. On February 19, 2015, at a White House summit with ministers from around the world to address preventing violent terrorism at its root, current UN Secretary General Ban Ki-moon echoed Annan when he challenged all nations to alter their military responses to terrorism. Instead of looking at the consequences of terrorism, he urged nations to focus on proactive preventive measures, which for him primarily involves good governance. It is a global challenge, for which he appealed for solidarity of nations.[16]

Development ethicists are mindful that unless we address the pervasive asymmetry of power, knowledge, and freedom so starkly evident in the world, construing solidarity as a participatory forum for representation would ring hollow. Accordingly, recognizing the importance of "localization" of participatory politics that has given people a voice and provided foci for organizations in response to globalism, such ethicists have called for a gradual and incremental power shift from collective action at the local and national level to collective and realistic reform in the global institutional architecture. For them, this broadly consequentialist approach addresses the vulnerabilities, needs, and dependencies of real people by promoting enabling institutions and practicable social realizations. This developmental approach requires the proactive policy of preventive non-intervention, which is the basis for my idea of a just peace (see, for example, Stiglitz 2002; Sen 1999, 2009).[17]

In a democratic global order that strives toward inclusion, participation, and human empowerment, no member nation would be considered an outlaw state meriting targeted preventive military strikes. Such a justice-based global order would vastly increase the prospect of democratic reforms within states in a peaceful manner, with initiatives generated by the states themselves. Because of the privilege of recognition and membership in a global body, individual states would feel incentivized not to compromise this position through a failure of granting democratic rights to citizens. It is a sad reality that stable democratic institutions are not easy to promote, build, or sustain in impoverished or inept states or in countries that do not have a history of democratic traditions. But this is as much a failure for these states as it is for the world at large.

In a collaborative global order, the rhetoric of military intervention for imposing democracy from without would fall flat. Also, the idea of a military strike against a group of subversive non-state actors in remote areas who are plotting attacks in populated areas—the oft-cited hypothetical scenario by today's just-war theorists for justifying preventive intervention (see, for example, Buchanan 2006)—would be an empty argument. If the actors are domestic terrorists, then the question of intervening in a sovereign state does not arise, and if they are known foreign terror groups, then in a collaborative global order the idea of intervention would go against the thrust of collaboration. There are numerous collaborative avenues for responding to such distant threats.

Scholars have been suggesting ways to demand accountability in the deployment of drones. Allen Buchanan and Robert Keohane (2015) have recently proposed an international collaborative framework, strikingly similar to the one I have presented here for a just-peace approach, for regulating the lethal use of drones by states to safeguard better compliance with the laws of war. This proposal is in accord with their earlier advocacy of the just-war permissibility of duly constrained preventive intervention (Buchanan and Keohane 2004), which I have critiqued above. Given the moral and legal conundrums of the very use of drone technologies in today's warfare, regardless of how they are being deployed, it is inconsistent to say that drones are acceptable if they are used properly in a just war. A war is just not only when its cause is just. It also needs to fulfill the other criteria of a just war. How a war is fought is one. Given grave doubts on the justness of the deployment of lethal drones, their use in an otherwise just war makes the entire mission unjust. This is a telling point against the proposal put forth by Buchanan and Keohane (2015), who are not against drones per se, but only against their improper use. This position is comparable to Strawser's (2012) touting of lethal drones as "moral predators" if they are used properly and only when the war itself is just, as well as to a recent call for establishing a "drone court" to approve American drone strikes (Guiora and Brand 2015).

Given all the roadblocks to the applicability of the just-war doctrine to today's unconventional warfare, it is imperative that we shift our discourse from the justifiability of preventive intervention under the rubric of just war to the imperative of preventive non-intervention via the proactive approach of a just peace.

Shifting the Discourse

In matters of individual and collective self-defense in a high-risk world, there is no fool-proof answer. But we need to see if some direction other than preventive intervention is at least as promising, if not more so. After all, anticipatory use of force can go horribly wrong, as recent military engagements in Iraq and elsewhere amply attest. It takes time and political will to build a collaborative and equitable world order that reflects a culture of just peace, and such work is never complete. But the policy of preventive intervention cannot guarantee total security either, even by the admissions of its own proponents (Doyle 2008; Buchanan and Keohane 2004; Buchanan 2006). It is high time that we shift our discourse from seeking security through just war to building security via just peace.

Unlike the principled anti-interventionist arguments of pacifists, my stance on just peace is anti-interventionist in a contingent sense. It is not necessarily against intervention per se (as for instance when intervention is the only option for preemptive reasons); it is against the way it usually takes place, or against its general feasibility in a complicated and interdependent world. But this limited provision for preemption, when construed on the premise of just peace but not just war, would not provide the moral mandate—and the crafty transition—to preventive military measures in the name of peace and security. A just-peace initiative allows measured military response as an option for self-defense or humanitarian reasons, either after an attack has occurred or when an attack is imminent, in the strictest sense of the imminence of act—that is, requiring interception when the offense is irrevocable. If there is any logic to just war, then it should be reserved for preemption only in this sense.

But in view of today's unconventional weapons in the hands of sworn enemies, it is understandable if the contemporary just-war theorists who advocate morally permissible preventive war find this strict construal of imminence too risky and thus unacceptable. After all, they could retort that the just-war doctrine is also a nuanced and balanced attempt to find peace in a measured way, especially with their proposed procedural and institutional safeguards. Thus, both sides may believe that theirs is the more reasonable and responsible approach.

However, these two approaches need not be considered to be diametrically opposed. Though both sides have their compelling arguments in this debate, neither approach is fool-proof or complete. There are striking similarities between them, though they give vastly different prescriptions. Needless to say, I would propose a gradual shifting of direction in favor of my vision. Here are my reasons:

If the proposed drone accountability regime is feasible, then the just peace I propose is feasible too. Of course, as I noted above, to build a global culture of just peace takes time and political will, and it is never complete. Buchanan and Keohane also note that "it would be entirely unrealistic to expect . . . a Drone Accountability Regime to appear suddenly" (2015, 34). What is more, they point out that their proposal "could have value even if never implemented" because it would draw attention to the current deficiencies in the system, which could motivate others to step in with their ideas. For them, "the process of deliberation on a Drone Accountability Regime would itself be a valuable, if small, step toward a better institutionalized world order" (35). Likewise, my proposal of just peace envisions the prospect of open deliberation in an inclusive and collaborative world. The practice of open dialogue between nations starts with the beginning of such dialogues at home.

The similarities between my approach and Buchanan and Keohane's proposal get even more striking. They imagine a "virtuous spiral" spinning out of their proposed scenario where the United States and other major nations can be "first movers" who would "take credible steps to induce others to reciprocate" (34) and identify incentives for others to join them. This is also how I envision the collaborative global arrangements promoting just peace. They claim that their proposed model is aspirational and that it relies on vigorous civil society mobilization, not on formal legal measures. My proposal for a "better institutionalized world order" is also aspirational. It is motivated by the urgent moral imperative of just peace against which existing arrangements can be measured.

In an earlier publication Buchanan (2006) suggested a global institutional initiative toward a responsible just-war provision for preventive war that is very similar to his proposed drone accountability regime. But he concluded there that if institutional safeguards could not be realized now or even in the future, then he would favor a just-war ban on such measures as preventive self-defense and forcible democratization.

However, the future is an indefinite time frame, and one may wonder how many horribly failed preventive military measures nations must go through, either for putative self-defense or for initiating political reforms abroad, before realizing that the future is now.

Concluding Note

Legitimizing preventive war creates the cycle of perpetual war. This situation could be mitigated by proactive non-intervention—a long and hard road that promises no quick results, but if we are looking for a fail-safe quick path to peace and security in today's murky and uncertain world, nothing can take us there. Preventive military interventions only make things worse. We should pay heed to Grotius, who said: "Human life exists under such conditions that complete security is never guaranteed to us" (1925, 184).

Understanding my idea of just peace can help us understand the seeming paradox of indifference and apathy leading to intervention and proactive engagement leading to non-intervention.[18]

NOTES

1. In an interview with David Speedie on the U.S. Global Engagement Program at the Carnegie Council for Ethics in International Affairs, November 20, 2014. Freeman was U.S. ambassador to Saudi Arabia during Operations Desert Shield and Desert Storm, principal deputy assistant secretary of state for African affairs, director of Chinese affairs at the U.S. Department of State with President Nixon (he went to China in 1972) and thus he has had extensive Middle Eastern, African, East Asian, and European diplomatic experience. See also Freeman 2014 and Bacevich 2005.
2. *PBS NewsHour*, aired on July 29, 2015.
3. This is happening with regard to America's expanding military engagement abroad. The Obama administration altered only the style and format of the Bush doctrine but kept the doctrinal substance intact.
4. Lucas defends the moral permissibility of preventive war and the use of lethal drones.
5. David Rodin notes that even the question of whether there exists a right of preemptive self-defense in international law is "by no means" a settled issue (2005, 145).
6. For a highly legalistic Christian pacifist like John Howard Yoder, not even preemption is justified in the just-war tradition. See Yoder 1984.
7. A recent article in the *New York Times* (Mazzetti 2014b) reports that although the Pentagon's armed drones have stopped flying in the skies above Yemen as a result of the ban on American military drone strikes imposed by the Yemeni government after a number of botched operations in recent years that killed Yemeni civilians, the CIA's drone war in Yemen continues. In Pakistan, the CIA remains in charge of drone operations, and may continue to do so long after American troops have left Afghanistan. And in Jordan, it is the CIA rather than the Pentagon that ran a program to arm and train Syrian rebels—a concession to the

Jordanian government, which would not allow an overt military presence in the country.

8   For law enforcement analogy for targeted killing, see McMahan (2012) and Finkelstein (2012). For a critique of McMahan and Finkelstein's arguments, see Kaplan (2014). See Luban (2003) for the hybrid war-law approach of the war on terror. Also, see Enemark (2014) for more on the legal and moral limbo involved in excessive reliance on killer drones.

9   In commenting on why he "insists the United States has been its own worst enemy in Pakistan," Imran Khan, who is a moderate and popular political leader in Pakistan, was quoted as saying: "It was all the killing of civilians—the drones and killing—that made anti-Americanism go up. Especially drones. I can't understand what human beings can sit thousands of miles away and kill people, women and children, here" (quoted in Nordland 2015).

10 As President Obama remarked at the National Defense University, May 23, 2013, "So this is a just war—a war waged proportionally, in last resort, and in self-defense." For a thoughtful critique of the United States' policy of drone warfare as just war, see Brunstetter 2012.

11 The Obama administration's secretive and unilateral policy of expanded drone warfare is part of this same strategy.

12 See Buchanan and Keohane 2004 for a detailed proposal of *ex post* and *ex ante* guidelines of accountability.

13 "Internationalism," of course, implies more cooperation and less unilateralism than "imperialism."

14 Commenting on the government failures that allow six million children under five years old to die of preventable diseases every year, the new United Nations Human Rights High Commissioner, Zeid Ra'ad Zeid al-Hussein, was quoted as asking: "If ISIL killed six million people a year, you wouldn't be able to talk about anything else, so why is it that we don't look aggressively at the right to better health?" (quoted in Cumming-Bruce 2015).

15 A 2008 Rand Corporation research paper titled "How Terrorist Groups End" (Jones 2008) sheds light on how terrorist groups ended in the past. In what Rand calls "the first systematic study" on the topic, researchers found that the use of military force did not undermine terror groups like al-Qaeda. A more effective U.S. approach, the study suggests, would involve empowering needy countries in effective policing and achieving peaceful political accommodation with the groups. The study suggested replacing the "war-on-terrorism orientation" with counterterrorism approaches that would promote cooperation and political engagement, as well as minimizing the use of U.S. military force. See also Macleod 2007.

16 Sarah Chayes, who has long worked in Afghanistan, argues in her important 2015 book, *Thieves of State: Why Corruption Threatens Global Security*, that the United States keeps repeating the same mistake in the Middle East—namely, overestimating the power of religious ideology and underappreciating the impact of poverty

and misgovernance. She observes that nothing feeds extremism more than in-your-face government corruption and injustice.

17 Sen and Stiglitz have been the two leading voices in articulating a conceptual framework for the slowly emerging trend of democratization of globalization, with a focus on the normative and institutional imperatives of human development and wellbeing.

18 I thank the anonymous readers of my earlier draft for their comments, and I especially thank Kerstin Fisk and Jennifer M. Ramos for their insightful ideas and unfailing help.

Portions of the essay draw on my "Deciding on Preventive War: Amartya Sen's Idea of Justice" (2015), "Technology, Information, and Modern Warfare: Challenges and Prospects in the 21st Century" (2014), "Enough about Just War, What about Just Peace?" (2013), "Veiled Politics: The Liberal Dilemma of Multiculturalism" (2012), and "Building Common Ground: Going Beyond the Liberal Conundrum" (2013).

BIBLIOGRAPHY

Bacevich, Andrew. 2005. *The New American Militarism: How Americans Are Seduced by War*. New York: Oxford University Press.

Berkowitz, Roger. 2014. "Drones and the Question of 'The Human.'" *Ethics & International Affairs* 28 (2): 159–169.

Blake, Michael. 2013. "Does International Law Make a Moral Difference? The Case of Preventive War." In *The Ethics of Preventive War*, edited by Deen Chatterjee, 65–86. New York: Cambridge University Press.

Brunstetter, Daniel R. 2012. "Can We Wage a Just Drone War?" *Atlantic*, July 19. http://www.theatlantic.com/technology/archive/2012/07/can-we-wage-a-just-drone-war/260055/.

Buchanan, Allen. 2006. "Institutionalizing the Just War." *Philosophy and Public Affairs* 34 (1): 1–11.

Buchanan, Allen. 2007. "Justifying Preventive War." In *Preemption: Military Action and Moral Justification*, edited by Henry Shue and David Rodin, 126–142. Oxford, UK: Oxford University Press.

Buchanan, Allen, and Robert Keohane. 2004. "The Preventive Use of Force: A Cosmopolitan Institutional Approach." *Ethics & International Affairs* 17 (1) 1–18.

Buchanan, Allen, and Robert Keohane. 2015. "Toward a Drone Accountability Regime." *Ethics & International Affairs* 29 (1): 15–37.

Burke, Anthony. 2005. "Against the New Internationalism." *Ethics & International Affairs* 19 (2): 73–90.

Burke, Anthony. 2007. *Beyond Security, Ethics and Violence: War against the Other*. London: Routledge.

Chatterjee, Deen. 2011. "Reciprocity, Closed Impartiality, and National Borders: Framing (and Extending) the Debate on Global Justice." *Social Philosophy Today* 27:199–217.

Chatterjee, Deen. 2012. "Veiled Politics: The Liberal Dilemma of Multiculturalism." *Monist* 95 (1): 127–150.

Chatterjee, Deen. 2013a. "Building Common Ground: Going Beyond the Liberal Conundrum." *Ethics & International Affairs* 27 (2): 119–127.

Chatterjee, Deen. 2013b. "Enough about Just War, What about Just Peace?" In *The Ethics of Preventive War*, edited by Deen Chatterjee, 214–239. New York: Cambridge University Press.

Chatterjee, Deen. 2015. "Deciding on Preventive War: Amartya Sen's Idea of Justice." *Philosophy and Social Criticism* 41 (1): 69–76.

Chatterjee, Deen, with Wayne McCormack. 2014. "Technology, Information, and Modern Warfare: Challenges and Prospects in the 21st Century." In *The Ethics of Information Warfare*, edited by Luciano Floridi and Mariarosaria Taddeo, 61–70. Dordrecht: Springer.

Chayes, Sarah. 2015. *Thieves of State: Why Corruption Threatens Global Security*. New York: W. W. Norton.

Chivers, C. J. 2007. "Dutch Soldiers Stress Restraint in Afghanistan." *New York Times*, April 6. http://www.nytimes.com/2007/04/06/world/asia/06afghan.html?pagewanted=all.

Coady, C. A. J. 2003. "War for Humanity: A Critique." In *Ethics and Foreign Intervention*, edited by Deen Chatterjee and Don E. Scheid, 274–295. New York: Cambridge University Press.

Cooper, Robert. 2002. "The New Liberal Imperialism." *Guardian*, April 7. http://www.theguardian.com/world/2002/apr/07/1.

Crane, Conrad. 2010. "The Bombing of Germany." In *American Experience*, produced by PBS Television.

Crawford, Neta. 2003. "The Slippery Slope to Preventive War." *Ethics & International Affairs* 17 (1): 30–36.

Cumming-Bruce, Nick. 2015. "U.N. Rights Chief Says He'll Shine a Light on Countries Big and Small." *New York Times*, January 30. http://www.nytimes.com/2015/01/31/world/un-rights-chief-to-shine-light-on-countries-big-and-small.html.

Doyle, Michael. 2008. *Striking First: Preemption and Prevention in International Conflict*. Edited, with an introduction by Stephen Macedo. Princeton, NJ: Princeton University Press.

Dumas, Lloyd. 2003. "Is Development an Effective Way to Fight Terrorism?" In *War after September 11*, edited by Verna Gehring, 65–74. Lanham, MD: Rowman and Littlefield.

Elshtain, Jean Bethke. 2013. "Prevention, Preemption, and Other Conundrums." In *The Ethics of Preventive War*, edited by Deen Chatterjee, 15–26. New York: Cambridge University Press.

Enemark, Christian. 2013. *Armed Drones and the Ethics of War: Military Virtue in a Post-Heroic Age*. New York: Routledge.

Enemark, Christian. 2014. "Drones, Risk, and Perpetual Force." *Ethics & International Affairs* 28 (3): 365–381.

Falk, Richard. 2014. "Why Drones Are More Dangerous than Nuclear Weapons." In *Drones and Targeted Killing: Legal, Moral and Geopolitical Issues*, edited by Marjorie Cohn, 29–50. Northampton, MA: Olive Branch Press.

Finkelstein, Claire. 2012. "Targeted Killing as Preemptive Action." In *Targeted Killings: Law and Morality in an Asymmetrical World*, edited by C. Finkelstein, J. D. Ohlin, and A. Altman, 156–182. Oxford, UK: Oxford University Press.

Freeman, Chas. 2014. "How Diplomacy Fails." Remarks to the Hammer Forum Review of the Diplomatic Lessons of 1914 for 2014. Los Angeles: The Hammer Museum, August 19.

Goldstein, Richard. 2014. "Theodore Van Kirk, 93, Enola Gay Navigator, Dies." *New York Times*, July 29. http://www.nytimes.com/2014/07/30/us/30vankirk.html.

Gould, Carol C. 2004. *Globalizing Democracy and Human Rights*. New York: Cambridge University Press.

Gould, Carol C. 2014. *Interactive Democracy: The Social Roots of Global Justice*. New York: Cambridge University Press.

Grotius, Hugo. 1925. *De Jure Belli ac Pacis* Translated by Francis W. Kelsey. Oxford, UK: Clarendon Press.

Guiora, Amos, and Jeffrey Brand. 2015. "Establishment of a Drone Court: A Necessary Restraint on Executive Power." In *Legitimacy and Drones: Investigating the Legality, Morality and Efficacy of UCAVs*, edited by Steven Barela, 323–358. Burlington, VT: Ashgate.

Hoffmann, Stanley. 1995. "The Crisis of Liberal Internationalism." *Foreign Policy* 98 (Spring): 159–177.

Ignatieff, Michael. 2003. "The American Empire: The Burden." *New York Times Magazine*, January 5. http://www.nytimes.com/2003/01/05/magazine/the-american-empire-the-burden.html?pagewanted=all.

Jones, Seth. 2008. "How Terrorist Groups End: Lessons for Countering al-Qaeda." Santa Monica, CA: RAND Corporation. http://www.rand.org/pubs/monographs/MG741-1.html.

Kagan, Robert. 2014. "Superpowers Don't Get to Retire." *New Republic*, May 26. http://www.newrepublic.com/article/117859/allure-normalcy-what-america-still-owes-world.

Kahn, Paul. 2003. "The Paradox of Riskless Warfare." In *War after September 11*, edited by Verna Gehring, 37–50. Lanham, MD: Rowman and Littlefield.

Kamm, F. M. 2014. "Taking Just War Seriously in Gaza." *Boston Review*, July 31. http://www.bostonreview.net/books-ideas/fm-kamm-taking-just-war-seriously-gaza.

Kant, Immanuel. 1983. *Perpetual Peace and Other Essays*. Translated by T. Humphrey. Indianapolis: Hackett Publishing.

Kaplan, Shawn. 2014. "Targeted Killing and the Paradox of Riskless Warfare." Unpublished manuscript, presented at the Ethics of War in the 21st Century Conference, Stockholm.

Landler, Mark. 2007. "German Judge Cites Koran, Stirring up Cultural Storm." *New York Times*, March 23. http://www.nytimes.com/2007/03/23/world/europe/23germany.html?pagewanted=all&_r=0.

Luban, David. 2003. "The War on Terrorism and the End of Human Rights." In *War after September 11*, edited by Verna Gehring, 51–64. Lanham, MD: Rowman and Litttlefield.

Luban, David. 2004. "Preventive War." *Philosophy and Public Affairs* 32 (3): 207–248.

Luban, David. 2007. "Preventive War and Human Rights." In *Preemption: Military Action and Moral Justification*, edited by Henry Shue and David Rodin, 171–201. Oxford, UK: Oxford University Press.

Lucas, George, Jr. 2013. "The Case for Preventive War." In *The Ethics of Preventive War*, edited by Deen Chatterjee, 46–64. New York: Cambridge University Press.

Macleod, Alistair. 2007. "The War against Terrorism and the 'War' against Terrorism." In *Intervention, Terrorism, and Torture*, edited by Steven Lee, 187–202. Dordrecht: Springer.

May, Larry. 2013. "Preventive War and Trials of Aggression." In *The Ethics of Preventive War*, edited by Deen Chatterjee, 101–120. New York: Cambridge University Press.

Mazzetti, Mark. 2014a. "Use of Drones for Killings Risks a War Without End, Panel Concludes in Report." *New York Times*, June 26. http://www.nytimes.com/2014/06/26/world/use-of-drones-for-killings-risks-a-war-without-end-panel-concludes-in-report.html.

Mazzetti, Mark. 2014b. "Delays in Efforts to Refocus C.I.A. from Drone War." *New York Times*, April 5. http://www.nytimes.com/2014/04/06/world/delays-in-effort-to-refocus-cia-from-drone-war.html.

McMahan, Jeff. 2012. "Targeted Killing: Murder, Combat or Law Enforcement?" In *Targeted Killings: Law and Morality in an Asymmetric World*, edited by C. Finkelstein, J. D. Ohlin, and A. Altman, 135–155. Oxford, UK: Oxford University Press.

Myers, Robert. 2006. "Notes on the Just War Theory: Whose Justice, Which Wars?" *Ethics & International Affairs* 10 (1): 115–130.

Nordland, Rod. 2015. "Imran Khan Sees His Vindication in a Calmer Pakistan." *New York Times*, November 19. http://www.nytimes.com/2015/11/20/world/asia/pakistan-imran-khan-pti.html?_r=0.

Obama, Barack. 2013. "Remarks by the President at the National Defense University." Washington, DC: White House Office of the Press Secretary, May 23. https://www.whitehouse.gov/the-press-office/2013/05/23/remarks-president-national-defense-university.

Pogge, Thomas. 2001. "Priorities of Global Justice." *Metaphilosophy* 32 (1–2): 6–24.

Rodin, David. 2005. *War and Self-Defense*. Oxford, UK: Oxford University Press.

Sandvik, Kristin Bergtora, and Kjersti Lohne. 2013. "The Promise and Perils of 'Disaster Drones.'" *Humanitarian Exchange Magazine* 58. http://www.odihpn.org/humanitarian-exchange-magazine/issue-58/the-promise-and-perils-of-disaster-drones.

Sen, Amartya. 1999. *Development as Freedom*. New York: Knopf and Oxford University Press.

Sen, Amartya. 2000. "Consequential Evaluation and Practical Reason." *Journal of Philosophy* 97 (9): 477–502.

Sen, Amartya. 2009. *The Idea of Justice*. Cambridge, MA: Harvard University Press.
Shane, Scott. 2012. "The Moral Case for Drones." *New York Times*, July 14. http://www.nytimes.com/2012/07/15/sunday-review/the-moral-case-for-drones.html
Shue, Henry. 1996. *Basic Rights, Subsistence, Affluence, and U.S. Foreign Policy*, 2nd ed. Princeton, NJ: Princeton University Press.
Singer, Peter W. 2012a. "Robot Ethics Won't Clean Up Combat." *Boston Globe*, February 5. https://www.bostonglobe.com/opinion/2012/02/05/robot-ethics-won-clean-combat/kb86gVVzkDpGhooCDUllaJ/story.html.
Singer, Peter W. 2012b. "Do Drones Undermine Democracy?" *New York Times*, January 21. http://www.nytimes.com/2012/01/22/opinion/sunday/do-drones-undermine-democracy.html?_r=0.
Slaughter, Anne-Marie. 2011. "Why Libya Skeptics Were Proved Badly Wrong." *Financial Times*, August 24. http://www.ft.com/intl/cms/s/0/18cb7f14-ce3c-11e0-99ec-00144feabdc0.html#axzz3VBdECcLZ.
Stiglitz, Joseph. 2002. *Globalization and Its Discontents*. New York: W. W. Norton.
Strawser, Bradley Jay. 2012. "Moral Predators: The Duty to Employ Uninhibited Aerial Vehicles." *Journal of Military Ethics* 9 (4): 342–368.
Walzer, Michael. 2006. *Just and Unjust Wars: A Moral Argument with Historical Illustrations*, 4th ed. New York: Basic Books.
Yoder, John Howard. 1984. *When War Is Unjust: Being Honest in Just-War Thinking*. Maryknoll, NY: Orbis Books.

13

Conclusions

JENNIFER M. RAMOS AND KERSTIN FISK

This volume has sought to address some of the important considerations regarding preventive uses of force in various contexts, including the war in Iraq and signature drone strikes in Pakistan. We have suggested that preventive force can be understood in terms of a continuum, where the scope and severity of preventive force may vary, but the basic logic remains the same: using force in order to stop a suspected, potential threat from emerging at an undefined point in the future.

We constructed this volume with three particular goals in mind. The first was to better understand the advantages and disadvantages of one of the most talked about instruments of U.S. foreign policy—armed drones. We believed (and still believe) that these could be clarified by bringing U.S. drone policy into a broader preventive force framework, as opposed to treating drone policy in isolation. Doing so recognizes a long tradition of academic scholarship on preventive motivations and helps provide theoretical underpinnings for analyzing drone policy. Second, we were interested in ascertaining the long-term impacts of the preventive use of force, including the seeming rise of preventive force as an international norm: Will it be a destabilizer in the international system or just a new status quo, a kind of new "balance of threat" (Walt 1987)? Finally, we sought to gain a firmer grasp of the major legal and ethical debates on targeted killing and their evolution. Out of this discussion, we hope to advance a constructive path forward, particularly on the question of whether or not a new regulatory framework is needed to govern drone technology.

To achieve these goals, we asked our contributors to address one or more of the following issues: the costs and benefits of employing preventive force, the political causes and consequences of preventive force, and the legal and ethical implications of preventive actions. In the following, we use these themes to guide our review of the volume's findings, before offering some thoughts on the future of preventive force.

The Ins and Outs of Preventive Force

Whether directly or indirectly, a good number of the chapters deal in some way with the costs and benefits of preventive force. From reputational costs to strategic advantages to political risk, the volume reflects a range of views of the cost-benefit analysis. Some, like our contributor Deen Chatterjee, believe that preventive action is never worth the cost, while others assert that in limited circumstances and for finite time periods may such force be deemed the best course of action.

Jennifer Taw concludes that preventive action is rarely worth the cost. There are simply too many unknowns to risk this type of action, including domestic ramifications and international fallout. However, she does concede that in a purely objective cost-benefit analysis, one type of preventive use of force short of war may make sense: drone strikes. Drones allow states a low-cost way to reduce potential terrorist threats, even ones that have yet to materialize. In her view, the benefits in this case significantly outweigh the costs. Miroslav Nincic concurs; he argues that preventive drone attacks can help minimize costs, particularly the human costs that citizens care so much about, though dangers to democratic accountability and a slippery slope from limited strikes to large-scale military force should slow our enthusiasm.

In his consideration of preventive war, Nincic finds that the public is less willing to support it than other types of war and considers the case that such action can achieve the desired goals with minimal costs (namely, casualties) less than convincing. Thomas Nichols argues that systemic transformations will nevertheless push states to engage in preventive actions, even war. His work implies that, as the norm of sovereignty recedes, and the imperatives of new types of threats like terrorism take hold, states may be more likely than not to downplay the costs and emphasize the benefits. He thus agrees with Taw's and Nincic's conclusions regarding the costs and benefits of preventive drone strikes, while going further to consider the implications across the preventive force continuum.

For his part, Stephan Sonnenberg pushes us to think about what we mean by cost, especially since drones are often touted as a "low-cost" weapon. He asks: Low-cost for whom? The issues raised by drones, from their legality to their impact on civilians—are of upmost importance to scholars like Stephan Sonnenberg, Christine Fair, and Daphne Eviatar. And as David Glazier points out, it is difficult to ascertain the exact human costs when different news websites and organizations like the *Long War Journal* or the Bureau of Investigative Journalism report different casualty rates. Highlighting these data challenges, Christine Fair critiques the assertions made by many human rights advocates regarding the negative effects of drones and offers suggestions for better ways to conduct research to assess the human cost of drone strikes.

Yet even if the outcome of preventive force leads to fewer lives lost than if an attack were to occur, is it ethical to perpetually be engaged in a state of war to the extent that it becomes habitual or a "normal" way of life (even if, via targeted strikes, it involves a relatively lower level of death and destruction)? Scholars including John Emery and Daniel Brunstetter and Ben Jones and John Parrish provide ethical perspectives on the gray areas that arise in the space between war and peace, between military force and law enforcement. As they examine current global issues such as targeted killing and the use of force short of war, there seems to be a consensus that drone technology is a step in the right direction for more "humane" warfare.[1] However, that does not necessarily mean that the policy and laws associated with armed drones are adequate. As we saw in the chapters by Sonnenberg, Eviatar, and Glazier, legal scholars debate both whether current laws are enough and which laws—domestic, international humanitarian law, international human rights law—are most applicable. Under the circumstances, it appears that until international law is more settled, states will continue to interpret legal questions as they see fit.

Given the risks and uncertainty, when might states be likely to see preventive force as their best option? Taw concludes that preventive action is the best option only when two conditions are met: the threat must be greater than the risks of the action, and the threat cannot be equally or better addressed by the alternatives. Yet there may be other situations when preventive action is chosen, even if it is not the best option. For example, a preventive use of force might be the *preferred* choice if a leader believes that he or she would benefit from a rally-round-the-flag effect, resulting in a surge in public approval. A leader may also be compelled to use force preventively when the anticipated political costs, if an attack were to occur on his or her watch, are considered significant. One could also assume that the political cost of being portrayed as "soft" on the state's presumed enemies may be enough to increase a leader's willingness to use anticipatory force, and as Taw notes, there is a psychological tendency to exaggerate threats, even distant and uncertain ones.

Of course, state leaders are not the only influential domestic players with clear incentives to use preventive force. Scholars have in particular noted militaries' tendency toward "offensive bias," which favors preventive grand strategy. For instance, Jack Snyder argues that a highly professionalized soldier is "necessarily preoccupied with the prospect of armed conflict" and has "a simplified, zero-sum view of international politics and the nature of war, in which wars are seen as difficult to avoid and almost impossible to limit" (1984, 118). The military's parochial interests can further encourage offensive bias, as Snyder also notes: "The prestige, self-image, and material health of military institutions will prosper if the military can convince civilians and themselves that wars can be short, decisive, and socially beneficial. One of the attractions of decisive, offensive strategies is that they hold out the promise of a demonstrable return on the nation's investment in military capability" (121). Private interests can also be influential. Some argue that the economic and political sway of Halliburton under the Bush administration may have helped to tip the scale toward war in Iraq (see, for example, Briody 2004). One might argue that the politically powerful defense industry (represented by such corporations as General Atomics, Lockheed Martin, Boeing, and Northrup Grumman) would have a similar interest in a state pursuing preventive force strategies to ensure the expansion of drone contracts.

Such domestic political incentives adversely affect the public's ability to restrain their leaders. As Eviatar's chapter details, the lack of transparency surrounding preventive drone strikes in Pakistan, Yemen, and Somalia means there are severe limitations on the public's knowledge of, and ability to assess, the counterterrorism "success" of the U.S. drone program. Fair's discussion of the serious challenges of reporting in remote regions of Pakistan further illustrates the extent of the gap between private and public knowledge, especially given hindrances to gathering adequate and reliable information.

Addressing the issue of transparency in drone policy from a different perspective, Emery and Brunstetter (Chapter 10) note that a lack of transparency "has opened the way for ever-expanding target lists" legitimized according to the logic of "perpetual imminence." Both Glazier and Sonnenberg argue that this has the effect of removing those accountability mechanisms that would otherwise restrain democratic leaders' willingness to wage and sustain wars of choice rather than necessity. Glazier notes in particular that the problem of executive overreach in the United States in recent years leads us farther away from the laws and principles upon which our security is based.

Indeed, many would argue that it is states' deference to reigning international norms and respect for the rule of law that enhance the stability of the international system. As many of the contributors point out, though, states are increasingly weighing in on their right to self-defense. With regard to preemptive and preventive force, Plaw and Reis argue that states are already asserting a broader interpretation of self-defense that moves beyond the "restrictionist" interpretation of international law to one that is more flexible in meeting the needs of the times—an approach these authors call "Caroline plus." Not surprisingly, scholars point to the role of the United States in setting a precedent for the use of preventive self-defense. As Sonnenberg observes, "When the first movers to adopt drones employ them in ways that offend other nations' understanding of IHL [international humanitarian law], the game-theory prediction suggests that those other nations would soon also abandon preexisting normative frameworks in favor of U.S. precedent" (see also Fisk and Ramos 2014). Such considerations and others regarding the future of preventive force are the subjects we turn to next.

Future of Preventive Force

As we mentioned in the introductory chapter, preventive force is nothing new in the international system. It is a long-standing, if infrequent, practice of states. Now, the question at hand seems to be whether or not such action will become a norm, an accepted standard of behavior for international interactions. Scholars have provided some evidence that a preventive force norm is evolving—at either end of the spectrum, from war (see, for example, Nichols, Chapter 4) to limited military strikes. At the same time, there seems to be considerable global backlash against preventive force at both the citizen and the state levels, making it difficult to know whether or not we have seen the climax of preventive force or if it the trend will continue.

If the trend toward state use of preventive force does continue, some suggest that the United Nations, or an international institutional authority, should be empowered to set standards for when preventive force is legitimate. Indeed, Whitley Kaufman (2005) argues that such legitimacy can be gained only through the approval of the United Nations Security Council and that such approval must be determined by criteria from the just war tradition as well as appeals to ethical norms. If such standards exist, states can be held accountable for their actions, even if that means they sacrifice only their international reputations. Allen Buchanan and Robert Keohane (2005) go so far as to suggest criteria by which preventive use of force should be evaluated both before *and after* such use of force is undertaken.

This suggestion, however, leaves out the role of increasingly powerful players in the international system: that of non-state actors. Thus far, our focus has been on states although non-state actors may be just as likely as state actors to find preventive logic compelling. Furthermore, Hezbollah and Hamas have already used drones in armed combat (New America Foundation, 2015). While their use of drones has not necessarily been preventive, it is easy to see how the precedent set by the United States could embolden these organizations to use drones in this way.

One thing to keep in mind, though, is that no matter the weapon or security force strategy, it goes only so far. We must recognize that there are limits to the use of force. Even if we can find scenarios in which preventive force might be legitimate or just, some might argue that the very nature of the inquiry itself already sets us up for a self-fulfilling prophecy of violence (Steele 2013). Scholars such as Deen Chatterjee (2013, 2015) urge us to consider alternatives such as preventive nonintervention. Resisting attempts to mold traditional just war thinking to current threat environments, Chatterjee shifts the focus to "just peace." In this view, how can we expect to build peace if our energy is spent on finding "better" ways to go to war? Trends in global justice and human rights norms suggest that this may, in fact, be a more humane and ultimately more secure path to the goals that just war is trying to achieve.

NOTE

1 Note that Sonnenberg asserts that efforts to make war more "humane" are deceiving because these will likely contribute to its frequency, not its restraint.

BIBLIOGRAPHY

Briody, Dan. 2004. *The Halliburton Agenda: The Politics of Oil and Money*. Hoboken, NJ: Wiley.

Buchanan, Allen, and Robert O. Keohane. 2004. "The Preventive Use of Force: A Cosmopolitan Institutional Proposal." *Ethics & International Affairs* 18 (1): 1–22.

Buchanan, Allen, and Robert O. Keohane. 2005. "Justifying Preventive Force." *Ethics & International Affairs* 19 (2): 109–112.

Chatterjee, Deen, ed. 2013. *The Ethics of Preventive War*. New York: Cambridge University Press.

Chatterjee, Deen. 2015. "Deciding on Preventive War: Amartya Sen's Idea of Justice." *Philosophy & Social Criticism* 41 (1): 69–76.

Fisk, Kerstin, and Jennifer M. Ramos. 2014. "Actions Speak Louder than Words: Preventive Self-Defense as a Cascading Norm." *International Studies Perspectives* 15 (2): 163–185.

Kaufman, Whitley. 2005. "What's Wrong with Preventive War? The Moral and Legal Basis for the Preventive Use of Force." *Ethics & International Affairs* 19 (3): 23–38.

New America Foundation. 2015. "World of Drones." Washington, DC. http://security-data.newamerica.net/world-drones.html.

Snyder, Jack. 1984. "Civil Military Relations and the Cult of the Offensive." *International Security* 9 (1): 108–146.

Steele, Brent J. 2013. *Alternative Accountabilities in Global Politics: The Scars of Violence*. New York: Routledge.

Walt, Stephen M. 1987. *The Origins of Alliances*. Ithaca, NY: Cornell University Press.

ABOUT THE CONTRIBUTORS

Daniel R. Brunstetter is Associate Professor of Political Science in the School of Social Sciences at the University of California, Irvine. His publications include articles in *Ethics & International Affairs*, the *Journal of Military Ethics*, the *Review of International Studies*, *Raisons Politiques*, and the *International Journal of Human Rights*, as well as a book, *Tensions of Modernity: Las Casas and His Legacy in the French Enlightenment*.

Deen Chatterjee is Senior Fellow of Law at the S. J. Quinney College of Law at the University of Utah and a Global Ethics Fellow at the Carnegie Council for Ethics in International Affairs in New York. He is the editor in chief of the two-volume *Encyclopedia of Global Justice* and the series editor of *Studies in Global Justice*. His publications include *The Ethics of Preventive War* and, with Don E. Scheid, *Ethics and Foreign Intervention*. His areas of specialization are justice and global initiatives, ethics of war and peace, and philosophy of religion and culture.

John Emery is a PhD candidate in Political Science at the University of California, Irvine. His research interests include ethics of technology in warfare, U.S. foreign policy, and issues of language, space, and time in terrorism and counterterrorism operations.

Daphne Eviatar is Senior Counsel for National Security at Human Rights First. She has written widely about law, national security, and human rights for such publications as the *New York Times*, the *Washington Post*, *Harper's Magazine*, *Politico*, and many others. She is the author of *Detained and Denied in Afghanistan* and a contributor to the book *Why Peace?* She writes a blog on the *Huffington Post*, is a founding editor of the legal blog *Just Security*, and serves on the advisory board of Rethink Media.

C. Christine Fair is Associate Professor at Georgetown University's Security Studies Program in the Edmund A. Walsh School of Foreign Service. Her research focuses on political and military affairs in South Asia. She has authored, co-authored, and co-edited several books including *Cops as Counterinsurgents: Policing Insurgencies*, edited with Sumit Ganguly, *Cuisines of the Axis of Evil and Other Irritating States*, and *Treading Softly on Sacred Ground: Counterinsurgency Operations on Sacred Space*, edited with Sumit Ganguly. She is also a Senior Fellow at the Combating Terrorism Center at West Point.

Kerstin Fisk is Assistant Professor of Political Science at Loyola Marymount University. Her research concerns issues of security, peace, and conflict, with emphases on foreign and defense policy and political violence. Her current projects investigate the implications of unmanned systems for foreign policy, international relations, and individual attitudes toward the use of force. She has published in *International Studies Perspectives* and *Civil Wars*.

David Glazier is Professor of Law at Loyola Law School, Los Angeles. He spent twenty-one years as a U.S. Navy surface warfare officer before retiring to attend law school, and now writes extensively on the application of the law of armed conflict to contemporary conflicts, particularly those involving "nontraditional" adversaries.

Ben Jones is a PhD candidate in Political Science at Yale University, with research interests in political theory, religion and politics, and public policy (especially in the area of criminal justice). His work on "The Paradox of Secular Apocalyptic Thought" examines ways in which Christian eschatology becomes secularized in political thought. His research has been featured in *Ethical Theory and Moral Practice* and *Politics, Groups, and Identities*.

Thomas M. Nichols is Professor in the National Security Affairs Department at the U.S. Naval War College in Newport, Rhode Island, and teaches for both the Harvard Extension and Summer Schools. Previously, he was a Fellow of the International Security Program and the Project on Managing the Atom at Harvard's John F. Kennedy School of

Government. He is currently a Senior Associate of the Carnegie Council on Ethics and International Affairs and a Fellow of the International History Institute at Boston University. He is also the author of several books, including *Eve of Destruction: The Coming of Age of Preventive War* and, most recently, *No Use: Nuclear Weapons and U.S. National Security*.

Miroslav Nincic is Professor of Political Science and International Relations at the University of California, Davis. Previously, he taught at New York University and the University of Michigan. He has written or co-authored eight books, edited two, and produced a large body of articles in the areas of international relations and U.S. foreign policy. His most recent books are *Renegade Regimes* and *The Logic of Positive Engagement*.

John M. Parrish is Professor and Chair of Political Science at Loyola Marymount University. He is the author of *Paradoxes of Political Ethics: From Dirty Hands to the Invisible Hand* and co-author, with Alex Tuckness, of *The Decline of Mercy in Public Life*.

Avery Plaw is Associate Professor of Political Science at the University of Massachusetts, Dartmouth, and is the Director of the University Honors Program. His research focuses on political theory and international relations. Plaw's book *Targeting Terrorists: A License to Kill?* was runner-up for the Canadian Political Science Association's Prize for Best Book of the Year in International Relations, and he has edited two volumes, *The Metamorphosis of War* and *Frontiers of Diversity: Explorations in Contemporary Pluralism*. He is currently working on a book on the ethics and legality of drone warfare.

Jennifer M. Ramos is Associate Professor of Political Science at Loyola Marymount University. Her research focuses on U.S. foreign policy, international security, and international norms. Her work has appeared in *Journal of Politics, International Studies Perspectives, Foreign Policy Analysis*, and other periodicals. Her recent books include *Changing Norms through Actions: The Evolution of Sovereignty*, and a co-edited volume with Richard L. Fox, *iPolitics: Citizens, Elections and Governing in the New Media Era*.

João Franco Reis is a PhD student and a Junior Researcher at the Institute of Political Studies at Portuguese Catholic University. He is a 2010 graduate of the University of Lisbon Law School. He was also part of the IMPACT Program at the University of Massachusetts, Dartmouth, in 2011 and a visiting student at St. Anthony's College at Oxford University in 2012.

Stephan Sonnenberg is a Clinical Supervising Attorney and Lecturer in Law with the International Human Rights and Conflict Resolution Clinic at Stanford Law School. He has authored or co-authored numerous advocacy reports, including the seminal 2012 report entitled "Living under Drones," which detailed the impact of the U.S. drones program on civilians living in Northwest Pakistan. He has also written on the issues of corporate social responsibility, human trafficking, and judicial remedies following instances of communal violence.

Jennifer Taw is Associate Professor of Government and International Relations at Claremont McKenna College. She is author of *Mission Revolution: The U.S. Military and Stability Operations* and co-author of *World Politics in a New Era*, 6th edition.

INDEX

ababil (holy swallows), 212
Abbot, Sebastian, 209–10, 216
abductions, 22n2
Abrams, Elliott, 99
accountability, 189–90, 305–7, 331, 333, 345
accuracy, targeting, 124–25, 126, 132
advocacy organizations, 214, 216, 217, 220n10
Afghanistan, 9, 39, 60–61, 82, 179, 206
aircraft, military, 157–58, 162
air warfare regulations, 162–63
al-Awlaki, Anwar, 142, 152, 154
Al Badawi, Jamal Ahmad Mohammad, 273
al-Harthi, Ali Qaed Senyan, 271, 308n5
al-Kibar, 49–50
al-Libi, Abu Anas, 276
allies, 2, 12–13, 186
al-Qaeda, 165n5, 202, 279n4; "associated forces" of, 146, 149, 174–77, 238, 273, 276–77; civilians compared to combatant in, 150; drone strikes on, 8, 9, 14, 146, 276–77; religious and societal values and, 41–42; territory of, 158; U.S. armed conflict distinction with, 174–77, 293; U.S. policy on targeting, 134, 149, 163, 174–77, 263–64, 301; U.S. sanction against, 50
al-Qaeda in the Arabian Peninsula (AQAP), 8, 14, 146, 177, 179, 270–71, 275
al-Quso, Fahd, 273
al-Radmi, Hamid, 275
al-Shabaab, 8, 41, 147, 279n5
Alston, Philip, 243
al-Wuhayshi, Nasir, 146
Amnesty International, 156, 209, 214–15, 217, 266

Anderson, Kenneth, 262
Annan, Kofi, 144, 239–40
Ansar al-Sharia, 8
anti-American sentiment, 13–14, 128–29, 171, 177, 180–83, 184, 189, 335n9
anti-drone sentiment, 123; Code Pink's, 218n5, 243; in Pakistan, 127, 146, 190n2, 204; research influenced by, 210–12, 216, 220n11; in U.S. and globally, 190n2
"anti-proliferation imperative," 7
AQAP. *See* al-Qaeda in the Arabian Peninsula
Arab League, 99
armament reduction, 118–19
armed conflict: co-belligerency in, 178; combatants targeted outside of, 4, 170, 172–73, 262, 295–302; distinction with al-Qaeda affiliates, 174–77, 293; interpretation of, 249–50; law enforcement compared to, 261–62, 272–73; for self-defense, 143–44, 146–47, 165n1, 202, 244–45, 249–50, 331–32; in Yemen and Pakistan, 179–80, 335n9. *See also* non-international armed conflict
armed conflict zones, 157–58, 309n7; criteria for force in and out of traditional, 172–73, 257, 263, 266, 278, 295–302, 298; laws governing, 4, 170, 172–73, 174–77, 258–59, 268–69; legality inside of compared to outside of, 174–77, 292–93, 295–303; problems with expanding, 299–300; Walzer on killing outside of, 258, 268, 284, 295–303. *See also* "in-between zones"

354 | INDEX

armed force: alternate use of, in Pakistan, 209, 212; criteria for, 17, 22n5, 116–17, 143, 150, 151–52, 153, 160, 161, 164, 172–73, 190n3, 252n9, 257, 263, 266, 269–70, 273–76, 278, 286, 292, 295–302, *298*, *303*, 308, 319; ethics and legality of, 120, 126–27, 143–45; IHRL on, outside of conflict zone, 4, 170, 173; under LEE, 292, 297–98, *298*, 303, 308; self-defense justification for, 143–44, 146–47, 165n1, 202, 244–45, 249–50; UN Charter and legality of, 120, 126–27, 137n4, 143–45, 323; by UN Security Council, 137n4, 143
armed forces (military), 148–49, 344
Article 51, 9, 22, 159, 202; criteria for force in, 22n5, 150, 190n3; "inherent right" interpretation of, 230, 233–34; "preventive force" interpretation of, 235–36; schools of thought on, 229, 230–36; "strict construction" interpretation of, 231–32; terrorism and interpretation of, 231–32, 234
Ashburton, Lord, 5, 159
al-Assad, Bashar, 52, 98, 101, 103
"associated forces," targeting: of al-Qaeda, 146, 149, 174–77, 238, 273, 276–77; international law on, 178; in World War II, 147
Authorization to Use Military Force (AUMF): legality of, 8–9, 163, 202, 293; specifications and limits in, 145–46, 161, 202; U.S. armed conflict in Yemen and, 179
Awakening Movement, 74
Axe, David, 215
Axelrod, Robert, 133, 134

Barzani, Massoud, 241
battlefields. *See* armed conflict zones
Benjamin, Medea, 154
Bergen, Peter, 9–10
Berger, Charles E., 176
Berra, Yogi, 90
Bethlehem, Daniel, 161, 166n14
Bhutto, Benazir, 205

bin Laden, Osama, 61, 91, 152, 205; death impact of, 175–76, 247–48; drone strikes targeting, 199, 242–43; killing associates of, 146, 277
Blair, Tony, 93, 103
Boko Haram, 41
Boyle, Michael, 13–14
Braun, Megan, 264, 265, 267, 268–69
Breedlove, Philip, 313
Brennan, John, 9, 170
Britain, 94–95, 103, 159, 234, 288. *See also* United Kingdom
Brodie, Bernard, 37
Brooks, Rosa, 267
Brunstetter, Daniel R., 264, 265, 267, 268–69
B'Tselem, 240
Buchanan, Allen, 331, 333
Budapest Memorandum (1994), 96–97
Burke, Anthony, 327
Bush, George H. W., 164, 188
Bush, George W., 73, *75*, 80–81; on counterterrorism, 11, 92, 145–46, 149, 183; ethical considerations of, 15; on Iraq War objectives, 6–7, 67, 74, 77; Obama's criticism of, 90, 257; on preemptive action, 144; on preventive force, 52, 56n9, 235; secret agreement with Pakistan and, 201; war on terror actions by, 147, 183, 184
Bush administration, 8, 165n4, 201–2, 218n4, 242, 293, 344
Bush Doctrine, 60, 235; critics of, 97, 324; legacy of, 92, 93, 257, 334n3; unilateralism in, 317, 322–23

Callen, David, 14
Cambodia, 159, 161
Cameron, David, 94, 100, 103, 107n13
Canada, 159, 234
capture and surrender, 131, 274–76, 277, 294, 320
Caroline, 5, 159, 234, 238, 248–49
Caroline doctrine, 5, 22n5, 234, 248–49, 252n9, 260

"Caroline plus," 21, 248–49, 250, 252nn9–10, 261, 345
Cartwright, James, 294
Casey-Maslen, Stuart, 122
Cassese, Antonio, 231
casualties: in cost and benefit analysis, 18, 62, 63, 65, 125; of flawed "precision" targeting, 125, 126, 132; in Iraq War compared to World War II, 72; minimizing Iraq War, 81–82; prohibition of foreseen civilian, 292, 294, 297, *298*, 304–5, 308n3; public expectations of Iraq War, 69, *69*; public opinion and, 18, 63, 65, *72*, 72–73, 76, 82–83; in threat assessment, 322–23. *See also* collateral damage; drone victims
Chayes, Sarah, 335n16
chemical weapons, 98–99, 101, 102, 119
Cheney, Dick, 322, 323
Chesney, Robert, 179
child witnesses, 214, 216, 220n15
China, 39, 40, 41, *44*, 49, 96
Christopher, Russell, 12
CIA, 9, 41, 125, 157, 179, 301; civilians killed by, 219n6, 308n6; drone strikes by, 161–63, 199, 271, 272, 273, 308nn5–6, 334n7; outside of international law, 319; Pakistani military agreement with, 200–201; target identities of, 206, 294
citizens, U.S., 142, 189, 308n6
citizenship, 154–55
civilians. *See* noncombatants
The Clash of Civilizations (Huntington), 42
climate change, 42, 55n5
Clinton, Bill, 7, 93
Clinton, Hillary, 90
Coady, C. A. J., 279, 288
Coady, Tony, 314
Code Pink, 218n5, 243
Cold War, 6, 19, 41, 89, 95, 97, 106
collaboration, 324, 330–31, 332, 333
collateral damage, 120, 142; proportionality and discrimination of, 116, 121, 151,

155–57, 164, 240, 270–71; from targeted killing, 146, 149, 156–57
Colombia, 241–42, 245, 246
combatants: distinction between non-combatant and, 149–50, 320, 322, 325; outside of armed conflict, 4, 170, 172–73, 262, 295–302; targeted killing of, 148–49, 210, 291–92
combat zones. *See* armed conflict zones
Commission on Human Rights, 173
communism, 41, 46
congressional elections. *See* elections
consent, 252n9; by neutral states, 160–61, 179–80, 245; NIAC, 244–45; Pakistani debate, 160, 179–80, 201–5, 242–43, 245
consequentialist ethics, 15–16, 291, 330
Constitution, U.S., 143, 154
Corten, Olivier, 236, 237
cost and benefit analysis, 124; accurate threat assessment for, 34, 37, 50–51, 53; casualties in, 18, 62, 63, 65, 125; of Israel's attacks in Syria and Iraq, 49–50; of preventive action/force, 47–48, 50–51, 72–73, 342–43; public opinion impacting, 34, 38, 51; public's, of war, 62, 63, 65, 81; for sanctions, 50; of targeting terrorists, 54–55; threats in, 38–42; types of benefits in, 36–38, 342–43; types of costs in, 35–36, 342–43; wealth and power of nation in, 35, 36, 38, 53–54, 324–25, 326–27
counterterrorism, 258; alternative approaches to, 335n15; Bush, G. W., on, 11, 92, 145–46, 149, 183; drones as tool for, 170–71, 257; ethical and legal dilemmas in, 268–69, 307–8, 309n9; Israeli example of, 239–40; law enforcement compared to armed conflict for, 261–62, 272–73; Obama on, 90–92, 107nn3–4, 146, 149; partisan conflict on, 107n5; self-defense justification in, 238–43; in Syria, 241
covert military operations, 218n4, 319

Crane, Conrad, 322
Crimea, 38–39, 40, 43, 50–51, 95–97, 188
Cuba, 39–40
Cuban Missile Crisis, 67
customary law, 236–37, 243–48, 251

Darwish, Kemal, 308n6
Davis, Raymond, 219n6
deaths, 23n14, 175–76, 209, 239, 247–48; in Pakistan, 156, 219n8, 220n12; in Syria, 98. *See also* casualties; drone victims
decapitations, 244
Declaration of St. Petersburg (1868), 117–18
defeat, 65
defense industry, 344
defensive war, 59, 60, 66
de Mazière, Thomas, 122
democracy, 3–4, 58, 326, 336n17, 346
Democrats, 62, 76–77, *78*, 107n5
Dempsey, Martin, 99–100
deontological ethics, 15–16, 291
DeRouen, Karl, 51
detentions, 22n2
development ethicists, 330
Dever, James, 234, 250
Dever, John, 234, 250
de Wijze, Stephen, 283, 284–86, 287, 289, 295
Dinstein, Yoram, 232, 234
diplomacy, 49, 313
"directly participating in hostilities" (DPH), 172
dirty hands framework: accountability in, 305–7; emergency ethics and, 287–90; examples of, 288–89, 302–5; moral conflict in, 285–86, 291, 305; as "policy," 289–90; targeted killing and, 283–87, 302–5, 307, 309n9
disease, 335n14
Doctor Zhivago, 41
domestic law, U.S., 142–43, 145, 163–64, 179, 343

domestic military action, 45
domestic terrorism, 299, 300
"double tap" strikes, 243, 252n6
Doyle, Michael, 258
DPH. *See* "directly participating in hostilities"
drone operators, 129–30, 131, 132, 218n1
drone policy, U.S., 162, 279n7; on al-Qaeda, 134, 149, 163, 174–77, 263–64, 301; analysis of, 2–3, 22n3, 116, 341; "associated forces" and, 146, 149, 171, 174; compared to Israeli policy, 157; consent in, 203; ethics and legality of, 121–22, 163–65, 257, 292–93, 294; international community adopting, 14–15, 134–35, 170, 188, 245–46, 345; international opinion and support of, 6, 13–14, 128–29, 306; law enforcement compared to "just war" view of, 260–61, 295–96; long-term risks with, 143; national interest primary in, 126–27; under Obama administration, 2, 7–8, 90, 92, 107n3, 131, 142, 146, 171, 174, 218n5, 242–43, 257, 258, 278, 309n10, 318, 335nn10–11; preventive war concerns with, 265–66; self-defense justification in, 246–47; target selection in, 270–71
drone regulation, 136, 188–89, 268–69, 331; global policy of, 19, 123, 252n8; hybrid law enforcement-just war paradigm for, 258–59, 266–67, 278–79, 335n8; under IHRL, 20, 258, 261; limitations of just war tradition win, 133, 262–66, 267; need for distinct, 115–16, 121–22, 135, 171, 266
drones, 2, 115, 165n3, 199, 220n14; accuracy reality of, 124–25; alternative names for, 22n1, 55n4; banning, 135; compared to historical war tactics, 264–65; compliance with international law, 19–20, 120–22, 126–27, 131–33, 142–43, 261–62, 267, 345; as counterterrorism tool, 170–71, 257; military markings law and,

162–63; as "moral predators," 318, 319, 331; peace impacted by, 131–32
drones, surveillance, 164, 215, 220n14
drone strikes, 1, 16; accountability with, 305–7, 331, 333, 345; allies alienated by, 12–13; on al-Qaeda, 8, 9, 14, 146, 276–77; alternatives to, 209, 212; on AQAP, 8, 146, 270–71, 275; AUMF and, 9, 163; on bin Laden, 199, 242–43; under Bush administration, 165n4, 201–2, 242; by CIA, 161–63, 199, 271, 272, 273, 308nn5–6, 334n7; civilians impacted by, 121, 126, 129, 132, 202, 208, 209–16, 271, 343; consent for, by neutral states, 160–61, 179–80, 245; data and analysis methods of, 20, 216–17; ethics and legality of, 4, 19–20, 120–22, 126–27, 129, 142–43, 147, 163–65, 252n9, 257–61; in FATA, 199, 200, 208, 212; indictment process for, 272–76; as last resort, 269–78; location requirements for employing, 157–61; as "low-cost" option, 2, 54, 105, 116, 127–28; under Obama administration, 7–8, 131, 165n4, 201–2, 218n4, 242; organizations authorized for, 161–63, 280n9; in Pakistan, 7–8, 20, 45, 115, 116, 120–22, 124–26, 128–29, 146, 179–80, 199, 200–205, 208, 242–43; peace efforts impacted by, 131–32; public opinion and support of, 10, 23nn13–14, 190n2, 243, 247, 306; risks of, 55, 129, 132; in Somalia, 8, 45, 115, 242; as sovereignty threat for target nation, 39; target transparency in, 8, 9–10, 345; territory of, 258, 268–69; terrorism increased by, 13–14, 156, 271; "tip-sheet" for avoiding, 125; trauma and devastation from, 125–26, 203, 215, 219n6, 306–7; types of covert, 218n4; unidentified victims in, 9, 23n12, 218n4, 270–71; in violation of international law, 121–22, 127, 162, 170; in Yemen, 7–8, 14, 23n11, 115, 137n7, 146,

242, 275, 296–97, 308nn5–6, 334n7. *See also* targeted killing

drone technology, 2, 22, 220n16, 220n18; arms race for developing, 134; casualties reduced with, 82; computer game compared to, 130–31; countries with, 132, 164; human rights view of, 116, 122–23, 128, 136; just war tradition and, 319–21; lack of capture ability in, 131; as structurally disruptive, 116, 122–23, 132–33, 136n1

drone victims, 92, 137n7, 146, 154; civilians as, 13, 125, 132, 142, 156, 170, 209–17, 218n4, 242, 270, 285, 287, 294, 304–5, 308n6, 335n9; unidentified, 9, 23n12, 218n4, 270–71

"dum-dum" bullets, 119, 137n3

Dunant, Henry, 117, 136n2

economic interests, 63, 64, 67, 75, 83
Ecuador, 242
elections, 79, 79–81, *80*
elite, policy, 62
Elshtain, Jean Bethke, 316
Emanuel, Rahm, 309n10
embassy bombings, U.S. (1998), 272, 276–77
emergency ethics, 287–90, 322
Emmerson, Ben, 202–3, 243
Enola Gay, 320
environmental issues, 64
Erdoğan, Recep Tayyip, 240–41
ethics and legality, 18, 21; armed conflict zones and, 172–73, 174–77, 258–59, 268–69, 292–93, 295–303; of armed force, 120, 126–27, 143–45; of authorized organizations for drone employment, 161–63; of CIA drones, 161–63; of contemporary self-defense practice, 248–51; counterterrorism dilemmas in, 268–69, 307–8, 309n9; in customary law interpretation, 236–37; deontoglogical compared to consequentialist, 15–16, 291;

ethics and legality (*cont.*)
developmental approach in, 330; distinction and proportionality in, 155–57, 265; of drone strikes, 4, 19–20, 120–22, 126–27, 129, 142–43, 147, 163–65, 252n9, 257–61; of interventions, 164, 188; of low-level fighters targeted, 16; peace over war in, 185; of preventive and preemptive force, 4, 37, 229, 230–36, 252n5, 257–58, 264, 315–26; of research on drone impacts, 210–12, 220n11; of strike locations, 157–61; of strikes in Pakistan, 127, 202–6; of targeted killing, 4, 16, 121, 143, 147–55, 187–89, 238, 259, 278, 284–85, 341; of target selection, 269–71; in tradition of war, 16–17, 283; of U.S. drone policy, 121–22, 163–65, 257, 292–93, 294; in U.S. foreign policy in World War II, 5–6. *See also* dirty hands framework; *jus ad bellum*; *jus ad vim*; *jus in bello*; just war ethic; law enforcement ethic

European Court of Human Rights, 190n7
European Union, 252n8
Eve of Destruction: The Coming Age of Preventive War (Nichols), 4, 18–19, 88, 89–90, 106, 107n1
expansionism, 46
extraterritorial force, 172, 180–83, 186, 233

Falk, Richard, 325
FARC. *See* Revolutionary Armed Forces of Colombia
FATA. *See* Federally Administered Tribal Areas and Frontier Regions
FCR. *See* Frontier Crimes Regulation
fear, 46, 213–14, 313
Feaver, Peter, 65
Federally Administered Tribal Areas and Frontier Regions (FATA), 120, 179, 219n8; drone strikes in, 199, 200, 208, 212; FCR control over, 207–10, 212;

media reports from, 209–10; organization of and restrictions in, 207–8, 219n9; Pakistani military attacks in, 214–15; terrorism in, 208–9

Feliciano, Florentino, 233
FFR. *See* Foundation for Fundamental Rights
Finkelstein, Claire, 274, 335n8
"foot soldiers." *See* low-level fighters
force. *See* armed force; preventive force
foreign policy, U.S., 4, 12–13, 22n5, 65, 107n12; economic interests in, 67; future of, 83; historical trends in, 5–10; on imminent threat, 6, 9–10, 296; interventionist, 105; after 9/11, 7, 8–9, 64, 68, 142–43; Obama's speeches on, 104–5, 262–63, 274; objectives of, 58, 62–63, 64, 68; public opinion and, 58, 62–63, 64; "tit-for-tat" strategy in, 134; of war preparedness, 313–14. *See also* domestic law; drone policy, U.S.
Foundation for Fundamental Rights (FFR), 210–12, 220n11
France, 46, 94–95, 103
Freeman, Chas, 313
Frontier Crimes Regulation (FCR), 207–10, 212

G-8, 40, 44
Gaddis, John Lewis, 12–13
Gallup surveys, 67, 68, *68*, 73, *78*
Gelpi, Christopher, 65
Geneva Conventions, 131–32, 147, 184
genocide, 98
Gentili, Alberico, 17
geographical spaces. *See* territory
Germany, 13, 122, 147, 288, 322
Glazier, David, 267
goals. *See* objectives
Gonzales, Alberto, 183–84
Greenwood, Christopher, 231
Grotius, Hugo, 17, 315, 317, 334

ground troops, 71, 83, 170, 189, 294
Guantánamo detentions, 90, 142, 147, 155, 163, 183
Gulf War (1991). *See* Operation Desert Storm

Hague Convention (1899), 118–19, 137n3
Halliburton, 344
Hamas, 132, 346
Hamdi v. Rumsfeld, 155
Haqqani, Husain, 204
Haqqani network, 8
Hay, Jon, 118–19
HCJ. *See* Israeli High Court of Justice
helicopters, 221n17
Heyns, Christof, 178, 243
Hezbollah, 43, 132, 239–40, 346
Hiroshima, 288, 320
Hitler, Adolf, 6, 184
Hollande, Francois, 102, 103, 107n13
holy swallows (*ababil*), 212
Houthi rebels, 45
Hudson, Leila, 14
"humane" war, 4, 116–20, 343, 346, 347n1
humanitarian disasters, 89, 93–94
human rights, 173; benefits of U.S. abiding by, 186–87; drone technology and, 116, 122–23, 128, 136; interventions for, 88–89, 323–24, 327–28; military and private interests compared to, 126–27, 133–34, 344; public opinion on, 63, 64; rise of global, awareness, 328–29; violations of, under Obama, 131. *See also* International Human Rights Law
Human Rights Committee, 190n7
Human Rights Watch, 122, 261, 266–67, 270, 271, 274–75
Huntington, Samuel, 42
Hussein, Saddam, 152; Iraq War justification and, 53, 61, 67; public opinion on, 74–75, 75; threat of, 1, 11, 52, 67–68, 73, 74

IAC. *See* International Armed Conflict
ICCPR. *See* International Covenant on Civil and Political Rights
ICESCR. *See* International Covenant on Economic, Social, and Cultural Rights
ICG. *See* International Crisis Group
ICJ. *See* International Court of Justice
ICRC. *See* International Committee of the Red Cross
IEDs. *See* improvised explosive devices
IHL. *See* International Humanitarian Law
IHRL. *See* International Human Rights Law
illegal immigration, 46, 55n6
imminent threat, 5, 56n9; anticipatory self-defense compared to, 12, 33, 144, 155, 248–49, 273–74, 314, 317; in criteria for targeting killing, 286, 293, 296, 297, 298, 303; defining, 249–50, 293, 296; emergency ethics and, 287–88; interpretation of, 186, 234, 250; lag and predictability with, 42, 43–44, 46, 53, 259, 260–61, 269; perpetual, 259, 260–61, 263, 264, 272–74, 276, 278, 345; requirement of, not met, 303; rogue states and terrorist as, 11; self-defense argument and, 233, 246, 260; U.S. foreign policy on, 6, 9–10, 296
imperialism, 326–27, 335n13
improvised explosive devices (IEDs), 82
IMU. *See* Islamic Movement of Uzbekistan
"in-between zones," 258, 284, 295–302, 298, *298*
Independents, 78
India, 40
indictment process. *See* law enforcement paradigm
infrastructure attacks, 22n2
"inherent right" school, 230, 233–34
institutionalist/restrictive approach, 236–37, 248

interests, 124; military and private, 126–27, 133–34, 344; national, as primary goal, 126–27, 206, 219n7; oil, 75, 83
internal factions, 39, 43, 45
International Armed Conflict (IAC), 172
International Bill of Human Rights, 173
international collaborative framework, 331–32
International Committee of the Red Cross (ICRC), 116, 117, 151–52, 153, 157, 175
international community, 16, 299; anti-American sentiment in, 13–14, 128–29, 171, 177; disapproval of preventive force in, 52–53, 190n2; drone regulation and, 19, 123, 252n8, 267; human rights awareness in, 328–29; on Iraq War, 12–13; preventive cooperation in, 4, 36; preventive force as trend in, 1–2, 14–15, 19, 106, 115–16; targeted killing adopted by, 1–2, 14–15, 127, 245–46, 345; threats to status in, 40–41, 44; U.S. drone policy adopted by, 14–15, 134–35, 170, 188, 245–46, 345; U.S. drone policy opinion within, 6, 13–14, 128–29, 306; U.S. reputation in, 6, 13–14, 128–29, 171, 186–89, 191n8, 293
International Court of Justice (ICJ), 231–32, 236, 237, 247–48, 251n2
International Covenant on Civil and Political Rights (ICCPR), 154, 173–74, 180–82, 186, 190n4, 190n7
International Covenant on Economic, Social, and Cultural Rights (ICESCR), 173
International Crisis Group (ICG), 202–3, 204, 207
International Humanitarian Law (IHL), 155, 157, 165n1; capture over kill in, 131; on civilians participating in hostilities, 149–52; on collateral damage and proportionality, 156; on combatants, 148–49; criteria for armed force in, 116–17, 143; drones' compatibility with, 120–22, 131–33, 261–62, 345; drones' regulation under, 135–36, 258; on "in-between" zones, 298; Israel campaign against Hizbullah and, 239–40; JWE link to, 291; signature strikes under, 154
International Human Rights Law (IHRL): drone regulation under, 20, 258, 261; extraterritorial application of, 172, 180–83, 186; on "in-between" zones, 298; LEE link to, 292; on lethal force outside of conflict zone, 4, 170, 173; "right to life" in, 143, 155, 173–74; U.S. interpretation of and adherence to, 180–83, 184, 189
internationalism. See liberal internationalism
international law, 17, 21, 171; CIA outside of, 319; drones compliance with, 19–20, 120–22, 126–27, 131–33, 142–43, 261–62, 267, 345; drone strikes in violation of, 121–22, 127, 162, 170; evolution of, 251, 299, 343; history of, on warfare, 118–20; on hostility in neutral territories, 158, 159–60; preventive force in, 5, 170, 230, 314, 317; sovereignty violations under, 251; targeted killing and, 238, 284; on targeting "associated forces," 178; terrorism and updating, 184; U.S. adherence to, 180–83, 184, 186–89, 190n4; on U.S. targeting al-Qaeda, 174–77. See also customary law
international peace, 137n4, 144, 230, 323–24
international sanctions. See sanctions, international
Interservices Intelligence Directorate (ISI), 200–201, 204
interventions: collaboration over, 330–31; future of U.S., 105; for human rights, 88–89, 323–24, 327–28; just war tradition on, 325, 330–31; lack of, in Syria, 88–89, 98–104, 106; legality of U.S. and Russian, 164, 188; in Libya,

88, 93–95, 106, 107n9, 327; Obama administration's advocacy for, 93, 94, 97; self-defense compared to humanitarian, 323–24; in Ukraine, 38–39, 40, 43, 50–51, 95–97, 106, 164, 188. *See also* non-interventionist platform; preventive non-intervention

Iran, 33, 41; Osirak attack's influence on, 36, 49–50; restricting access as threat, 40, *44*; sanctions and power in, 39–40

Iraq: ground troops to, 83; nuclear proliferation in, 7, 49–50; possession of WMD by, 15–16; sovereignty threat and, 39; Turkey's alliance with, 241

Iraqi Survey group, 74

Iraq Wars, 1, 4, 5; arguments for and against, 6–7, 11–15; casualties in, 72, 81–82; cost absorption by U.S., 53–54; effectiveness and purpose in, 73–76, *75*; elections impacted by, *79*, 79–81, *80*; ethical debate on, 15–16; Hussein as justification for, 53, 61, 67; international view of, 12–13; Obama's support of, 90–92; objectives of, 6–7, 67–68, 74, 77; partisan opinion and support, 76–77, *78*; public opinion and support of, 65, *68*, 68, *69*, 69, 70–83, *71*, *72*, *75*, *81*. *See also* Operation Desert Storm; Operation Iraqi Freedom

ISI. *See* Interservices Intelligence Directorate

ISIS. *See* Islamic State in Iraq and Syria

Islamic militancy, 206, 207–8

Islamic Movement of Uzbekistan (IMU), 8

Islamic State in Iraq and Syria (ISIS), 83, 175

Israel, 132, 164; counterterrorism offensive (2006) by, 239–40; drone policy in, 157, 245–46; self-defense claim of, 232; sovereignty threats and, 39; strike on Iraq nuclear facility by, 6, 33, 36, 49–50; "Wrath of God" operation, 244

Israeli High Court of Justice (HCJ), 150, 153, 157, 246

Jacobson, Gary C., 80–81
Japan, 147, 288
Jervis, Robert, 52
jihad, 176
Johnsen, Gregory D., 176
Johnston, Patrick, 14, 244
Jordan, 334n7
journalists. *See* research and reporting
juggernaut, 47, 50–51
jus ad bellum (justice of going to war): consent and, 160, 161; criteria of, 120, 126, 127, 129, 143, 147, 164, 252n9, 263, 301, 320–21; defining, 308n1; LEE with, 259; limitations of, in drone regulation, 262–66, 267; preventive and preemptive force impact on, 229, 252n5, 257
jus ad vim (just use of force), 17, 21, 277; critics of, 279n3; defining, 258; parameters of, 259, 261–62, 267
jus in bello (just conduct during war), 252n10, 267, 319; criteria for, 129, 161, 320–21; defining, 120, 143, 308n1; drone regulation and limitations of, 133, 262–66; target selection under, 164, 269–70, 289
Just and Unjust Wars (Walzer), 287, 322
just conduct during war. *See jus in bello*
justice of going to war. *See jus ad bellum*
just peace strategy, 22, 313; collaboration for, 333; defining, 314–15; democracy over reactive intervention, 326, 346; liberalism in, 327–28; non-state actors and, 330–31; preventive non-intervention in, 327–28, 329–30, 334
just use of force. *See jus ad vim*
just war ethic (JWE): criteria for force under, 297–98, *298*; outside of armed conflict zones, 292–93, 295–301; principles of, 291–92; targeted killing under, 297, *298*, 302

just war tradition, 15, 16, 21, 283; criteria challenges in, 321–26, 331; drone regulation limitations in, 133, 262–66; drone technology and, 319–21; forefathers of, 23n17; on interventions, 325, 330–31; law enforcement and, 260–62, 266–67, 290–96; Obama references to, 262–63; preventive force in, 258, 316–19, 332; proportionality and discrimination in, 17, 160, 161, 164, 190n3, 252n9, 297, 298, 301, 319; weaknesses of, 316–17

JWE. *See* just war ethic

Kaag, John, 265
Kagan, Robert, 327
Kahn, Paul, 325
Kamm, F. M., 320–21
Kant, Immanuel, 315, 316, 327
Kashmiri, Ilyas, 125
Kaufman, Whitley, 15, 258, 346
Kellogg-Briand Pact (1928), 119–20
Kennan, George, 37
Kennedy, Edward, 12
Kenya, 272, 276–77
Keohane, Robert, 331, 332
Kerry, John, 99–103
Khan, Imran, 335n9
Khrushchev, Nikita, 96
"kill list": Obama's, 91, 106, 147–48, 155; reliance on, for identity of target, 153, 272
Kittrick, Jan, 232
Koh, Harold, 9, 142
Korean War, 65, 72
Kosovo, 95, 96
Krasner, Stephen, 88
Kreps, Sarah, 265
Ku Klux Klan, 300
Kurdistan Workers' Party (PKK), 240–41
Kuwait, 67

lag and predictability, 42, 43–44, 46, 53, 259, 260–61, 269

Lalman, David, 51
Landay, Jonathan, 206
landmines, 135–36
Larson, Eric, 65
The Last Refuge: Yemen, al Qaeda and America's War in Arabia (Johnsen), 176
last resort, 313–14; criteria for force as, 269–70, 273–76, 278; example cases of, principle, 276–77; law enforcement paradigm and, 271–72, 277–78
law enforcement ethic (LEE), 291; criteria for force in, 292, 297–98, 298, 303, 308; for domestic terrorism, 300; targeted killing outside combat zones and, 295–301; targeted killing under, 293–94, 297, 298, 301–2, 303
law enforcement paradigm, 279n4; armed conflict compared to, 261–62, 272–73; drone regulation hybrid of just war ethic with, 258–59, 266–67, 278–79, 335n8; indictment process in, 272–76, 277; "just war" tradition and, 260–62, 266–67, 290–96; last resort under, 271–72, 277–78; presumed innocence and due process in, 271, 292; targeted killing regulated under, 284, 287, 335n8
law of armed conflict (LOAC). *See* International Humanitarian Law
League of Nations, 119
Lebanon, 239–40
LEE. *See* law enforcement ethic
legality. *See* ethics and legality
Lemke, Douglas, 40–41
Levy, Jack, 3–4
liberal internationalism, 326–27
liberalism, 327–28
Libya intervention, 88, 93–95, 106, 107n9, 327
Lincoln, Abraham, 286
"Living under Drones," 120–21, 122–23, 126, 132, 202, 210
LOAC (law of armed conflict). *See* International Humanitarian Law

Long War Journal, 156, 177–78, 209
low-level fighters, 9–10, 14, 16, 152, 244
Lucas, George, Jr., 316

Mahsud, Hakimullah, 153
Man, the State, and War (Waltz), 33
Manual on International Law Applicable to Air and Missile Warfare (2009), 162–63
Marine Corps, 157
Markey, Daniel, 203
Mazzetti, Mark, 200
McDonough, Denis, 101, 102
McDougal, Myres, 233
media. *See* research and reporting
Mehsud, Baitullah, 125, 146, 157, 178
Mehsud, Hakimullah, 178
A Memory of Solferino (Dunant), 117
de Mesquita, Bruce Bueno, 51
Mexico, 39
military. *See* armed forces
Mintz, Alex, 51
Mohammad, Nek, 146, 200–201
"moral predators," 318, 319, 331
morals. *See* dirty hands framework; ethics and legality
Morgenthau, Hans J., 37, 55n7
Mueller, John, 52, 63, 65, 72
multilateralism, 45, 317, 323, 324
Musharraf, Pervez, 127, 201, 203

Nagasaki, 288
nationality, 154–55
national power. *See* power, national
National Security Strategy (2002), 9, 10–11, 12, 235, 317
NATO, 93, 96, 98, 219n6, 313
natural gas, 40, 41
Navy, U.S., 158
necessity principle, 292, 304; in *Caroline* affair, 5, 159, 238; neutral states and, 160–61; self-defense claim and, 5, 116–17, 143, 156, 159–61, 164, 190n3, 234, 248–50, 252n9; unidentified targets and, 271
neutral states: consent of, 160–61, 179–80, 245; "unwilling or unable" standard with, 158–60, 164, 246, 268, 279n7
New America Foundation, 9–10, 201, 209, 210
NGOs, 240, 243, 261
NIAC. *See* non-international armed conflict
Nicaragua, 242
Nicaragua v. United States of America, 231–32, 233, 251n2
Nicholas II (czar), 118–19
Nichols, Thomas M., 4, 18–19, 88, 89–90, 106, 107n1
Nigeria, 41
Nincic, Miroslav, 34, 38
9/11: "al Qaeda core" and, 174; AUMF specifications and, 145–46, 161, 202; imminent threat prior to, 9; public opinion on terrorism after, 63, 69; security strategies after, 10–11, 23n10, 53, 283; targeted killings after, 244, 273; U.S. foreign policy after, 7, 8–9, 64, 68, 142–43
Nixon administration, 159–60
noncombatants: casualties in World War II, 322; CIA killing, 219n6, 308n6; distinction between combatant and, 149–50, 320, 322, 325; drone strikes impact on, 121, 126, 129, 132, 202, 208, 209–16, 271, 343; as drone victims, 13, 125, 132, 142, 156, 170, 209–17, 218n4, 242, 270, 285, 287, 294, 304–5, 308n6, 335n9; emergency ethics in killing, 289, 322; foreseen casualties of, 292, 294, 297, 298, 304–5, 308n3; IHL on, participating in hostilities, 149–52; with military authority, 152–53; participating in hostilities, 149–52; protection of, 21, 149, 284, 291–92, 294, 301; risk assessment for, 297, 298, 300–301, 320, 323, 343; targeting, in armed conflict, 172, 294

non-international armed conflict (NIAC), 20, 293; consent for, 244–45; defining, 308n4; target requirements in, 172, 174–75, 178
non-interventionist platform: pacifist compared to contingent, 332; prevention of war through, 314–15
nonlethal force, 22n2, 297, *298*
non-preventive war, 60, 61, 66
non-state actors: deterrence of, 11–12; drone proliferation to, 121; ICJ rulings on self-defense against, 247–48; just peace theory and, 330–31; NIAC on, 172; policies of force on, 229, 231, 232, 233, 237, 249; threats from, 21, 43, *43*, *48*, 235, 246, 346
North Korea, 7, 39–40
North Waziristan, Pakistan, 120–21, 125–26, 211, 212, 220n13, 294
NSA, 125
nuclear facilities, 6, 7, 33, 36, 49–50
nuclear proliferation: in Iraq and Syria, 7, 49–50; public opinion on, 63, *64*; threat of, *47*, 60

Obama, Barack, 257, 283; on accountability, 189–90; on counterterrorism, 90–92, 107nn3–4, 146, 149; ethical considerations of, 15, 283; foreign policy speeches of, 104–5, 262–63, 274; "kill list" of, 91, 106, 147–48, 155; on Syrian situation, 98, 101, 102–3
Obama administration, 293; ban on landmines under, 136; drone policy under, 2, 7–8, 90, 92, 107n3, 131, 142, 146, 171, 174, 218n5, 242–43, 257, 258, 278, 309n10, 318, 335nn10–11; drone strikes under, 7–8, 131, 165n4, 201–2, 218n4, 242; human rights violations and, 131; imminent threat definition under, 9; intervention advocacy in, 93, 94, 97; Iraq Wars' impact on, 80–81; legal paradigm under, 136, 185, 259; Libya intervention (2001) and, 93–95; reputation concerns in, 13; response to Crimea conflict, 96–97; self-defense justification under, 202, 263, 321; Syria intervention debate under, 88–89, 99–104; targeted killing under, 8, 91–92, 263–64, 315
Obama Doctrine, 95, 104
objectives, 36; effectiveness assessment based on, 65–66; Iraq War, 6–7, 67–68, 74, 77; of preventive force, 35, 343–44; public opinion and support of, 58, 62–63, *64*, 67, 73–76, *75*, 81; U.S. foreign policy, 58, 62–63, *64*, 68
O'Connell, Mary Ellen, 160–61
oil interests, *75*, 83
Olmert, Ehud, 239
Omar, Mullah, 152, 153
"1 percent doctrine," 322, 323
On the Law of War and Peace (Grotius), 17
Operation Desert Storm (1991), 61, 67, *68*, *69*, 69, 81
Operation Enduring Freedom, 61–62
Operation Iraqi Freedom (2003), 22n22, 59, 61; impact of costs of, 72; objectives of, 67; partisan support of, 76–77, *78*; public opinion and support of, *68*, 68, *69*, 69, 81
Operation Odyssey Dawn (2001), 93, 95
opinio juris, 236–37, 251
Orange Revolution (2004), 97
Osiel, Mark, 243
Osirak nuclear facility, 6, 33, 36, 49–50
Ottawa Mine Ban Treaty, 135–36
Owens, Colin, 14

Pace, Peter, 99
Pak Institute for Peace Studies, 219n8
Pakistan, 91, 153, 171–72, 334n7; ability and willingness to act on targets, 205–9; anti-American sentiment in, 13, 189, 335n9; consent debate, 160, 179–80, 201–5, 242–43, 245; death toll in, 156,

219n8, 220n12; drone program impact in, 199–200; drone protest in, 127, 146, 190n2, 204; drone strikes in, 7–8, 20, 45, 115, 116, 120–22, 124–26, 128–29, 146, 179–80, 199, 200–205, 208, 242–43; Islamic militancy in, 206, 207–8; legality of drone strikes in, 127, 202–6; low-level targets in, 244; national security policy in, 205; as neutral territory, 158; population displacement in, 220n13; reporting and research in, 209–16, 219n9; sovereignty issues with, 201–5, 206, 207, 219n7; terrorism in, 14, 268; U.S. armed conflict in, 179–80, 335n9. *See also* Federally Administered Tribal Areas and Frontier Regions; North Waziristan, Pakistan

Pakistani military, 200–201, 207, 209, 214–15, 220n12, 220n13, 220n17

Pakistani Taliban (TTP), 13–14, 146, 153, 157, 177–78, 180, 206, 207

Palestine, 240

Panetta, Leon, 131

partisanship, 76–77, 78, 92, 107n5

Paust, Jordan, 243

peace, 185; deterrence compared to humanist views on, 118–19; drones use impacting, efforts, 131–32; preventive non-intervention for, 21–22, 313–14, 315, 327–28, 329, 332, 346. *See also* international peace; just peace strategy

perpetual imminence, 259, 260–61, 263, 264, 272–74, 276, 278, 345

personality strikes, 218n4, 294

PKK. *See* Kurdistan Workers' Party

poliheuristic theory, 51

political cause and consequence, 18, 41, 236–37; in accountability for drone strikes, 305–7, 345; dirty hands in, 283–84; preventive force/war and, 58, 344; public opinion influencing, 58; rationality debate of, 51–52; threat assessment and, 36, 38, 51–52

Politics among Nations (Morgenthau), 37, 55n7

polls, 67, 68, *68*, 73, *78*, 83n5, 84nn9–21, 85n25

Post-Traumatic Stress Disorder (PTSD), 129–30, 214

poverty and inequality, 329–30, 335nn14–16

power, national, 55n7; asymmetry of, 324–25, 326–27, 330; in cost and benefit analysis, 35, 36, 38, 53–54, 324–25, 326–27; threats to, 39–40, *44*, 47

Power, Samantha, 93, 97, 101

"precision" targeting, 124–25, 126, 132

Predator system, 82, 220n16, 220n18

predictability. *See* lag and predictability

preemptive force, 58, 135, 269; legal impact of, 229, 252n5, 257; options for, 47–48; preventive compared to, 5–6, 12, 59–60, 105, 259, 260, 317, 318; self-defense argument for, 144, 234, 248–49, 314, 345; WMDs and, 322

preemptive war, 59–60, 105, 317

presidential elections. *See* elections

presumed innocence and due process, 271, 292

preventive action, 55n1, 89; cost and benefit analysis of, 47–48, 50–51, 72–73, 342–43; against humanitarian disasters, 93; normalizing of, 92, 334; for sovereignty threats, 46, *47*; under UN Security Council, 314, 346

preventive force, 34, 58, 123, 135; action compared to, 55n1; alternatives to, 47–48; Bush, G. W., on, 52, 56n9, 235; capture and surrender considerations in, 274–76; criteria for and examples of, 43–44, 45–47, 279n8, 302–5; in defense of international status, 41, *44*; ethics and legality of, 4, 37, 229, 230–36, 252n5, 257–58, 264, 315–26; forms of, 1, 2, 33, 341; global trend and future of, 1–2, 14–15, 19, 106, 115–16, 345–46;

preventive force (*cont.*)
  goals of, 35, 343–44; history of, 5–10; increasing war not lessening, 16, 22n3, 321–26, 334; international community's disapproval of, 52–53, 190n2; in international law, 5, 170, 230, 314, 317; in just war tradition, 258, 316–19, 332; as last resort, 269–78; limitations and guidelines for, 21, 257–58, 346; perpetual imminent threat and, 264; preemptive compared to, 5–6, 12, 59–60, 105, 259, 260, 317, 318; psychology of choosing, 52, 54; public opinion and support of, 3–4, 66, 69; threat assessment for, 35, 52–53, 343–44. *See also* armed force; preventive war
"preventive force" school, 230, 235–36
preventive interventions. *See* interventions
preventive non-intervention, 20, 334; peace secured through, 21–22, 313–14, 315, 327–28, 329, 332, 346; for poverty and inequality, 329–30; self-defense and, 332–33
preventive war, 18, 23n6, 52, 56n9, 83n2; critics against, 37–38; empirical research of, 60–61; example of, 59–60; future possibilities of, 83; impact of, 72–73; increasing warfare, 321–26, 334; in Libya, 93–95; nature and purpose of, 59–60; non-preventive compared to, 61, 66; partisanship and, 76–77, *78*; political cause and consequence of, 58, 344; preemptive compared to, 59–60, 105; public support of, 75–76, *77*, 81; rise of, 88; Syrian case and, 103; with U.S. drone policy, 265–66
Price, Bryan, 244
prisoner of war, 165n6
proportionality and discrimination, 263–65, 271; *Caroline* incident and, 5, 159, 248–49; of collateral damage, 116, 121, 151, 155–57, 164, 240, 270; just force criteria and, 17, 160, 161, 164, 190n3, 252n9, 297, 298, 301, 319
prospect theory, 63
protest. *See* anti-drone sentiment; public opinion and support
psychological issues, 52, 54, 129–30
PTSD. *See* Post-Traumatic Stress Disorder
public opinion and support, 23n14; casualties impacting, 18, 63, 65, 72, 72–73, 76, 82–83; of drone strikes, 10, 23nn13–14, 190n2, 243, 247, 306; effectiveness influencing, 65–66, 73–74, 81; on foreign policy objectives, 58, 62–63, 64; impacting cost and benefit analysis, 34, 38, 51; of Iraq Wars, 65, 68, *68*, 69, 69, 70–83, *71*, *72*, *75*, 81; of Israeli self-defense against Hizbullah, 239–40; of overthrowing foreign governments, 70, *70*; policy elites influencing, 62; polls of, 67, 68, *68*, 73, *78*, 83n5, 84nn9–21, 85n25; of preventive force and war, 3–4, 66, 69, 75–76, *77*, 81; self and domestic interest in, 124; terrorism after 9/11 and, 63, 69; in Vietnam War, 69–70, 72, 76; war's objectives and, 67, 73–76, *75*, 81
Putin, Vladimir, 96, 164, 188, 191n8

Qaddafi, Moammar, 94
al-Qaeda, 165n5, 202, 279n4; "associated forces" of, 146, 149, 174–77, 238, 273, 276–77; civilians compared to combatant in, 150; drone strikes on, 8, 9, 14, 146, 276–77; religious and societal values and, 41–42; territory of, 158; U.S. armed conflict distinction with, 174–77, 293; U.S. policy on targeting, 134, 149, 163, 174–77, 263–64, 301; U.S. sanction against, 50
Qatar, 238

R2P (responsibility to protect), 93
raids, 22n2

Rand Corporation, 188, 335n15
Ratner, Steven, 301
Rawls, John, 291
reactive war, 59, 69, 81
Reagan, Ronald, 164, 188
realist/extensive approach, 236, 237, 245, 248
Reaper system, 82, 220n16, 220n18
Red Cross Societies, 136n2
regulation. *See* drone regulation; warfare regulation
Rehman, Zubair, 214–15, 217
Reifler, Jason, 65
religious values, 41–42, 44, 46, 335n16
remotely piloted aircraft (RPA). *See* drones
remotely piloted vehicles (RPVs). *See* drones
reporting. *See* research and reporting
Reprieve, 210–12, 220n11
Republicans, 62, 76–77, 78, 107n5
reputation. *See* status, national
research and reporting, 60–61, 207, 243, 309n11, 344; child witnesses in, 214, 216, 220n15; on counterterrorism alternatives, 335n15; ethics and legality in, 210–12, 220n11; false reports in, 214–15, 216–17, 219n9; fear impacting, 213–14; methods for improving, 216–17; in Pakistan, 209–10, 219n9; role in public opinion, 62
resources, 40, 44, 48, 50
responsibility to protect. *See* R2P
retaliation, 35, 36, 213
Revolutionary Armed Forces of Colombia (FARC), 241–42
Reyes, Raúl, 241–42
Rice, Condoleezza, 11, 53
Rice, Susan, 93, 101
"right to life," 143, 155, 173–74
Rio Group, 242
risks, 38, 295; dangers of removing, from warfare, 318, 319–21, 325, 326–27; drone policy and long-term, 143; of drone strikes, 55, 129, 132; to noncombatants, 297, 298, 300–301, 320, 323, 343; over potential loss, 52, 54, 342; public willingness for, 63; retaliation and escalation, 35, 36
robots, land, 82
rogue states, 10–11, 89, 107n2
Roosevelt, Eleanor, 173
Root, Elihu, 33
Roth, Kenneth, 271
RPA (remotely piloted aircraft). *See* drones
RPVs (remotely piloted vehicles). *See* drones
Rumsfeld, Donald, 199
Russia, 13, 41, 118–19; G-8 expulsion of, 40, 44; Syrian intervention and, 100, 102; targeted killing (2004) by, 238, 245; threats to Ukraine's sovereignty, 38–39, 40, 43, 50–51, 95–97, 106, 164, 188
Russian Federation, 96
Ruys, Tom, 233, 234, 235

sanctions, international, 39–40, 41, 44, 46, 50
Sarbahi, Anoop, 14
Sartre, Jean-Paul, 286
Saudi Arabia, 67
Savych, Bogdan, 65
Scales, Robert, 99
Schiff, Adam, 8–9
Schmitt, Michael N., 234
"security nation," 315
security strategies, 1, 12, 205, 235, 315, 317, 332; based on rogue states and terrorists, 9–11; after 9/11, 10–11, 23n10, 53, 283
self-defense: armed conflict for, 143–44, 146–47, 165n1, 202, 244–45, 249–50, 331–32; Caroline doctrine on justified, 22n5, 234, 248–49, 252n9; cases of force justified by, 238–43;

self-defense (cont.)
    collaborative framework for, 332; humanitarian intervention compared to, 323–24; ICJ rulings on, 247–48; imminent threat and argument for, 233, 246, 260; imminent threat compared to anticipatory, 12, 33, 144, 155, 248–49, 273–74, 314, 317; "inherent right" view of, 230, 233–34; Israeli, against Hizbullah, 239–40; legality and merits of contemporary practice of, 248–51; limits to, actions, 185–86, 190n3; necessity principle of, 5, 116–17, 143, 156, 159–61, 164, 190n3, 234, 248–50, 252n9; neutral territories and criteria of, 159; NIAC without appeal to, 244; Obama administration's justification of, 202, 263, 321; pattern of force based on, 237–38, 245; preemptive force justified by, 144, 234, 248–49, 314, 345; "preventive force" view of, 235–36; preventive non-intervention and, 332–33; schools of thought on, meaning, 229, 230–36, 250–51; sovereignty violations in, 242, 244, 248; "strict construction" view of, 231–32; UN Charter on, 144, 185–86, 202, 229–31, 233; U.S. drone policy justification of, 246–47
self-defense law. See Article 51; customary law
Serbia, 93
al-Shabaab, 8
Sharif, Nawaz, 204
Shue, Henry, 279n8
signature strikes, 121, 153–54, 218n4, 243, 252n6, 294; examples of, 270–71; *jus ad vim* ruling out, 259; permanent imminence and, 264, 272
Silverstein, Gordon, 107n12
Silverstone, Scott, 6, 7
Singer, Peter W., 127, 318, 319
Singh, Ritika, 203

Slaughter, Anne-Marie, 93, 327
Sloane, Robert D., 184
Smith, Megan, 14
socioeconomic values, 41–42
Somalia, 8, 41, 45, 115, 146, 242, 268
South Korea, 13, 39
sovereignty, 158; *Caroline* incident and, 159; consent and issues of, 161; drone strikes and threats to, 39; international law violations for, 251; Pakistani, debate, 201–5, 206, 207, 219n7; preventive action for threats to, 46, 47; self-defense justification for violations of, 242, 244, 248; threats to Ukraine's, 38–39, 40, 43, 50–51, 95–97, 106, 164, 188
sovereignty norm, 88, 89–90, 103, 106
Special Weapons Observation Reconnaissance system (SWORD), 82
status, national: threats to, 40–41, 44; of U.S., 6, 13–14, 128–29, 171, 186–89, 191n8, 293
Stern, Jessica, 13
Stevenson, John R., 160
Strait of Hormuz, 40, 44
strategic areas, 40, 44, 48, 50
"strict construction" school, 230, 231–32
stringers, 209–10, 216
structurally disruptive military technology, 116, 122–23, 132–33, 136n1
Sudan, 39, 41, 98
Sunni insurgency, 73, 74
surrender. See capture and surrender
surveillance drones, 164, 215, 220n14
SWORD. See Special Weapons Observation Reconnaissance system
Syria, 39, 41, 83, 241; chemical attacks by, 98–99, 101, 102; debates on intervention in, 88–89, 98–104, 106; nuclear proliferation in, 49–50

TADS (terrorist attack disruption strikes), 23n11

Taiwan, 39
Taliban, 8, 202; noncombatant compared to combatants in, 150; Pakistan peace negotiations with, 180; territory of, 158; U.S. policy on, 61, 149, 163, 174–77. *See also* Pakistani Taliban
Talon robot, 82
Tanzania, 272, 276–77
targetable political perpetrators, 43, 43–44, 45, 53
targeted killing, 185, 243; after 9/11, 244, 273; of American citizens, 142; of civilians participating in hostilities, 149–52; collateral damage from, 146, 149, 156–57; Colombia's, 241–42; of combatants, 148–49, 210, 291–92; without consent of state, 161; criteria not met for, 302–5; defining, 286–87; dirty hands framework and, 283–87, 302–5, 307, 309n9; distinction and proportionality, 155–57, 164, 270–71; ethics and legality of, 4, 16, 121, 143, 147–55, 187–89, 238, 259, 278, 284–85, 341; extrajudicial, 137n7, 144; "in-between zones" for, 258, 284, 295–302, 298; international community adopting policy of, 1–2, 14–15, 127, 245–46, 345; Israeli, of Palestinians, 240; under JWE and LEE, 293–302, 298, 303; under law enforcement paradigm, 284, 287, 335n8; legitimate targets for, 172, 174–75, 178, 269–72, 289, 294, 297, 298, 303–4; of low-level fighters, 9–10, 14, 16, 152, 244; of Mohammad, N., 146, 200–201; NIAC requirements for, 172, 174–75, 178; of noncombatants, 172, 294; under Obama administration, 8, 91–92, 263–64, 315; precision flaws in, 124–25, 126, 132; public reports of, 309n11; by Russia, 238, 245; of senior military/terrorist leadership, 152–54, 244, 295–96; of unidentified individuals, 121, 153–54, 206, 218n4, 270–71,

303–4. *See also* dirty hands framework; "double tap" strikes; "kill list"; personality strikes; signature strikes
"Targeted Killing: A 'Dirty Hands' Analysis" (de Wijze), 284–85
technology, military, 82, 317, 318–20. *See also* drone technology
Tehreek-e-Taliban-e-Pakistan. *See* Pakistani Taliban
territory: of conflict parties, 158; of drone strikes, 258, 268–69; of neutral states, 158–61; of terrorists, 158, 232, 300–301. *See also* extraterritorial force
terrorism, 11; Article 51 on, 231–32, 234; as crime compared to war, 145, 261–62; domestic compared to global, 299; drone strikes increasing, 13–14, 156, 271; force against, as self-defense, 240; international law outmoded with, 184; in Pakistan, 14, 208–9, 268; perpetual compared to lagged imminence with, 260–61; public opinion after 9/11 on, 63, 69; Russian policy on, 238; vulnerability to, 89. *See also* counterterrorism
terrorism, domestic, 299, 300
terrorist attack disruption strikes. *See* TADS
terrorists, 14, 268; cost and benefit of targeting, 54–55; decapitations of, 244; NIAC target requirements and, 172; non-recognized territory of, 158, 232; risk-taking profile of, 11–12; security strategy based on, 9–10; sovereignty threats and, 39; targeting criteria and, 152–54, 244, 295–96; territory of, 232, 300–301; threat from, 11, 45, 48, 54–55, 89; U.S. primary, targets, 172. *See also* non-state actors
Tesón, Fernando, 261–62
Tet offensive, 72
Thailand, 39
third party state. *See* neutral states

threat, 55n2; climate change as, 42, 55n5; elimination as primary goal, 36; expansionism as, 46; Hussein as, 1, 11, 52, 67–68, 73, 74; illegal immigration as, 46, 55n6; from internal factions, 39, 43, 45; to international status, 40–41, 44; in Iraq War, 53, 61, 67; lag and predictability of, 42, 43–44, 46, 53, 259, 260–61, 269; to nations' power, 39–40, 44, 47; nations' range of, 38; from non-state actors, 21, 43, 43, 48, 235, 246, 346; of nuclear proliferation, 47, 60; restriction of resources and areas as, 40, 44, 48, 50; rogue states as, 89; self-defense based on, 235–36; to sovereignty, 38–39, 40, 43, 46, 47, 50–51, 95–97, 106, 159, 164, 188; from targetable political perpetrators, 43; from terrorists, 11, 45, 48, 54–55, 89; types of, 43, 43–44, 44; to values and ideologies, 41–42, 44, 45–46, 47; "windows of opportunity" for eliminating, 274–76; of WMDs, 70, 75, 75, 76–77, 322. *See also* imminent threat

threat assessment, 297, 298; civilian casualties in, 322–23; for cost and benefit analysis, 34, 37, 50–51, 53; difficulty in, 60; exaggeration in, 34, 51–52, 54; political cause and consequence and, 36, 38, 51–52; for preventive force decision, 35, 52–53, 343–44; public opinion and credibility of, 66

Times Square bombing (2010), 146

"tit-for-tat" strategy, 133, 134

"Title 50" operations, 201

Totten, Mark, 11–12, 23n10, 269

trade, 40, 44

trauma, 203, 215, 219n6, 306–7. *See also* Post-Traumatic Stress Disorder

treaties, 119, 135–36, 173–74, 185, 236

Truman, Harry S., 6, 37

TTP. *See* Pakistani Taliban

Turkey, 240–41, 246

UAV (unmanned aerial vehicle). *See* drones

UDHR. *See* Universal Declaration of Human Rights

Ukraine: protest to Russian actions in, 96–97; sovereignty threat in, 38–39, 40, 43, 50–51, 95–97, 106, 164, 188; after Soviet Union collapse, 95–96. *See also* Crimea

UN. *See* United Nations

UN Charter: human rights law in, 173; multilateralism in, 324; purpose of, 230; regulations on warfare in, 120, 126–27, 137n4, 143–45, 323; self-defense rights in, 144, 185–86, 202, 229–31, 233. *See also* Article 51

unilateralism, 45, 317, 322–23, 324–25, 335n13

United Kingdom, 13, 96, 132, 164

United Nations (UN), 13, 96, 266, 346

United States (U.S.), 96, 190n2; adherence to international law, 180–83, 184, 186–89, 190n4; armed conflict distinction with al-Qaeda, 174–77, 293; armed conflict in Yemen and Pakistan, 179–80, 335n9; *Caroline* incident as sovereignty threat to, 159; drone employment within, 158; ICJ ruling against, 251n2; as imperial democracy, 326; Iraq War's cost absorption by, 53–54; military power of, 164; Pakistani security interests of, 206, 219n7; on preventive force, 106; reputation in international community of, 6, 13–14, 128–29, 171, 186–89, 191n8, 293; sanctions against al-Qaeda, 50; as "security nation," 315; Syria intervention debate in, 88–89, 99–104; threats to international status of, 41, 44; threats to power of, 40; war against communism, 46. *See also* anti-American sentiment

Universal Declaration of Human Rights (UDHR), 173, 180

unmanned aerial vehicle (UAV). *See* drones
UN Security Council, 6, 96–97, 103, 190n7; armed force by, 137n4, 143; enforcement of international peace by, 230, 323–24; inefficacy of, 106; preventive action under, 314, 346
"unwilling or unable" standard, 158–60, 164, 246, 268, 279n7
U.S. *See* United States
U.S.S. *Cole* bombing, 271, 272–73

values and ideologies, 41–42, *44*, 45–46, *47*, 100, 335n16
Van Kirk, Theodore, 320
Van Schaack, Beth, 181–82, 183, 187
VCLT. *See* Vienna Convention on the Law of Treaties
Venezuela, 13, 40, 242
victory, 65, 69, *69*, 72, *72*
Vienna Convention on the Law of Treaties (VCLT), 185
Vietnam, 41, 157, 159
Vietnam War, 65, 69–70, 72, 76, 81, 85n22

Walsh, James Igoe, 14
Waltz, Kenneth, 33
Walzer, Michael, 17, 265, 269, 278, 308n5; on dirty hands framework, 305–6; on emergency ethics, 287–90, 322; on killing outside armed conflict zones, 258, 268, 284, 295–303
war crimes, 130, 151, 157, 163
warfare, 9, 59, 185, 290; advanced technology and future of, 317, 318–19; attempts at humanizing, 116–17, 120, 173; battlefields distinction in, 157, 292, 309n7; dangers of risk-less, 318, 319–21, 325, 326–27; economic considerations as barrier to, 123–24; history of international law on, 118–20; "humane," 4, 116–20, 343, 346, 347n1; hybrid law enforcement, framework, 258–59, 266–67, 278–79, 335n8; Kellogg-Briand Pact banning, 119–20; objectives and costs of, 62–63, 64, 65; preventive force/war increasing, 16, 22n3, 321–26, 334; psychology of "distancing" in, 130; public opinion and support of, 62–63, 65–83, *68*, *71*, *72*, *75*, *77*; tradition of ethics in, 16–17. *See also* armed conflict; armed conflict zones; defensive war; *jus ad bellum*; *jus in bello*; "just war" tradition; preventive war; war on terror
warfare regulation: for air conflict, 162–63; attempts at, 116–17, 123; in UN Charter, 120, 126–27, 137n4, 143–45, 323; on weapon types, 118–20. *See also* drone regulation; International Humanitarian Law; Kellogg-Briand Pact
war on terror, 59, 146, 283, 314; in Afghanistan, 60–61; Bush, G. W., actions based on, 147, 183, 184; drones use in, 2, 199; international law and, 184; number and scope of targets in, 8–9; perpetuity of, 317; risks in, 295
war zones. *See* armed conflict zones; "in-between zones"
weapons, 118–20, 135–36, 136n1, 137n3, 217
weapons, chemical, 98–99, 101, 102, 119
weapons of mass destruction. *See* WMDs
Webster, Daniel, 5, 159, 234
Westphalian sovereignty, 88
Whetham, David, 12
Wikileaks, 127
Williams, Bernard, 285
Williams, Brian Glyn, 165
WMDs (weapons of mass destruction), 10, 53, 318; Iraqi possession of, 15–16; preemptive action on threat of, 322; public opinion on threat of, 70, *75*, *75*, 76–77; search team for, 74

World War I, 119
World War II, 46, 157, 320; casualties in, 72, 132; combatant and noncombatant treatment in, 322; drone use compared to tactics in, 264–65; emergency ethics in, 288; international efforts to humanize war after, 116–17, 120, 173; Kellogg-Briand Pact and, 119–20; targeting "associated forces" in, 147; U.S. foreign policy in, 5–6
"Wrath of God" operation, 244

Yandarbiyev, Zelimkhan, 238

Yemen, 45, 171–72, 179, 190n5, 279n4; anti-American sentiment in, 177; compensation for drone victims in, 154; consent for drone strikes from, 160, 245; drone strikes in, 7–8, 14, 23n11, 115, 137n7, 146, 242, 275, 296–97, 308nn5–6, 334n7; strike on low-level fighters, 10; terrorism in, 14, 268

Zaller, John, 62
Zenko, Micah, 92, 165
zones of war. *See* armed conflict zones; "in-between zones"

www.ingramcontent.com/pod-product-compliance
Lightning Source LLC
Chambersburg PA
CBHW020350080526
44584CB00014B/963